机器学习算法 第2版(影印版)
Machine Learning Algorithms, 2nd Edition

Giuseppe Bonaccorso 著

南京　东南大学出版社

图书在版编目(CIP)数据

机器学习算法：英文/(意)朱塞佩·博纳科尔索 (Giuseppe Bonaccorso) 著. ——2 版(影印本). ——南京：东南大学出版社,2019.3

书名原文：Machine Learning Algorithms, 2nd Edition

ISBN 978-7-5641-8291-5

Ⅰ.①机… Ⅱ.①朱… Ⅲ.①机器学习-算法-英文 Ⅳ.①TP181

中国版本图书馆 CIP 数据核字(2019)第 025733 号

图字：10-2018-489 号

© 2018 by PACKT Publishing Ltd.

Reprint of the English Edition, jointly published by PACKT Publishing Ltd and Southeast University Press, 2019. Authorized reprint of the original English edition, 2018 PACKT Publishing Ltd, the owner of all rights to publish and sell the same.

All rights reserved including the rights of reproduction in whole or in part in any form.

英文原版由 PACKT Publishing Ltd 出版 2018。

英文影印版由东南大学出版社出版 2019。此影印版的出版和销售得到出版权和销售权的所有者——PACKT Publishing Ltd 的许可。

版权所有，未得书面许可，本书的任何部分和全部不得以任何形式重制。

机器学习算法 第 2 版(影印版)

出版发行：东南大学出版社
地　　址：南京四牌楼 2 号　　邮编：210096
出 版 人：江建中
网　　址：http://www.seupress.com
电子邮件：press@seupress.com
印　　刷：常州市武进第三印刷有限公司
开　　本：787 毫米×980 毫米　　16 开本
印　　张：32.5
字　　数：636 千字
版　　次：2019 年 3 月第 1 版
印　　次：2019 年 3 月第 1 次印刷
书　　号：ISBN 978-7-5641-8291-5
定　　价：108.00 元

本社图书若有印装质量问题，请直接与营销部联系。电话(传真)：025-83791830

To my family and to all the people who always believed in me and encouraged me in this long journey!

– *Giuseppe Bonaccorso*

mapt.io

Mapt is an online digital library that gives you full access to over 5,000 books and videos, as well as industry leading tools to help you plan your personal development and advance your career. For more information, please visit our website.

Why subscribe?

- Spend less time learning and more time coding with practical eBooks and Videos from over 4,000 industry professionals

- Improve your learning with Skill Plans built especially for you

- Get a free eBook or video every month

- Mapt is fully searchable

- Copy and paste, print, and bookmark content

PacktPub.com

Did you know that Packt offers eBook versions of every book published, with PDF and ePub files available? You can upgrade to the eBook version at www.PacktPub.com and as a print book customer, you are entitled to a discount on the eBook copy. Get in touch with us at service@packtpub.com for more details.

At www.PacktPub.com, you can also read a collection of free technical articles, sign up for a range of free newsletters, and receive exclusive discounts and offers on Packt books and eBooks.

Contributors

About the author

Giuseppe Bonaccorso is an experienced team leader/manager in AI, machine/deep learning solution design, management, and delivery. He got his MScEng in electronics in 2005 from the University of Catania, Italy, and continued his studies at the University of Rome Tor Vergata and the University of Essex, UK. His main interests include machine/deep learning, reinforcement learning, big data, bio-inspired adaptive systems, cryptocurrencies, and NLP.

> *I want to thank the people who have been close to me and have supported me, especially my parents, who never stopped encouraging me.*

About the reviewer

Doug Ortiz is an experienced enterprise cloud, big data, data analytics, and solutions architect who has architected, designed, developed, re-engineered, and integrated enterprise solutions. Other expertise includes Amazon Web Services, Azure, Google Cloud, business intelligence, Hadoop, Spark, NoSQL databases, and SharePoint, to name a few.

He is the founder of Illustris, LLC and is reachable at `dougortiz@illustris.org`.

> *Huge thanks to my wonderful wife, Milla, Maria, Nikolay, and our children for all their support.*

Packt is searching for authors like you

If you're interested in becoming an author for Packt, please visit `authors.packtpub.com` and apply today. We have worked with thousands of developers and tech professionals, just like you, to help them share their insight with the global tech community. You can make a general application, apply for a specific hot topic that we are recruiting an author for, or submit your own idea.

Table of Contents

Preface — 1

Chapter 1: A Gentle Introduction to Machine Learning — 7
  Introduction – classic and adaptive machines — 8
    Descriptive analysis — 11
    Predictive analysis — 12
  Only learning matters — 13
    Supervised learning — 14
    Unsupervised learning — 17
    Semi-supervised learning — 19
    Reinforcement learning — 21
    Computational neuroscience — 23
  Beyond machine learning – deep learning and bio-inspired adaptive systems — 24
  Machine learning and big data — 26
  Summary — 27

Chapter 2: Important Elements in Machine Learning — 29
  Data formats — 29
    Multiclass strategies — 33
      One-vs-all — 33
      One-vs-one — 34
  Learnability — 34
    Underfitting and overfitting — 36
    Error measures and cost functions — 39
    PAC learning — 43
  Introduction to statistical learning concepts — 44
    MAP learning — 45
    Maximum likelihood learning — 46
  Class balancing — 51
    Resampling with replacement — 52
    SMOTE resampling — 54
  Elements of information theory — 57
    Entropy — 57
    Cross-entropy and mutual information — 59
    Divergence measures between two probability distributions — 61
  Summary — 62

Chapter 3: Feature Selection and Feature Engineering — 65
  scikit-learn toy datasets — 66

Table of Contents

Creating training and test sets	67
Managing categorical data	69
Managing missing features	72
Data scaling and normalization	74
Whitening	76
Feature selection and filtering	78
Principal Component Analysis	81
Non-Negative Matrix Factorization	88
Sparse PCA	90
Kernel PCA	92
Independent Component Analysis	95
Atom extraction and dictionary learning	99
Visualizing high-dimensional datasets using t-SNE	102
Summary	104
Chapter 4: Regression Algorithms	**105**
Linear models for regression	105
A bidimensional example	107
Linear regression with scikit-learn and higher dimensionality	112
R2 score	116
Explained variance	117
Regressor analytic expression	118
Ridge, Lasso, and ElasticNet	119
Ridge	119
Lasso	122
ElasticNet	124
Robust regression	125
RANSAC	126
Huber regression	128
Bayesian regression	130
Polynomial regression	134
Isotonic regression	138
Summary	141
Chapter 5: Linear Classification Algorithms	**143**
Linear classification	144
Logistic regression	147
Implementation and optimizations	150
Stochastic gradient descent algorithms	153
Passive-aggressive algorithms	157
Passive-aggressive regression	163
Finding the optimal hyperparameters through a grid search	167
Classification metrics	170
Confusion matrix	172

Precision	176
Recall	176
F-Beta	177
Cohen's Kappa	178
Global classification report	180
Learning curve	180
ROC curve	182
Summary	186

Chapter 6: Naive Bayes and Discriminant Analysis — 187
Bayes' theorem — 188
Naive Bayes classifiers — 190
Naive Bayes in scikit-learn — 191
- Bernoulli Naive Bayes — 191
- Multinomial Naive Bayes — 194
  - An example of Multinomial Naive Bayes for text classification — 196
- Gaussian Naive Bayes — 199
Discriminant analysis — 203
Summary — 208

Chapter 7: Support Vector Machines — 209
Linear SVM — 209
SVMs with scikit-learn — 214
- Linear classification — 215
Kernel-based classification — 217
- Radial Basis Function — 218
- Polynomial kernel — 219
- Sigmoid kernel — 219
- Custom kernels — 219
- Non-linear examples — 220
v-Support Vector Machines — 225
Support Vector Regression — 228
- An example of SVR with the Airfoil Self-Noise dataset — 232
Introducing semi-supervised Support Vector Machines (S3VM) — 236
Summary — 243

Chapter 8: Decision Trees and Ensemble Learning — 245
Binary Decision Trees — 246
- Binary decisions — 247
- Impurity measures — 250
  - Gini impurity index — 250
  - Cross-entropy impurity index — 250
  - Misclassification impurity index — 252
- Feature importance — 252
Decision Tree classification with scikit-learn — 252
Decision Tree regression — 260

       Example of Decision Tree regression with the Concrete Compressive
       Strength dataset   261
  Introduction to Ensemble Learning   267
    Random Forests   268
       Feature importance in Random Forests   271
    AdaBoost   273
    Gradient Tree Boosting   277
    Voting classifier   280
  Summary   284

Chapter 9: Clustering Fundamentals   285
  Clustering basics   285
  k-NN   288
  Gaussian mixture   294
    Finding the optimal number of components   298
  K-means   301
    Finding the optimal number of clusters   308
       Optimizing the inertia   308
       Silhouette score   310
       Calinski-Harabasz index   314
       Cluster instability   316
  Evaluation methods based on the ground truth   319
    Homogeneity   319
    Completeness   320
    Adjusted Rand Index   321
  Summary   322

Chapter 10: Advanced Clustering   323
  DBSCAN   324
  Spectral Clustering   328
  Online Clustering   331
    Mini-batch K-means   332
    BIRCH   334
  Biclustering   337
  Summary   340

Chapter 11: Hierarchical Clustering   343
  Hierarchical strategies   343
  Agglomerative Clustering   344
    Dendrograms   346
    Agglomerative Clustering in scikit-learn   349
    Connectivity constraints   354
  Summary   359

Chapter 12: Introducing Recommendation Systems   361
  Naive user-based systems   362

Implementing a user-based system with scikit-learn	363
Content-based systems	365
Model-free (or memory-based) collaborative filtering	367
Model-based collaborative filtering	370
Singular value decomposition strategy	371
Alternating least squares strategy	373
ALS with Apache Spark MLlib	374
Summary	378

Chapter 13: Introducing Natural Language Processing — 379
NLTK and built-in corpora — 380
    Corpora examples — 381
The Bag-of-Words strategy — 382
    Tokenizing — 384
        Sentence tokenizing — 384
        Word tokenizing — 385
    Stopword removal — 386
        Language detection — 387
    Stemming — 388
    Vectorizing — 389
        Count vectorizing — 389
            N-grams — 391
        TF-IDF vectorizing — 391
Part-of-Speech — 393
    Named Entity Recognition — 395
A sample text classifier based on the Reuters corpus — 396
Summary — 397

Chapter 14: Topic Modeling and Sentiment Analysis in NLP — 399
Topic modeling — 399
    Latent Semantic Analysis — 400
    Probabilistic Latent Semantic Analysis — 407
    Latent Dirichlet Allocation — 413
Introducing Word2vec with Gensim — 418
Sentiment analysis — 422
    VADER sentiment analysis with NLTK — 426
Summary — 427

Chapter 15: Introducing Neural Networks — 429
Deep learning at a glance — 429
    Artificial neural networks — 430
MLPs with Keras — 434
    Interfacing Keras to scikit-learn — 443
Summary — 445

Chapter 16: Advanced Deep Learning Models — 447
Deep model layers — 447

Table of Contents

    Fully connected layers — 448
        Convolutional layers — 449
        Dropout layers — 450
        Batch normalization layers — 451
        Recurrent Neural Networks — 451
An example of a deep convolutional network with Keras — 452
An example of an LSTM network with Keras — 456
A brief introduction to TensorFlow — 462
    Computing gradients — 464
    Logistic regression — 467
    Classification with a multilayer perceptron — 471
    Image convolution — 474
Summary — 476

Chapter 17: Creating a Machine Learning Architecture — 477
Machine learning architectures — 477
    Data collection — 479
    Normalization and regularization — 480
    Dimensionality reduction — 480
    Data augmentation — 481
    Data conversion — 483
    Modeling/grid search/cross-validation — 483
    Visualization — 484
    GPU support — 484
    A brief introduction to distributed architectures — 488
Scikit-learn tools for machine learning architectures — 491
    Pipelines — 491
    Feature unions — 495
Summary — 496

Other Books You May Enjoy — 497

Index — 501

Preface

This book is an introduction to the world of machine learning, a topic that is becoming more and more important, not only for IT professionals and analysts but also for all the data scientists and engineers who want to exploit the enormous power of techniques such as predictive analysis, classification, clustering, and natural language processing. In order to facilitate the learning process, all theoretical elements are followed by concrete examples based on Python.

A basic but solid understanding of this topic requires a foundation in mathematics, which is not only necessary to explain the algorithms, but also to let the reader understand how it's possible to tune up the hyperparameters in order to attain the best possible accuracy. Of course, it's impossible to cover all the details with the appropriate precision. For this reason, some topics are only briefly described, limiting the theory to the results without providing any of the workings. In this way, the user has the double opportunity to focus on the fundamental concepts (without too many mathematical complications) and, through the references, examine in depth all the elements that generate interest.

The chapters can be read in no particular order, skipping the topics that you already know. Whenever necessary, there are references to the chapters where some concepts are explained. I apologize in advance for any imprecision, typos or mistakes, and I'd like to thank all the Packt editors for their collaboration and constant attention.

Who this book is for

This book is for machine learning engineers, data engineers, and data scientists who want to build a strong foundation in the field of predictive analytics and machine learning. Familiarity with Python would be an added advantage and will enable you to get the most out of this book.

What this book covers

Chapter 1, *A Gentle Introduction to Machine Learning*, introduces the world of machine learning, explaining the fundamental concepts of the most important approaches to creating intelligent applications and focusing on the different kinds of learning methods.

Preface

Chapter 2, *Important Elements in Machine Learning*, explains the mathematical concepts regarding the most common machine learning problems, including the concept of learnability and some important elements of information theory. This chapter contains theoretical elements, but it's extremely helpful if you are learning this topic from scratch because it provides an insight into the most important mathematical tools employed in the majority of algorithms.

Chapter 3, *Feature Selection and Feature Engineering*, describes the most important techniques for preprocessing a dataset, selecting the most informative features, and reducing the original dimensionality.

Chapter 4, *Regression Algorithms*, describes the linear regression algorithm and its optimizations: Ridge, Lasso, and ElasticNet. It continues with more advanced models that can be employed to solve non-linear regression problems or to mitigate the effect of outliers.

Chapter 5, *Linear Classification Algorithms*, introduces the concept of linear classification, focusing on logistic regression, perceptrons, stochastic gradient descent algorithms, and passive-aggressive algorithms. The second part of the chapter covers the most important evaluation metrics, which are used to measure the performance of a model and find the optimal hyperparameter set.

Chapter 6, *Naive Bayes and Discriminant Analysis*, explains the Bayes probability theory and describes the structure of the most diffused Naive Bayes classifiers. In the second part, linear and quadratic discriminant analysis is analyzed with some concrete examples.

Chapter 7, *Support Vector Machines*, introduces the SVM family of algorithms, focusing on both linear and non-linear classification problems thanks to the employment of the kernel trick. The last part of the chapter covers support vector regression and more complex classification models.

Chapter 8, *Decision Trees and Ensemble Learning*, explains the concept of a hierarchical decision process and describes the concepts of decision tree classification, random forests, bootstrapped and bagged trees, and voting classifiers.

Chapter 9, *Clustering Fundamentals*, introduces the concept of clustering, describing the Gaussian mixture, K-Nearest Neighbors, and K-means algorithms. The last part of the chapter covers different approaches to determining the optimal number of clusters and measuring the performance of a model.

Chapter 10, *Advanced Clustering*, introduces more complex clustering techniques (DBSCAN, Spectral Clustering, and Biclustering) that can be employed when the dataset structure is non-convex. In the second part of the chapter, two online clustering algorithms (mini-batch K-means and BIRCH) are introduced.

Chapter 11, *Hierarchical Clustering*, continues the explanation of more complex clustering algorithms started in the previous chapter and introduces the concepts of agglomerative clustering and dendrograms.

Chapter 12, *Introducing Recommendation Systems*, explains the most diffused algorithms employed in recommender systems: content- and user-based strategies, collaborative filtering, and alternating least square. A complete example based on Apache Spark shows how to process very large datasets using the ALS algorithm.

Chapter 13, *Introduction to Natural Language Processing*, explains the concept of the Bag-of-Words strategy and introduces the most important techniques required to efficiently process natural language datasets (tokenizing, stemming, stop-word removal, tagging, and vectorizing). An example of a classifier based on the Reuters dataset is also discussed in the last part of the chapter.

Chapter 14, *Topic Modeling and Sentiment Analysis in NLP*, introduces the concept of topic modeling and describes the most important algorithms, such as latent semantic analysis (both deterministic and probabilistic) and latent Dirichlet allocation. The second part of the chapter covers the problem of word embedding and sentiment analysis, explaining the most diffused approaches to address it.

Chapter 15, *Introducing Neural Networks*, introduces the world of deep learning, explaining the concept of neural networks and computational graphs. In the second part of the chapter, the high-level deep learning framework Keras is presented with a concrete example of a Multi-layer Perceptron.

Chapter 16, *Advanced Deep Learning Models*, explains the basic functionalities of the most important deep learning layers, with Keras examples of deep convolutional networks and recurrent (LSTM) networks for time-series processing. In the second part of the chapter, the TensorFlow framework is briefly introduced, along with some examples that expose some of its basic functionalities.

Chapter 17, *Creating a Machine Learning Architecture*, explains how to define a complete machine learning pipeline, focusing on the peculiarities and drawbacks of each step.

Preface

To get the most out of this book

To fully understand all the algorithms in this book, it's important to have a basic knowledge of linear algebra, probability theory, and calculus.

All practical examples are written in Python and use the scikit-learn machine learning framework, **Natural Language Toolkit** (**NLTK**), Crab, langdetect, Spark (PySpark), Gensim, Keras, and TensorFlow (deep learning frameworks). These are available for Linux, macOS X, and Windows, with Python 2.7 and 3.3+. When a particular framework is employed for a specific task, detailed instructions and references will be provided. All the examples from chapters 1 to 14 can be executed using Python 2.7 (while TensorFlow requires Python 3.5+); however, I highly suggest using a Python 3.5+ distribution. The most common choice for data science and machine learning is Anaconda (https://www.anaconda.com/download/), which already contains all the most important packages.

Download the example code files

You can download the example code files for this book from your account at www.packtpub.com. If you purchased this book elsewhere, you can visit www.packtpub.com/support and register to have the files emailed directly to you.

You can download the code files by following these steps:

1. Log in or register at www.packtpub.com.
2. Select the **SUPPORT** tab.
3. Click on **Code Downloads & Errata**.
4. Enter the name of the book in the **Search** box and follow the onscreen instructions.

Once the file is downloaded, please make sure that you unzip or extract the folder using the latest version of:

- WinRAR/7-Zip for Windows
- Zipeg/iZip/UnRarX for Mac
- 7-Zip/PeaZip for Linux

The code bundle for the book is also hosted on GitHub at https://github.com/PacktPublishing/Machine-Learning-Algorithms-Second-Edition. In case there's an update to the code, it will be updated on the existing GitHub repository.

We also have other code bundles from our rich catalog of books and videos available at https://github.com/PacktPublishing/. Check them out!

Download the color images

We also provide a PDF file that has color images of the screenshots/diagrams used in this book. You can download it here: https://www.packtpub.com/sites/default/files/downloads/MachineLearningAlgorithmsSecondEdition_ColorImages.pdf.

Conventions used

There are a number of text conventions used throughout this book.

`CodeInText`: Indicates code words in text, database table names, folder names, filenames, file extensions, pathnames, dummy URLs, user input, and Twitter handles. Here is an example: "scikit-learn provides the `SVC` class, which is a very efficient implementation that can be used in most cases."

A block of code is set as follows:

```
from sklearn.svm import SVC
from sklearn.model_selection import cross_val_score

svc = SVC(kernel='linear')
print(cross_val_score(svc, X, Y, scoring='accuracy', cv=10).mean())
0.93191356542617032
```

Bold: Indicates a new term, an important word, or words that you see onscreen. For example, words in menus or dialog boxes appear in the text like this:

Warnings or important notes appear like this.

Tips and tricks appear like this.

Get in touch

Feedback from our readers is always welcome.

General feedback: Email `feedback@packtpub.com` and mention the book title in the subject of your message. If you have questions about any aspect of this book, please email us at `questions@packtpub.com`.

Errata: Although we have taken every care to ensure the accuracy of our content, mistakes do happen. If you have found a mistake in this book, we would be grateful if you would report this to us. Please visit `www.packtpub.com/submit-errata`, selecting your book, clicking on the Errata Submission Form link, and entering the details.

Piracy: If you come across any illegal copies of our works in any form on the Internet, we would be grateful if you would provide us with the location address or website name. Please contact us at `copyright@packtpub.com` with a link to the material.

If you are interested in becoming an author: If there is a topic that you have expertise in and you are interested in either writing or contributing to a book, please visit `authors.packtpub.com`.

Reviews

Please leave a review. Once you have read and used this book, why not leave a review on the site that you purchased it from? Potential readers can then see and use your unbiased opinion to make purchase decisions, we at Packt can understand what you think about our products, and our authors can see your feedback on their book. Thank you!

For more information about Packt, please visit `packtpub.com`.

A Gentle Introduction to Machine Learning

In the last few years, machine learning has become one of the most important and prolific IT and artificial intelligence branches. It's not surprising that its applications are becoming more widespread day by day in every business sector, always with new and more powerful tools and results. Open source, production-ready frameworks, together with hundreds of papers published every month, are contributing to one of the most pervasive democratization processes in IT history. But why is machine learning so important and valuable?

In this chapter, we are going to discuss the following:

- The difference between classic systems and adaptive ones
- The general concept of learning, proving a few examples of different approaches
- Why bio-inspired systems and computational neuroscience allowed a dramatic improvement in performances
- The relationship between big data and machine learning

Introduction – classic and adaptive machines

Since time immemorial, human beings have built tools and machines to simplify their work and reduce the overall effort needed to complete many different tasks. Even without knowing any physical law, they invented levers (formally described for the first time by Archimedes), instruments, and more complex machines to carry out longer and more sophisticated procedures. Hammering a nail became easier and more painless thanks to a simple trick, and so did moving heavy stones or wood using a cart. But, what's the difference between these two examples? Even if the latter is still a simple machine, its complexity allows a person to carry out a composite task without thinking about each step. Some fundamental mechanical laws play a primary role in allowing a horizontal force to contrast gravity efficiently, but neither human beings, nor horses or oxen, knew anything about them. The primitive people simply observed how a genial trick (the wheel) could improve their lives.

The lesson we've learned is that a machine is never efficient or trendy without a concrete possibility to use it with pragmatism. A machine is immediately considered useful and destined to be continuously improved if its users can easily understand what tasks can be completed with less effort or automatically. In the latter case, some intelligence seems to appear next to cogs, wheels, or axles. So, a further step can be added to our evolution list: automatic machines, built (nowadays, we'd say programmed) to accomplish specific goals by transforming energy into work. Wind or watermills are some examples of elementary tools that are able to carry out complete tasks with minimal (compared to a direct activity) human control.

In the following diagram, there's a generic representation of a classical system that receives some input values, processes them, and produces output results:

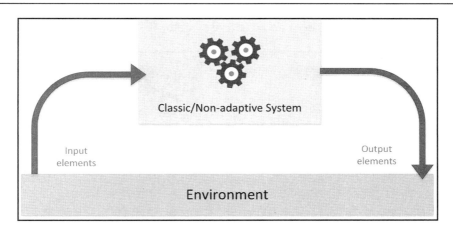

Interaction diagram of a classic/non-adaptive system

But again, what's the key to the success of a mill? It's not hasty at all to say that human beings have tried to transfer some intelligence into their tools since the dawn of technology. Both the water in a river and the wind show a behavior that we can simply call flowing. They have a lot of energy to give us free of any charge, but a machine should have some awareness to facilitate this process. A wheel can turn around a fixed axle millions of times, but the wind must find a suitable surface to push on. The answer seems obvious, but you should try to think about people without any knowledge or experience; even if implicitly, they started a brand new approach to technology. If you prefer to reserve the word intelligence to more recent results, it's possible to say that the path started with tools, moved first to simple machines, and then moved to smarter ones.

Without further intermediate (but no less important) steps, we can jump into our epoch and change the scope of our discussion. Programmable computers are widespread, flexible, and more and more powerful instruments; moreover, the diffusion of the internet allowed us to share software applications and related information with minimal effort. The word-processing software that I'm using, my email client, a web browser, and many other common tools running on the same machine, are all examples of such flexibility. It's undeniable that the IT revolution dramatically changed our lives and sometimes improved our daily jobs, but without **machine learning** (and all its applications), there are still many tasks that seem far out of the computer domain. Spam filtering, **Natural Language Processing** (**NLP**), visual tracking with a webcam or a smartphone, and predictive analysis are only a few applications that revolutionized human-machine interaction and increased our expectations. In many cases, they transformed our electronic tools into actual cognitive extensions that are changing the way we interact with many daily situations. They achieved this goal by filling the gap between human perception, language, reasoning, and model and artificial instruments.

Here's a schematic representation of an adaptive system:

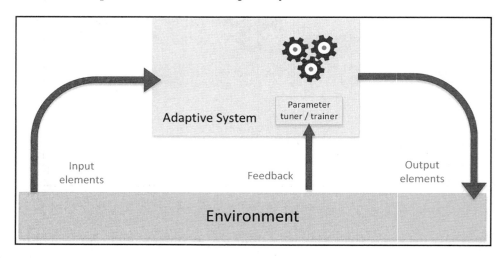

Interaction diagram of an adaptive system

Such a system isn't based on static or permanent structures (model parameters and architectures), but rather on a continuous ability to adapt its behavior to external signals (datasets or real-time inputs) and, like a human being, to predict the future using uncertain and fragmentary pieces of information.

Before moving on with a more specific discussion, let's briefly define the different kinds of system analysis that can be performed. These techniques are often structured as a sequence of specific operations whose goal is increasing the overall domain knowledge and allowing answering specific questions, however, in some cases, it's possible to limit the process to a single step in order to meet specific business needs. I always suggest to briefly consider them all, because many particular operations make sense only when some conditions are required. A clear understanding of the problem and its implications is the best way to make the right decisions, also taking into consideration possible future developments.

Descriptive analysis

Before trying any machine learning solution, it's necessary to create an abstract description of the context. The best way to achieve this goal is to define a mathematical model, which has the advantage of being immediately comprehensible by anybody (assuming the basic knowledge). However, the goal of **descriptive analysis** is to find out an accurate description of the phenomena that are observed and validate all the hypothesis. Let's suppose that our task is to optimize the supply chain of a large store. We start collecting data about purchases and sales and, after a discussion with a manager, we define the generic hypotheses that the sales volume increases during the day before the weekend. This means that our model should be based on a periodicity. A descriptive analysis has the task to validate it, but also to discover all those other particular features that were initially neglected.

At the end of this stage, we should know, for example, if the time series (let's suppose we consider only a variable) is periodic, if it has a trend, if it's possible to find out a set of standard rules, and so forth. A further step (that I prefer to consider as whole with this one) is to define a **diagnostic model** that must be able to connect all the effects with precise causes. This process seems to go in the opposite direction, but its goal is very close to the descriptive analysis one. In fact, whenever we describe a phenomenon, we are naturally driven to finding a rational reason that justifies each specific step. Let's suppose that, after having observed the periodicity in our time series, we find a sequence that doesn't obey this rule. The goal of diagnostic analysis is to give a suitable answer (that is, the store is open on Sunday). This new piece of information enriches our knowledge and specializes it: now, we can state that the series is periodic only when there is a day off, and therefore (clearly, this is a trivial example) we don't expect an increase in the sales before a working day. As many machine learning models have specific prerequisites, a descriptive analysis allows us to immediately understand whether a model will perform poorly or if it's the best choice considering all the known factors. In all of the examples we will look at, we are going to perform a brief descriptive analysis by defining the features of each dataset and what we can observe. As the goal of this book is to focus on adaptive systems, we don't have space for a complete description, but I always invite the reader to imagine new possible scenarios, performing a *virtual* analysis before defining the models.

Predictive analysis

The goal of machine learning is almost related to this precise stage. In fact, once we have defined a model of our system, we need to infer its future states, given some initial conditions. This process is based on the discovery of the rules that underlie the phenomenon so as to *push them forward* in time (in the case of a time series) and observe the results. Of course, the goal of a predictive model is to minimize the error between actual and predictive value, considering all possible interfering factors.

In the example of the large store, a good model should be able to forecast a peak before a day off and a normal behavior in all the other cases. Moreover, once a predictive model has been defined and trained, it can be used as a fundamental part of a decision-based process. In this case, the prediction must be turned into a suggested **prescription**. For example, the object detector of a self-driving car can be extremely accurate and detect an obstacle on time. However, which is the best action to perform in order to achieve a specific goal? According to the prediction (position, size, speed, and so on), another model must be able to pick the action that minimizes the risk of damage and maximizes the probability of a safe movement. This is a common task in reinforcement learning, but it's also extremely useful whenever a manager has to make a decision in a context where there are many factors. The resultant model is, hence, a pipeline that is fed with raw inputs and uses the single outcomes as inputs for subsequent models. Returning to our initial example, the store manager is not interested in discovering the hidden oscillations, but in the right volumes of goods that he has to order every day. Therefore, the first step is predictive analysis, while the second is a prescriptive one, which can take into account many factors that are discarded by the previous model (that is, different suppliers can have shorter or longer delivery times or they can apply discounts according to the volume).

So, the manager will probably define a goal in terms of a function to maximize (or minimize), and the model has to find the best amount of goods to order so as to fulfill the main requirement (that, of course, is the availability, and it depends on the sales prediction). In the remaining part of this book, we are going to discuss many solutions to specific problems, focusing on the predictive stage. But, in order to move on, we need to define what learning means and why it's so important in more and more different business contexts.

Only learning matters

What does learning exactly mean? Simply, we can say that learning is the ability to change according to external stimuli and remember most of our previous experiences. So, machine learning is an engineering approach that gives maximum importance to every technique that increases or improves the propensity for changing adaptively. A mechanical watch, for example, is an extraordinary artifact, but its structure obeys stationary laws and becomes useless if something external is changed. This ability is peculiar to animals and, in particular, to human beings; according to Darwin's theory, it's also a key success factor for the survival and evolution of all species. Machines, even if they don't evolve autonomously, seem to obey the same law.

Therefore, the main goal of machine learning is to study, engineer, and improve mathematical models that can be trained (once or continuously) with context-related data (provided by a generic environment) to infer the future and to make decisions without complete knowledge of all influencing elements (external factors). In other words, an agent (which is a software entity that receives information from an environment, picks the best action to reach a specific goal, and observes the results of it) adopts a statistical learning approach, trying to determine the right probability distributions, and use them to compute the action (value or decision) that is most likely to be successful (with the fewest errors).

I do prefer using the term **inference** instead of **prediction**, but only to avoid the weird (but not so uncommon) idea that machine learning is a sort of modern magic. Moreover, it's possible to introduce a fundamental statement: an algorithm can extrapolate general laws and learn their structure with relatively high precision, but only if they affect the actual data. So, the term *prediction* can be freely used, but with the same meaning adopted in physics or system theory. Even in the most complex scenarios, such as image classification with convolutional neural networks, every piece of information (geometry, color, peculiar features, contrast, and so on) is already present in the data and the model has to be flexible enough to extract and learn it permanently.

In the following sections, we will give you a brief description of some common approaches to machine learning. Mathematical models, algorithms, and practical examples will be discussed in later chapters.

Supervised learning

A supervised scenario is characterized by the concept of a teacher or supervisor, whose main task is to provide the agent with a precise measure of its error (directly comparable with output values). With actual algorithms, this function is provided by a training set made up of couples (input and expected output). Starting from this information, the agent can correct its parameters so as to reduce the magnitude of a global loss function. After each iteration, if the algorithm is flexible enough and data elements are coherent, the overall accuracy increases and the difference between the predicted and expected values becomes close to zero. Of course, in a supervised scenario, the goal is training a system that must also work with samples that have never been seen before. So, it's necessary to allow the model to develop a generalization ability and avoid a common problem called **overfitting**, which causes *overlearning* due to an excessive capacity (we're going to discuss this in more detail in the following chapters, however, we can say that one of the main effects of such a problem is the ability to only correctly predict the samples used for training, while the error rate for the remaining ones is always very high).

In the following graph, a few training points are marked with circles, and the thin blue line represents a perfect generalization (in this case, the connection is a simple segment):

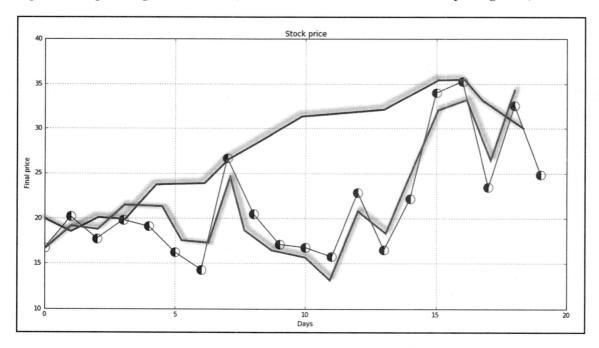

Example of regression of a stock price with different interpolating curves

Two different models are trained with the same datasets (corresponding to the two larger lines). The former is unacceptable because it cannot generalize and capture the fastest dynamics (in terms of frequency), while the latter seems a very good compromise between the original trend, and has a residual ability to generalize correctly in a predictive analysis.

Formally, the previous example is called **regression** because it's based on continuous output values. Instead, if there is only a discrete number of possible outcomes (called **categories**), the process becomes a **classification.** Sometimes, instead of predicting the actual category, it's better to determine its probability distribution. For example, an algorithm can be trained to recognize a handwritten alphabetical letter, so its output is categorical (in English, there'll be 26 allowed symbols). On the other hand, even for human beings, such a process can lead to more than one probable outcome when the visual representation of a letter isn't clear enough to belong to a single category. This means that the actual output is better described by a discrete probability distribution (for example, with 26 continuous values normalized so that they always sum up to 1).

In the following graph, there's an example of classification of elements with two features. The majority of algorithms try to find the best separating hyperplane (in this case, it's a linear problem) by imposing different conditions. However, the goal is always the same: reducing the number of misclassifications and increasing the noise-robustness. For example, look at the triangular point that is closest to the plane (its coordinates are about *[5.1 - 3.0]*). If the magnitude of the second feature were affected by noise and so the value were quite smaller than *3.0*, a slightly higher hyperplane could wrongly classify it. We're going to discuss some powerful techniques to solve these problems in later chapters:

A Gentle Introduction to Machine Learning

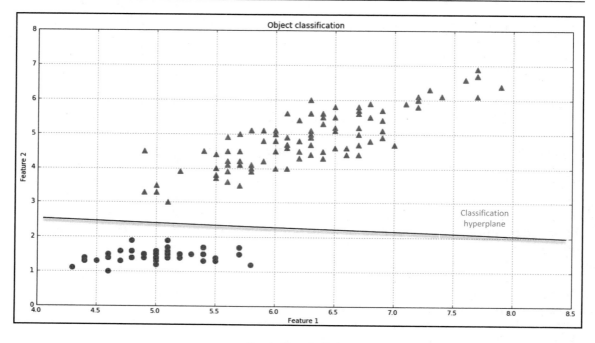

Example of linear classification

Common supervised learning applications include the following:

- Predictive analysis based on regression or categorical classification
- Spam detection
- Pattern detection
- NLP
- Sentiment analysis
- Automatic image classification
- Automatic sequence processing (for example, music or speech)

Unsupervised learning

This approach is based on the absence of any supervisor and therefore of absolute error measures. It's useful when it's necessary to learn how a set of elements can be grouped (clustered) according to their similarity (or distance measure). For example, looking at the previous graph, a human being can immediately identify two sets without considering the colors or the shapes. In fact, the circular dots (as well as the triangular ones) determine a coherent set; it is separate from the other one much more than how its points are internally separated. Using a metaphor, an ideal scenario is a sea with a few islands that can be separated from each other, considering only their mutual position and internal cohesion. Clearly, unsupervised learning provides an implicit descriptive analysis because all the pieces of information discovered by the clustering algorithm can be used to obtain a complete insight of the dataset. In fact, all objects share a subset of features, while they are different under other viewpoints. The aggregation process is also aimed to extend the characteristics of some points to their neighbors, assuming that the similarity is not limited to some specific features. For example, in a recommendation engine, a group of users can be clustered according to the preference expressed for some books. If the chosen criteria detected some analogies between users A and B, we can share the non-overlapping elements between the users. Therefore, if A has read a book that can be suitable for B, we are implicitly authorized to recommend it. In this case, the decision is made by considering a goal (sharing the features) and a descriptive analysis. However, as the model can (and should) manage unknown users too, its purpose is also predictive.

In the following graph, each ellipse represents a cluster and all the points inside its area can be labeled in the same way. There are also boundary points (such as the triangles overlapping the circle area) that need a specific criterion (normally a trade-off distance measure) to determine the corresponding cluster. Just as for classification with ambiguities (P and malformed R), a good clustering approach should consider the presence of outliers and treat them so as to increase both the internal coherence (visually, this means picking a subdivision that maximizes the local density) and the separation among clusters.

For example, it's possible to give priority to the distance between a single point and a centroid, or the average distance among points belonging to the same cluster and different ones. In this graph, all boundary triangles are close to each other, so the nearest neighbor is another triangle. However, in real-life problems, there are often boundary areas where there's a partial overlap, meaning that some points have a high degree of uncertainty due to their feature values:

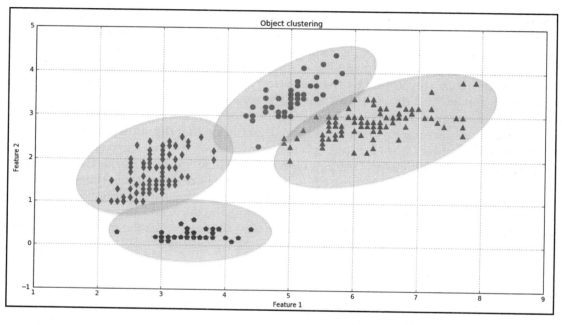

Example of clustering with a bidimensional dataset split into four *natural* clusters

Another interpretation can be expressed by using probability distributions. If you look at the ellipses, they represent the area of multivariate Gaussians bound between a minimum and maximum variance. Considering the whole domain, a point (for example, a blue star) could potentially belong to all clusters, but the probability given by the first one (lower-left corner) is the highest, and so this determines the membership. Once the variance and mean (in other words, the shape) of all Gaussians become stable, each boundary point is automatically captured by a single Gaussian distribution (except in the case of equal probabilities). Technically, we say that such an approach maximizes the likelihood of a Gaussian mixture given a certain dataset. This is a very important statistical learning concept that spans many different applications, so it will be examined in more depth in the next chapter, Chapter 2, *Important Elements in Machine Learning*. Moreover, we're going to discuss some common clustering methodologies, considering both strong and weak points, and compare their performances for various test distributions.

Other important techniques involve the use of both labeled and unlabeled data. This approach is therefore called semi-supervised and can be adopted when it's necessary to categorize a large amount of data with a few complete (labeled) examples or when there's the need to impose some constraints to a clustering algorithm (for example, assigning some elements to a specific cluster or excluding others).

Commons unsupervised applications include the following:

- Object segmentation (for example, users, products, movies, songs, and so on)
- Similarity detection
- Automatic labeling
- Recommendation engines

Semi-supervised learning

There are many problems where the amount of labeled samples is very small compared with the potential number of elements. A direct supervised approach is infeasible because the data used to train the model couldn't be representative of the whole distribution, so therefore it's necessary to find a trade-off between a supervised and an unsupervised strategy. Semi-supervised learning has been mainly studied in order to solve these kinds of problems. The topic is a little bit more advanced and won't be covered in this book (the reader who is interested can check out *Mastering Machine Learning Algorithms, Bonaccorso G., Packt Publishing*). However, the main goals that a semi-supervised learning approach pursues are as follows:

- The propagation of labels to unlabeled samples considering the graph of the whole dataset. The samples with labels become *attractors* that extend their influence to the neighbors until an equilibrium point is reached.
- Performing a classification training model (in general, **Support Vector Machines** (**SVM**); see `Chapter 7`, *Support Vector Machines*, for further information) using the labeled samples to enforce the conditions necessary for a good separation while trying to exploit the unlabeled samples as *balancers*, whose influence must be mediated by the labeled ones. Semi-supervised SVMs can perform extremely well when the dataset contains only a few labeled samples and dramatically reduce the burden of building and managing very large datasets.

- Non-linear dimensionality reduction considering the graph structure of the dataset. This is one of most challenging problems due to the constraints existing in high-dimensional datasets (that is, images). Finding a low-dimensional distribution that represents the original one minimizing the discrepancy is a fundamental task necessary to visualize structures with more than three dimensions. Moreover, the ability to reduce the dimensionality without a significant information loss is a key element whenever it's necessary to work with simpler models. In this book, we are going to discuss some common linear techniques (such as **Principal Component Analysis** (**PCA**) that the reader will be able to understand when some features can be removed without impacting the final accuracy but with a training speed gain.

It should now be clear that semi-supervised learning exploits the ability of finding out separating hyperplanes (classification) together with the auto-discovery of structural relationships (clustering). Without a loss of generality, we could say that the real supervisor, in this case, is the data graph (representing the relationships) that corrects the decisions according to the underlying informational layer. To better understand the logic, we can imagine that we have a set of users, but only 1% of them have been labeled (for simplicity, let's suppose that they are uniformly distributed). Our goal is to find the most accurate labels for the remaining part. A clustering algorithm can rearrange the structure according to the similarities (as the labeled samples are uniform, we can expect to find unlabeled neighbors whose center is a labeled one). Under some assumptions, we can propagate the center's label to the neighbors, repeating this process until every sample becomes stable. At this point, the whole dataset is labeled and it's possible to employ other algorithms to perform specific operations. Clearly, this is only an example, but in real life, it's extremely common to find scenarios where the cost of labeling millions of samples is not justified considering the accuracy achieved by semi-supervised methods.

Chapter 1

Reinforcement learning

Even if there are no actual supervisors, reinforcement learning is also based on feedback provided by the environment. However, in this case, the information is more qualitative and doesn't help the agent in determining a precise measure of its error. In reinforcement learning, this feedback is usually called **reward** (sometimes, a negative one is defined as a penalty), and it's useful to understand whether a certain action performed in a state is positive or not. The sequence of most useful actions is a policy that the agent has to learn in order to be able to always make the best decision in terms of the highest immediate and cumulative (future) reward. In other words, an action can also be imperfect, but in terms of a global policy, it has to offer the highest total reward. This concept is based on the idea that a rational agent always pursues the objectives that can increase his/her wealth. The ability to *see* over a distant horizon is a distinctive mark for advanced agents, while short-sighted ones are often unable to correctly evaluate the consequences of their immediate actions and so their strategies are always sub-optimal.

Reinforcement learning is particularly efficient when the environment is not completely deterministic, when it's often very dynamic, and when it's impossible to have a precise error measure. During the last few years, many classical algorithms have been applied to deep neural networks to learn the best policy for playing Atari video games and to teach an agent how to associate the right action with an input representing the state (usually, this is a screenshot or a memory dump).

A Gentle Introduction to Machine Learning

In the following diagram, there's a schematic representation of a deep neural network that's been trained to play a famous Atari game:

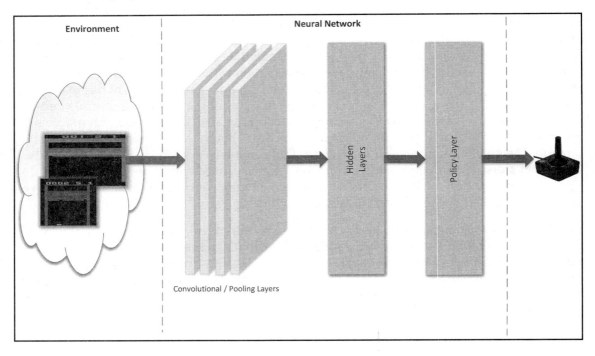

The generic structure of a deep reinforcement learning architecture

As input, there are one or more subsequent screenshots (this can often be enough to capture the temporal dynamics as well). They are processed using different layers (discussed briefly later) to produce an output that represents the policy for a specific state transition. After applying this policy, the game produces a feedback (as a reward-penalty), and this result is used to refine the output until it becomes stable (so the states are correctly recognized and the suggested action is always the best one) and the total reward overcomes a predefined threshold.

We're going to discuss some examples of reinforcement learning in the `Chapter 15`, *Introducing Neural Networks*, and `Chapter 16`, *Advanced Deep Learning Models*, dedicated to introducing deep learning and TensorFlow. However, some common examples are as follows:

- Automatic robot control
- Game solving
- Stock trade analysis based on feedback signals

Computational neuroscience

It's not surprising that many machine learning algorithms have been defined and refined thanks to the contribution of research in the field of computational neuroscience. On the other hand, the most diffused adaptive systems are the animals, thanks to their nervous systems, that allow an effective interaction with the environment. From a mechanistic viewpoint, we need to assume that all the processes working inside the gigantic network of neurons are responsible for all computational features, starting from low-level perception and progressing until the highest abstractions, like language, logic reasoning, artistic creation, and so forth.

At the beginning of 1900, Ramón y Cajal and Golgi discovered the structure of the nerve cells, the neurons, but it was necessary to fully understand that their behavior was purely computational. Both scientists drew sketches representing the input units (dendrites), the body (soma), the main channel (axon) and the output gates (synapses), however neither the dynamic nor the learning mechanism of a cell group was fully comprehended. The neuroscientific community was convinced that learning was equivalent to a continuous and structural change, but they weren't able to define exactly what was changing during a learning process. In 1949, the Canadian psychologist Donald Hebb proposed his famous rule (a broader discussion can be found in *Mastering Machine Learning Algorithms*, Bonaccorso G., Packt Publishing, 2018 and *Theoretical Neuroscience*, Dayan P., Abbott L. F., The MIT Press, 2005) that is focused on the synaptic plasticity of the neurons. In other words, the changing element is the number and the nature of the output gates which connect a unit to a large number of other neurons. Hebb understood that if a neuron produces a spike and a synapse propagates it to another neuron which behaves in the same way, the connection is strengthened, otherwise, it's weakened. This can seem like a very simplistic explanation, but it allows you to understand how elementary neural aggregates can perform operations such as detecting the borders of an object, denoising a signal, or even finding the dimensions with maximum variance (PCA).

The research in this field has continued until today, and many companies, together with high-level universities, are currently involved in studying the computational behavior of the brain using the most advanced neuroimaging technologies available. The discoveries are sometimes surprising, as they confirm what was only imagined and never observed. In particular, some areas of the brain can easily manage supervised and unsupervised problems, while others exploit reinforcement learning to predict the most likely future perceptions. For example, an animal quickly learns to associate the sound of steps to the possibility of facing a predator, and learns how to behave consequently. In the same way, the input coming from its eyes are manipulated so as to extract all those pieces of information that are useful to detect the objects. This denoising procedure is extremely common in machine learning and, surprisingly, many algorithms achieve the same goal that an animal brain does! Of course, the complexity of human minds is beyond any complete explanation, but the possibility to double-check these intuitions using computer software has dramatically increased the research speed. At the end of this book, we are going to discuss the basics of deep learning, which is the most advanced branch of machine learning. However, I invite the reader to try to understand all the dynamics (even when they seem very abstract) because the underlying logic is always based on very simple and natural mechanisms, and your brain is very likely to perform the same operations that you're learning while you read!

Beyond machine learning – deep learning and bio-inspired adaptive systems

During the last few years, thanks to more powerful and cheaper computers, many researchers started adopting complex (deep) neural architectures to achieve goals that were unimaginable only two decades ago. Since 1957, when Rosenblatt invented the first perceptron, interest in neural networks has grown more and more. However, many limitations (concerning memory and CPU speed) prevented massive research and hid lots of potential applications of these kinds of algorithms.

In the last decade, many researchers started training bigger and bigger models, built with several different layers (that's why this approach is called **deep learning**), in order to solve new challenging problems. The availability of cheap and fast computers allowed them to get results in acceptable timeframes and to use very large datasets (made up of images, texts, and animations). This effort led to impressive results, in particular for classification based on photo elements and real-time intelligent interaction using reinforcement learning.

The idea behind these techniques is to create algorithms that work like a brain, and many important advancements in this field have been achieved thanks to the contribution of neurosciences and cognitive psychology. In particular, there's a growing interest in pattern recognition and associative memories whose structure and functioning are similar to what happens in the neocortex. Such an approach also allows simpler algorithms called **model-free**; these aren't based on any mathematical-physical formulation of a particular problem, but rather on generic learning techniques and repeating experiences.

Of course, testing different architectures and optimization algorithms is rather simpler (and it can be done with parallel processing) than defining a complex model (which is also more difficult to adapt to different contexts). Moreover, deep learning showed better performance than other approaches, even without a context-based model. This suggests that, in many cases, it's better to have a less precise decision made with uncertainty than a precise one determined by the output of a very complex model (often not so fast). For animals, this is often a matter of life and death, and if they succeed, it is thanks to an implicit renounce of some precision.

Common deep learning applications include the following:

- Image classification
- Real-time visual tracking
- Autonomous car driving
- Robot control
- Logistic optimization
- Bioinformatics
- Speech recognition and **Natural Language Understanding** (NLU)
- **Natural Language Generation** (NLG) and speech synthesis

Many of these problems can also be solved by using classic approaches that are sometimes much more complex, but deep learning outperformed them all. Moreover, it allowed extending their application to contexts initially considered extremely complex, such as autonomous cars or real-time visual object identification.

This book covers, in detail, only some classical algorithms; however, there are many resources that can be read both as an introduction and for a more advanced insight.

Many interesting results have been achieved by the Google DeepMind team (https://deepmind.com) and I suggest that you visit their website to learn more about their latest research and goals. Another very helpful resource is OpenAI (https://openai.com/), where there's also a virtual gym with many reinforcement learning environments ready to use.

Machine learning and big data

Another area that can be exploited using machine learning is big data. After the first release of Apache Hadoop, which implemented an efficient MapReduce algorithm, the amount of information managed in different business contexts grew exponentially. At the same time, the opportunity to use it for machine learning purposes arose and several applications such as mass collaborative filtering became a reality.

Imagine an online store with 1 million users and only 1,000 products. Consider a matrix where each user is associated with every product by an implicit or explicit ranking. This matrix will contain 1,000,000 x 1,000 cells, and even if the number of products is very limited, any operation performed on it will be slow and memory-consuming. Instead, using a cluster, together with parallel algorithms, such a problem disappears, and operations with a higher dimensionality can be carried out in a very short time.

Think about training an image classifier with 1 million samples. A single instance needs to iterate several times, processing small batches of pictures. Even if this problem can be performed using a streaming approach (with a limited amount of memory), it's not surprising to wait even for a few days before the model begins to perform well. Adopting a big data approach instead, it's possible to asynchronously train several local models, periodically share the updates, and re-synchronize them all with a master model. This technique has also been exploited to solve some reinforcement learning problems, where many agents (often managed by different threads) played the same game, providing their periodical contribution to a *global* intelligence.

Not every machine learning problem is suitable for big data, and not all big datasets are really useful when training models. However, their conjunction in particular situations can lead to extraordinary results by removing many limitations that often affect smaller scenarios. Unfortunately, both machine learning and big data are topics subject to continuous hype, hence one of the tasks that an engineer/scientist has to accomplish is understanding when a particular technology is really helpful and when its burden can be heavier than the benefits. Modern computers often have enough resources to process datasets that, a few years ago, were easily considered big data. Therefore, I invite the reader to carefully analyze each situation and think about the problem from a business viewpoint as well. A Spark cluster has a cost that is sometimes completely unjustified. I've personally seen clusters of two medium machines running tasks that a laptop could have carried out even faster. Hence, always perform a descriptive/prescriptive analysis of the problem and the data, trying to focus on the following:

- The current situation
- Objectives (what do we need to achieve?)

- Data and dimensionality (do we work with batch data? Do we have incoming streams?)
- Acceptable delays (do we need real-time? Is it possible to process once a day/week?)

Big data solutions are justified, for example, when the following is the case:

- The dataset cannot fit in the memory of a high-end machine
- The incoming data flow is huge, continuous, and needs prompt computations (for example, clickstreams, web analytics, message dispatching, and so on)
- It's not possible to split the data into small chunks because the acceptable delays are minimal (this piece of information must be mathematically quantified)
- The operations can be parallelized efficiently (nowadays, many important algorithms have been implemented in distributed frameworks, but there are still tasks that cannot be processed by using parallel architectures)

In the chapter dedicated to recommendation systems, Chapter 12, *Introduction to Recommendation Systems*, we're going to discuss how to implement collaborative filtering using Apache Spark. The same framework will also be adopted for an example of Naive Bayes classification.

> If you want to know more about the whole Hadoop ecosystem, visit http://hadoop.apache.org. Apache Mahout (http://mahout.apache.org) is a dedicated machine learning framework, and Spark (http://spark.apache.org), one the fastest computational engines, has a module called **Machine Learning Library** (**MLlib**) which implements many common algorithms that benefit from parallel processing.

Summary

In this chapter, we introduced the concept of adaptive systems; they can learn from their experiences and modify their behavior in order to maximize the possibility of reaching a specific goal. Machine learning is the name given to a set of techniques that allow you to implement adaptive algorithms to make predictions and auto-organize input data according to their common features.

The three main learning strategies are supervised, unsupervised, and reinforcement. The first one assumes the presence of a teacher that provides a precise feedback on errors. The algorithm can, therefore, compare its output with the right one and correct its parameters accordingly. In an unsupervised scenario, there are no external teachers, so everything is learned directly from the data. An algorithm will try to find out all of the features that are common to a group of elements so that it's able to associate new samples with the right cluster. Examples of the former type are provided by all the automatic classifications of objects into a specific category according to some known features, while common applications of unsupervised learning are the automatic groupings of items with a subsequent labeling or processing. The third kind of learning is similar to supervised, but it only receives an environmental feedback about the quality of its actions. It doesn't know exactly what is wrong or the magnitude of its error, but receives generic information that helps it in deciding whether to continue to adopt a policy or to pick another one.

In the next chapter, Chapter 2, *Important Elements in Machine Learning*, we're going to discuss some fundamental elements of machine learning, with particular focus on the mathematical notation and the main definitions that we'll need in the rest of the chapters. We'll also discuss important statistical learning concepts and some theory about learnability and its limits.

2
Important Elements in Machine Learning

In this chapter, we're going to discuss some important elements and approaches that span through all machine learning topics and also create a philosophical foundation for many common techniques. First of all, it's useful to understand the mathematical foundation of data formats and prediction functions. In most algorithms, these concepts are treated in different ways, but the goal is always the same. More recent techniques, such as deep learning, extensively use energy/loss functions, just like the one described in this chapter, and even if there are slight differences, a good machine learning result is normally associated with the choice of the best loss function and the use of the right algorithm to minimize it.

In particular, we will be discussing the following topics:

- The generic structure of a machine learning problem and the data that is employed
- Properties of machine learning models and their impact on performances
- Class balancing
- Elements of statistical learning (**Maximum A Posteriori** (**MAP**) and **maximum likelihood estimation** (**MLE**) estimations)
- Introduction to information theory with a focus on the most important machine learning tools

Data formats

In both supervised and unsupervised learning problems, there will always be a dataset, defined as a finite set of real vectors with m features each:

$$X = \{\bar{x}_1, \bar{x}_2, \ldots, \bar{x}_n\} \ \ where \ \ \bar{x}_i \in \mathbb{R}^m$$

Considering that our approach is always probabilistic, we need to assume each X as drawn from a statistical multivariate distribution, D, that is commonly known as a **data generating process** (the probability density function is often denoted as $p_{data}(x)$). For our purposes, it's also useful to add a very important condition upon the whole dataset X: we expect all samples to be **independent and identically distributed (i.i.d)**. This means that all variables belong to the same distribution, D, and considering an arbitrary subset of k values, it happens that the following is true:

$$p(\bar{x}_1, \bar{x}_2, \ldots, \bar{x}_k) = \prod_{i=1}^{k} p(\bar{x}_i)$$

It's fundamental to understand that all machine learning tasks are based on the assumption of working with well-defined distributions (even if they can be partially unknown), and the actual datasets are made up of samples drawn from it. In the previous chapter, Chapter 1, *A Gentle Introduction to Machine Learning*, we defined the concept of learning considering the interaction between an agent and an unknown situation. This is possible because of the ability to learn a representation of the distribution and not the dataset itself! Hence, from now on, whenever a finite dataset is employed, the reader must always consider the possibility of coping with new samples that share the same distribution.

The corresponding output values can be both numerical-continuous or categorical. In the first case, the process is called **regression**, while in the second, it is called **classification**. Examples of numerical outputs are as follows:

$$Y = \{y_1, y_2, \ldots, y_n\} \quad where \quad y_i \in (0,1) \quad or \quad y_i \in \mathbb{R}^+$$

When the label can assume a finite number of values (for example, it's binary or bipolar), the problem is discrete (also known as categorical, considering that each label is normally associated with a well-defined class or category), while it's called continuous when $y_i \in \mathcal{R}$.

Other categorical examples are as follows:

$$y_i \in \{red, black, white, green\} \quad or \quad y_i \in \{0, 1, 2, 3\}$$

We define a generic **regressor**, a vector-valued function, *r(•)*, which associates an input value to a continuous output and generic **classifier**, and a vector-valued function *c(•)*, whose predicted output is categorical (discrete). If they also depend on an internal parameter vector that determines the actual instance of a generic predictor, the approach is called **parametric learning**:

$$\begin{cases} \tilde{y} = r(\bar{x}; \bar{\theta}) \\ \tilde{y} = c(\bar{x}; \bar{\theta}) \end{cases}$$

The vector θ is a summary of all model parameters which, in general, are the only elements we are going to learn. In fact, the majority of models assume a standard structure that cannot be modified (even if there are some particular dynamic neural networks that allows adding or removing computational units), and the adaptability relies only on the range of possible parameters.

On the other hand, **non-parametric learning** doesn't make initial assumptions about the family of predictors (for example, defining a generic parameterized version of *r(•)* and *c(•)*). A very common non-parametric family is called **instance-based learning** and makes real-time predictions (without pre-computing parameter values) based on a hypothesis that's only determined by the training samples (instance set). A simple and widespread approach adopts the concept of neighborhoods (with a fixed radius). In a classification problem, a new sample is automatically surrounded by classified training elements and the output class is determined considering the preponderant one in the neighborhood. In this book, we're going to talk about another very important algorithm family belonging to this class: **kernel-based Support Vector Machines**.

The internal dynamics and the interpretation of all elements are peculiar to every single algorithm, and for this reason, we prefer not to talk about thresholds or probabilities and try to work with an abstract definition. A generic parametric training process must find the best parameter vector that minimizes the regression/classification error given a specific training dataset, and it should also generate a predictor that can correctly generalize when unknown samples are provided.

Another interpretation can be expressed in terms of additive noise:

$$\begin{cases} \tilde{y} = r(\bar{x}; \bar{\theta}) + n(\mu; \sigma^2) \\ \tilde{y} = c(\bar{x}; \bar{\theta}) + n(\mu; \sigma^2) \end{cases}$$
$$\text{where } \mu = E[n] = 0 \text{ and } \sigma^2 = E[n^2] \ll 1$$

Important Elements in Machine Learning

For our purposes, we can expect zero-mean and low-variance Gaussian noise to be added to a perfect prediction. A training task must increase the signal-noise ratio by optimizing the parameters. Of course, whenever such a term doesn't have a null mean (independently from the other X values), it probably implies that there's a hidden trend that must be taken into account (maybe a feature that has been prematurely discarded). On the other hand, high noise variance means that X is extremely corrupted and its measures are not reliable.

In unsupervised learning, we normally only have an input set X with *m*-length vectors, and we define the clustering function *cl(•)* (with *n* target clusters) with the following expression:

$$k_t = cl(\bar{x}; \bar{\theta}) \ where \ k_t \in \{0, 1, \ldots, n-1\}$$

As explained in the previous chapter, Chapter 1, *A Gentle Introduction to Machine Learning*, a clustering algorithm tries to discover similarities among samples and group them accordingly; therefore *cl(•)* will always output a label between *0* and *n-1* (or, alternatively, between *1* and *n*), representing the cluster that best matches the sample *x*. As *x* is assumed to be drawn from the same data generating process used during the training phase, we are mathematically authorized to accept the result as reliable in the limit of the accuracy that has been achieved. On the other hand (this is true in every machine learning problem), if *x* is drawn from a completely different distribution, any prediction will be indistinguishable from a random one. This concept is extremely important and the reader must understand it (together with all the possible implications). Let's suppose that we classify images of Formula 1 cars and military planes with a final accuracy of 95%. This means that only five photos representing actual cars or planes are misclassified. This is probably due to the details, the quality of the photo, the shape of the objects, the presence of noise, and so on. Conversely, if we try to classify photos of SUVs and large cargo planes, all of the results are meaningless (even if they can be correct). This happens because the classifier will seldom output a classification probability of 50% (meaning that the uncertainty is maximum), and the final class will always be one of the two. However, the *awareness* of the classifier is not very different from that of an oracle that tosses a coin. Therefore, whenever we need to work with specific samples, we must be sure to train the model with elements drawn from the same distribution. In the previous example, it's possible to retrain the classifier with all types of cars and planes, trying, at the same time, to reach the same original accuracy.

In most scikit-learn models, there is a `coef_` instance variable which contains all of the trained parameters. For example, in a single parameter linear regression (we're going to widely discuss it in the following chapters), the output will be as follows:

```
model = LinearRegression()
model.fit(X, Y)
print(model.coef_)
array([ 9.10210898])
```

Multiclass strategies

Until now, we've assumed that both regression and classification operate on *m*-length vectors but produce a single value or single label (in other words, an input vector is always associated with only one output element). When the number of output classes is greater than one, there are two main possibilities to manage a classification problem:

- One-vs-all
- One-vs-one

In both cases, the choice is transparent, and the output returned to the user will always be the final value or class. However, it's important to understand the different dynamics in order to optimize the model and to always pick the best alternative (scikit-learn can manage these options automatically, so as to avoid an explicit choice when not necessary).

One-vs-all

This is probably the most common strategy and is widely adopted by scikit-learn for most of its algorithms. If there are *n* output classes, *n* classifiers will be trained in parallel considering there is always a separation between an actual class and the remaining ones. This approach is relatively lightweight (at most, *n-1* checks are needed to find the right class, so it has an *O(n)* complexity) and, for this reason, it's normally the default choice, so there's no need for further actions.

One-vs-one

The alternative to one-vs-all is training a model for each pair of classes. The complexity is no longer linear (it's $O(n^2)$ indeed), and the right class is determined by a majority vote. In general, this choice is more expensive and should only be adopted when a full dataset comparison is not preferable.

If you want to learn more about multiclass strategies implemented by scikit-learn, visit http://scikit-learn.org/stable/modules/multiclass.html.

Learnability

A parametric model can be split into two parts: a static structure and a dynamic set of parameters. The former is determined by the choice of a specific algorithm and is normally immutable (except in the cases when the model provides some remodeling functionalities), while the latter is the objective of our optimization. Considering n unbounded parameters, they generate an n-dimensional space (imposing bounds results in a subspace without relevant changes in our discussion) where each point, together with the immutable part of the estimator function, represents a learning hypothesis H (associated with a specific set of parameters):

$$H = \{\bar{\theta}_1, \bar{\theta}_2, \ldots, \bar{\theta}_n\}$$

The goal of a parametric learning process is to find the best hypothesis whose corresponding prediction error is at minimum and the residual generalization ability is enough to avoid overfitting. In the following diagram, we can see an example of a dataset whose points must be classified as red (**Class A**) or blue (**Class B**). Three hypotheses are shown: the first one (the middle line starting from the left) misclassifies 3 samples, while the lower and upper ones misclassify 14 and 24 samples, respectively:

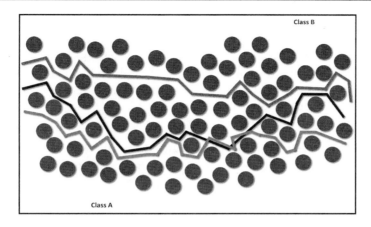

Example of classifiers based on 3 different hypotheses

Of course, the first hypothesis is almost optimal and should be selected; however, it's important to understand an essential concept that can determine a potential overfitting. Think about an n-dimensional binary classification problem. We say that dataset X is linearly separable (without transformations) if a hyperplane exists that divides the space into two subspaces containing only elements belonging to the same class. Removing the constraint of linearity, we have infinite alternatives by using generic hypersurfaces. However, a parametric model only adopts a family of non-periodic and approximate functions whose ability to oscillate and fit the dataset is determined (sometimes in a very complex way) by the number of parameters.

Consider the example shown in the following diagram:

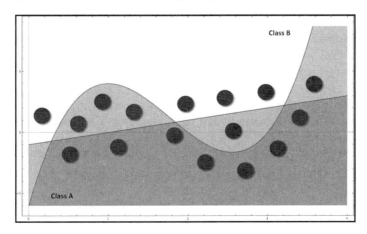

Example of a linear and a non-linear classifier

The blue classifier is linear while the red one is cubic. At a glance, the non-linear strategy seems to perform better, because it can capture more expressivity, thanks to its concavities. However, if new samples are added following the trend defined by the last four ones (from the right), they'll be completely misclassified. In fact, while a linear function is globally better but cannot capture the initial oscillation between **0** and **4**, a cubic approach can fit this data almost perfectly but, at the same time, loses its ability to keep a global linear trend. Therefore, there are two possibilities:

- If we expect future data to be exactly distributed as training samples, a more complex model can be a good choice in order to capture small variations that a lower-level one will discard. In this case, a linear (or lower-level) model will drive to underfitting, because it won't be able to capture an appropriate level of expressivity.
- If we think that future data can be locally distributed differently but keeps a global trend, it's preferable to have a higher residual misclassification error as well as a more precise generalization ability. Using a bigger model focusing only on training data can drive to overfitting.

Underfitting and overfitting

The purpose of a machine learning model is to approximate an unknown function that associates input elements to output ones (for a classifier, we call them classes). However, a training set is normally a representation of a global distribution, but it cannot contain all possible elements, otherwise, the problem could be solved with a one-to-one association. In the same way, we don't know the analytic expression of a possible underlying function, therefore, when training, it's necessary to think about fitting the model but keeping it free to generalize when an unknown input is presented. In this regard, it's useful to introduce the concept of the **representational capacity** of a model, as the ability to learn a small/large number of possible distributions over the dataset. Clearly, a low capacity is normally associated with simpler models that, for example, cannot solve non-linear problems, while a high capacity, that is both a function of the underlying model and of the number of parameters, leads to more complex separation hyperplanes. Considering the last example in the previous section, it's easy to understand that the linear classifier is equivalent to the equation of a straight line:

$$y = mx + q$$

In this case, there are two parameters, *m* and *q*, and the curve can never change its slope (which is defined by *m*). Conversely, the second classifier could be imagined as a cubic equation:

$$y = ax^3 + bx^2 + cx + d$$

Now, we have four parameters and two powers of the input value. These conditions allow modeling a function that can change its slope twice and can be adapted to more complex scenarios. Obviously, we could continue this analysis by considering a generic polynomial function:

$$y = \sum_{n=0}^{p-1} a_n x^n$$

The complexity (and, hence, the capacity) is proportional to the degree *p*. Joining polynomials and non-linear functions, we can obtain extremely complex representations (such as the ones achieved using neural networks) that can be flexible enough to capture the details of non-trivial datasets. However, it's important to remember that increasing the capacity is normally an irreversible operation. In other words, a more complex model will always be more complex, even when a simpler one would be preferable. The learning process can *stretch* or *bend* the curve, but it will never be able to remove the slope changes (for a more formal explanation, please check *Mastering Machine Learning Algorithms, Bonaccorso G., Packt Publishing, 2018*). This condition leads to two different potentials dangers:

- **Underfitting**: It means that the model isn't able to capture the dynamics shown by the same training set (probably because its capacity is too limited).
- **Overfitting**: The model has an excess capacity and it's not longer able to generalize effectively, considering the original dynamics provided by the training set. It can associate almost perfectly all the known samples to the corresponding output values, but when an unknown input is presented, the corresponding prediction error can be very high.

Important Elements in Machine Learning

In the following graph, there are examples of interpolation with low capacity (underfitting), normal capacity (normal fitting), and excessive capacity (overfitting):

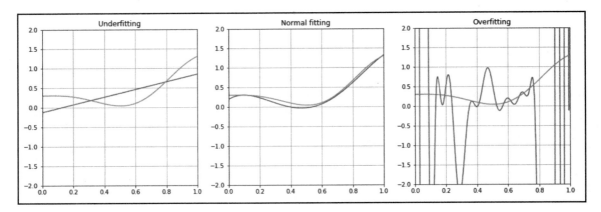

Example of underfitting (left), overfitting (right), and normal fitting (center)

An underfitted model usually has a high **bias**, which is defined as the difference between the expected value of the estimation of the parameters θ and the true ones:

$$Bias[\tilde{\theta}] = E[\tilde{\theta}] - \bar{\theta}$$

When the bias is null, the model is defined as **unbiased**. On the other hand, the presence of a bias means that the algorithm is not able to learn an acceptable representation of θ. In the first example of the previous graph, the straight line only has a negligible error in the neighborhoods of two points (about 0.3 and 0.8) and, as it's not able to change the slope, the bias will force the error to increase in all the other regions. Conversely, overfitting is often associated with a high **variance**, which is defined as follows:

$$Variance[\tilde{\theta}] = E\left[\left(\tilde{\theta} - E[\tilde{\theta}]\right)^2\right]$$

A high variance is often the result of a high capacity. The model now has the ability to oscillate changing its slope many times, but it can't behave as a simpler one anymore. The right-hand example in the previous graph shows an extremely complex curve that will probably fail to classify the majority of never-seen samples. Underfitting is easier to detect considering the prediction error, while overfitting may prove to be more difficult to discover as it could be initially considered the result of a perfect fitting. In fact, in a classification task, a high-variance model can easily learn the structure of the dataset employed in the training phase, but, due to the excess complexity, it can become *frozen* and hyperspecialized. It often means that it will manage never-seen samples with less accuracy, as their features cannot be recognized as variants of the samples belonging to a class. Every small modification is captured by the model, which can now adapt its separation surface with more freedom, so the similarities (which are fundamental for the generalization ability) are more difficult to detect. Cross-validation and other techniques that we're going to discuss in the following chapters can easily show how our model works with test samples that are never seen during the training phase. That way, it would be possible to assess the generalization ability in a broader context (remember that we're not working with all possible values, but always with a subset that should reflect the original distribution) and make the most reasonable decisions. The reader must remember that the real goal of a machine learning model is not to overfit the training set (we'll discuss this in the next chapter, Chapter 3, *Feature Selection and Feature Engineering*), but to work with never-seen samples, hence it's necessary to pay attention to the performance metrics before moving to a production stage.

Error measures and cost functions

In general, when working with a supervised scenario, we define a non-negative error measure e_m which takes two arguments (expected y_i and predicted output $\tilde{y}_i$) and allows us to compute a total error value over the whole dataset (made up of N samples):

$$Error_H = \sum_{i=1}^{N} e_m(\tilde{y}_i, y_i) \quad where \quad e_m \geq 0 \quad \forall \, \tilde{y}_i, y_i$$

This value is also implicitly dependent on the specific hypothesis H through the parameter set, and therefore optimizing the error implies finding an optimal hypothesis (considering the hardness of many optimization problems, this is not the absolute best one, but an acceptable approximation). In many cases, it's useful to consider the **mean square error** (**MSE**):

$$Error_H = \frac{1}{N} \sum_{i=1}^{N} (\tilde{y}_i - y_i)^2$$

Its initial value represents a starting point over the surface of an *n*-variable function. A generic training algorithm has to find the global minimum or a point quite close to it (there's always a tolerance to avoid an excessive number of iterations and a consequent risk of overfitting). This measure is also called **loss function** (or **cost function**) because its value must be minimized through an optimization problem. When it's easy to determine an element that must be maximized, the corresponding loss function will be its reciprocal.

Another useful loss function is called **zero-one-loss**, and it's particularly efficient for binary classifications (also for the one-versus-rest multiclass strategy):

$$L_{0/1_H}(\tilde{y}_i, y_i) = \begin{cases} 0 \ if \ \tilde{y}_i = y_i \\ 1 \ if \ \tilde{y}_i \neq y_i \end{cases}$$

This function is implicitly an indicator and can be easily adopted in loss functions based on the probability of misclassification.

A helpful interpretation of a generic (and continuous) loss function can be expressed in terms of potential energy:

$$Energy_H = \frac{1}{2} \sum_{i=1}^{N} e_m(\tilde{y}_i, y_i)^2$$

Chapter 2

The predictor is like a ball upon a rough surface: starting from a random point where energy (that is, error) is usually rather high, it must move until it reaches a stable equilibrium point where its energy (relative to the global minimum) is null. In the following diagram, there's a schematic representation of some different situations:

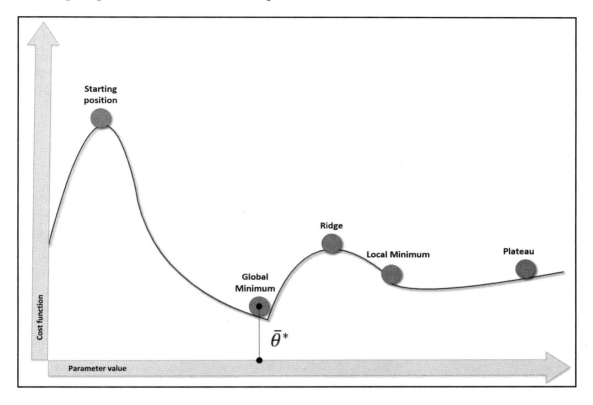

Energy curve with some peculiar points

Just like in the physical situation, the starting point is stable without any external perturbation, so to start the process, it's necessary to provide the initial kinetic energy. However, if such an energy is strong enough, then after descending over the slope, the ball cannot stop in the global minimum. The residual kinetic energy can be enough to overcome the ridge and reach the right-hand valley (plateau). If there are no other energy sources, the ball gets trapped in the plain valley and cannot move anymore. There are many techniques that have been engineered to solve this problem and avoid local minima. Another common problem (in particular, when working with deep models) is represented by the saddle points, which are characterized by a null gradient and positive semidefinite Hessian matrix. This means that the point is neither a local minimum, nor a maximum, but it can behave like a minimum moving in a direction and, like a maximum, moving in another one. In the following diagram, we can see a 3-dimensional example of this:

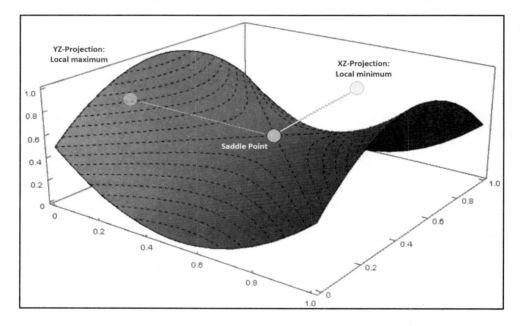

Example of a saddle point

Every situation must always be carefully analyzed to understand what level of residual energy (or error) is acceptable, or whether it's better to adopt a different strategy. In particular, all the models that are explicitly based on a loss/cost function can normally be trained using different optimization algorithms that can overcome the problems that a simpler solution isn't able to do. We're going to discuss some of them in the following chapters.

PAC learning

In many cases, machine learning seems to work seamlessly, but is there any way to formally determine the learnability of a concept? In 1984, the computer scientist L. Valiant proposed a mathematical approach to determine whether a problem is learnable by a computer. The name of this technique is **probably approximately correct** (**PAC**). The original formulation (you can read it in *A Theory of the Learnable*, Valiant L., *Communications of the ACM*, Vol. 27, No. 11, Nov. 1984) is based on a particular hypothesis, however, without a considerable loss of precision, we can think about a classification problem where an algorithm A has to learn a set of concepts. In particular, a concept is a subset of input patterns X that determine the same output element. Therefore, learning a concept (parametrically) means minimizing the corresponding loss function restricted to a specific class, while learning all possible concepts (belonging to the same universe) means finding the minimum of a global loss function.

However, given a problem, we have many possible (sometimes, theoretically infinite) hypotheses, and a probabilistic trade-off is often necessary. For this reason, we accept good approximations with high probability based on a limited number of input elements that are produced in polynomial time. Therefore, an algorithm A can learn the class C of all concepts (making them PAC learnable) if it's able to find a hypothesis H with a procedure $O(n^k)$ so that A, with a probability p, can classify all patterns correctly with a maximum allowed error m_e. This must be valid for all statistical distributions on X and for a number of training samples which must be greater than or equal to a minimum value depending only on p and m_e.

The constraint to computation complexity is not a secondary matter. In fact, we expect our algorithms to learn efficiently in a reasonable time when the problem is quite complex. An exponential time could lead to computational explosions when the datasets are too large or the optimization starting point is very far from an acceptable minimum. Moreover, it's important to remember the so-called **curse of dimensionality**, which is an effect that often happens in some models where training or prediction time is proportional (not always linearly) to the dimensions, so when the number of features increases, the performance of the models (that can be reasonable when the input dimensionality is small) gets dramatically reduced. Moreover, in many cases, in order to capture the full expressivity, it's necessary to have a very large dataset. However, without enough training data, the approximation can become problematic (this is called the **Hughes phenomenon**). For these reasons, looking for polynomial-time algorithms is more than a simple effort, because it can determine the success or the failure of a machine learning problem. For these reasons, in the following chapters, we're going to introduce some techniques that can be used to efficiently reduce the dimensionality of a dataset without a problematic loss of information.

Introduction to statistical learning concepts

Imagine that you need to design a spam-filtering algorithm, starting from this initial (over-simplistic) classification based on two parameters:

Parameter	Spam emails (X_1)	Regular emails (X_2)
p_1 - Contains > 5 blacklisted words	80	20
p_2 - Message length < 20 characters	75	25

We have collected 200 email messages (X) (for simplicity, we consider p_1 and p_2 as mutually exclusive) and we need to find a couple of probabilistic hypotheses (expressed in terms of p_1 and p_2), to determine the following:

$$p(Spam|h_{p_1}, h_{p_2})$$

We also assume the conditional independence of both terms (it means that h_{p1} and h_{p2} contribute in conjunction to spam in the same way as they would alone).

For example, we could think about rules (hypotheses) like so: "If there are more than five blacklisted words" or "If the message is less than 20 characters in length" then "the probability of spam is high" (for example, greater than 50%). However, without assigning probabilities, it's difficult to generalize when the dataset changes (like in a real-world antispam filter). We also want to determine a partitioning threshold (such as green, yellow, and red signals) to help the user in deciding what to keep and what to trash.

As the hypotheses are determined through the dataset X, we can also write (in a discrete form) the following:

$$p(Spam|X) = \sum_i p(Spam|h_{p_i}) p(h_{p_i}|X)$$

In this example, it's quite easy to determine the value of each term. However, in general, it's necessary to introduce the Bayes formula (which will be discussed in Chapter 6, *Naive Bayes and Discriminant Analysis*):

$$p(h_{p_i}|X) \propto p(X|h_{p_i}) p(h_{p_i})$$

The proportionality is necessary to avoid the introduction of the marginal probability *P(X)*, which only acts as a normalization factor (remember that, in a discrete random variable, the sum of all possible probability outcomes must be equal to 1).

In the previous equation, the first term is called **A Posteriori** (which comes after) probability, because it's determined by a marginal **Apriori** (which comes first) probability multiplied by a factor called **likelihood**. To understand the philosophy of such an approach, it's useful to take a simple example: tossing a fair coin. Everybody knows that the marginal probability of each face is equal to 0.5, but who decided that? It's a theoretical consequence of logic and probability axioms (a good physicist would say that it's never 0.5 because of several factors that we simply discard). After tossing the coin 100 times, we observe the outcomes and, surprisingly, we discover that the ratio between heads and tails is slightly different (for example, 0.46). How can we correct our estimation? The term called likelihood measures how much our actual experiments confirm the Apriori hypothesis and determine another probability (A Posteriori), which reflects the actual situation. The likelihood, therefore, helps us in correcting our estimation dynamically, overcoming the problem of a fixed probability.

In Chapter 6, *Naive Bayes and Discriminant Analysis*, which is dedicated to naive Bayes algorithms, we're going to discuss these topics deeply and implement a few examples with scikit-learn, however, it's useful to introduce two statistical learning approaches that are very diffused.

MAP learning

When selecting the right hypothesis, a Bayesian approach is normally one of the best choices, because it takes into account all the factors and, as we'll see, even if it's based on conditional independence, such an approach works perfectly when some factors are partially dependent. However, its complexity (in terms of probabilities) can easily grow because all terms must always be taken into account. For example, a real coin is a very short cylinder, so, in tossing a coin, we should also consider the probability of even (when the coin lies on its border). Let's say, it's 0.001. It means that we have three possible outcomes: *P(head) = P(tail) = (1.0 - 0.001)/2.0* and *P(even) = 0.001*. The latter event is obviously unlikely, but, in Bayesian learning, it must be considered (even if it'll be squeezed by the strength of the other terms).

An alternative is picking the most probable hypothesis in terms of A Posteriori probability:

$$h_{MAP} : p(h_{MAP}|X) = max_i \{p(h_{p_i}|X)\}$$

This approach is called MAP and it can really simplify the scenario when some hypotheses are quite unlikely (for example, in tossing a coin, a MAP hypothesis will discard *P(even)*). However, it still does have an important drawback: it depends on Apriori probabilities (remember that maximizing the A Posteriori implies also considering the Apriori). As Russel and Norvig (*Artificial Intelligence: A Modern Approach*, Russel S., Norvig P., Pearson) pointed out, this is often a delicate part of an inferential process, because there's always a theoretical background that can lead to a particular choice and exclude others. In order to rely only on data, it's necessary to have a different approach.

Maximum likelihood learning

We have defined likelihood as a filtering term in the Bayes formula. In general, it has the form of the following:

$$L(h_{p_i}|X) = p(X|h_{p_i})$$

Here, the first term expresses the actual likelihood of a hypothesis, given a dataset X. As you can imagine, in this formula, there are no more Apriori probabilities, so, maximizing it doesn't imply accepting a theoretical preferential hypothesis, nor considering unlikely ones. A very common approach, known as **Expectation Maximization** (**EM**), which is used in many algorithms (we're going to see an example in logistic regression), is split into two main parts:

- Determining a log-likelihood expression based on model parameters (they will be optimized accordingly). This is normally a proxy that depends on a set of parameters computed at time t θ_t.

- Maximizing it until the residual error is small enough. This operation is performed with respect to θ_t. Hence, at the next iteration, the proxy is computed with the new set of parameters.

This is a special case of a famous general algorithm of the same name (EM). A complete explanation is complex and it's beyond the scope of this book (it can be found in *Mastering Machine Learning Algorithms*, Bonaccorso G., Packt Publishing, 2018), however, it's easy to catch the main concepts.

A log-likelihood (normally called **L**) is a useful trick that can simplify gradient calculations. A generic likelihood expression is as follows:

$$L(h_i|X) = \prod_k p(X|h_i)$$

As all parameters are summarized inside h_i, the gradient is a complex expression that isn't very manageable. However, our goal is maximizing the likelihood, but it's easier to minimize its reciprocal:

$$max_i L(h_i|X) = min_i \frac{1}{L(h_i|X)} = min_i \frac{1}{\prod_k p(X|h_i)}$$

This can be turned into a very simple expression by applying a natural logarithm (which is monotonic):

$$max_i \, log \, L(h_i|X) = min_i \, (-log \, L(h_i|X)) = min_i \sum_k -log \, p(X|h_i)$$

The last term is a summation that can be easily derived and used in most of the optimization algorithms. Contrary to direct algorithms, EM works in an iterative fashion, alternating a step where the likelihood is computed using the current parameter estimation and another one where these parameters are chosen so as to maximize the expected log-likelihood. At the end of this process, we can find a set of parameters that provide the **maximum likelihood** (**ML**) without any strong statement about prior distributions. This approach can seem very technical, but its logic is really simple and intuitive. To understand how it works, I propose a simple exercise, which is part of the Gaussian mixture technique which is also discussed in *Mastering Machine Learning Algorithms, Bonaccorso G., Packt Publishing, 2018*.

Let's consider 100 points drawn from a Gaussian distribution with zero mean and a standard deviation equal to 2.0 (the Gaussian noise is made up of independent samples):

```
import numpy as np

np.random.seed(1000)

nb_samples = 100
X_data = np.random.normal(loc=0.0, scale=np.sqrt(2.0), size=nb_samples)
```

Important Elements in Machine Learning

The plot is shown in the following graph:

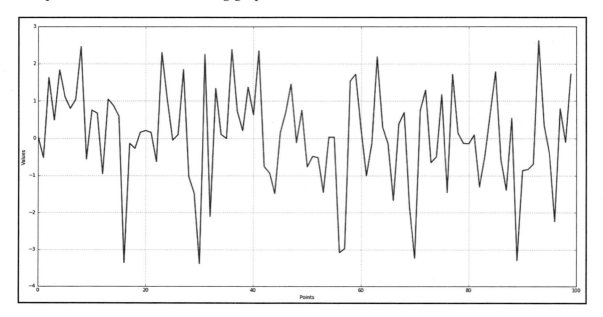

Points sampled from a Gaussian distribution

In this case, there's no need for a deep exploration (we know how they are generated), however, after restricting the hypothesis space to the Gaussian family (the most suitable considering only the graph), we'd like to find the best value for mean and variance. First of all, we need to compute the log-likelihood (which is rather simple thanks to the exponential function):

$$L(\mu, \sigma^2 | X) = \log p(X|\mu, \sigma^2) = \sum_i \log \frac{1}{\sqrt{2\pi\sigma^2}} e^{-\frac{(x_i-\mu)}{2\sigma^2}}$$

A simple Python implementation is provided next (for ease of use, there's only a single array which contains both mean (0) and variance (1)):

```
def negative_log_likelihood(v):
    l = 0.0
    f1 = 1.0 / np.sqrt(2.0 * np.pi * v[1])
    f2 = 2.0 * v[1]

    for x in X_data:
        l += np.log(f1 * np.exp(-np.square(x - v[0]) / f2))

    return -l
```

Then, we need to find its minimum (in terms of mean and variance) with any of the available methods (gradient descent or another numerical optimization algorithm). For example, using the `scipy` minimization function, we can easily get the following:

```
from scipy.optimize import minimize

minimize(fun=negative_log_likelihood, x0=np.array([0.0, 1.0]))

     fun: 181.6495875832933
hess_inv: array([[ 0.01959881, -0.00035322],
       [-0.00035322,  0.09799955]])
     jac: array([ 1.90734863e-06, 0.00000000e+00])
 message: 'Optimization terminated successfully.'
    nfev: 40
     nit: 8
    njev: 10
  status: 0
 success: True
       x: array([ 0.08052498, 2.21469485])
```

Important Elements in Machine Learning

A graph of the `negative_log_likelihood` function is plotted next. The global minimum of this function corresponds to an optimal likelihood given a certain distribution. It doesn't mean that the problem has been completely solved because the first step of this algorithm is determining an expectation, which must always be realistic. The likelihood function, however, is quite sensitive to wrong distributions because it can easily get close to zero when the probabilities are low. For this reason, ML learning is often preferable to MAP learning, which needs Apriori distributions, and can fail when they are not selected in the most appropriate way:

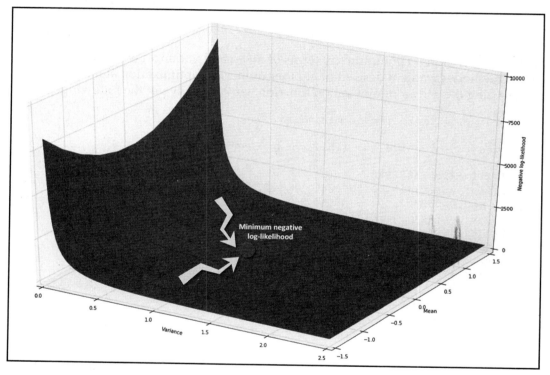

Trajectories from two different starting points and the MLE target

This approach has been applied to a specific distribution family (which is indeed very easy to manage), but it also works perfectly when the model is more complex. Of course, it's always necessary to have an initial awareness about how the likelihood should be determined, because more than one feasible family can generate the same dataset. In all of these cases, **Occam's razor** is the best way to proceed: the simplest hypothesis should be considered first. If it doesn't fit, an extra level of complexity can be added to our model. As we'll see, in many situations, the easiest solution is the winning one, and increasing the number of parameters or using a more detailed model can only add noise and a higher possibility of overfitting.

SciPy (https://www.scipy.org) is a set of high-end scientific and data-oriented libraries that are available for Python. It includes NumPy, Pandas, and many other useful frameworks. If you want to read more about Python scientific computing, refer to *Learning SciPy for Numerical and Scientific Computing Second Edition, Rojas S. J. R. G., Christensen E. A., Blanco-Silva F. J., Packt Publishing, 2017*.

Class balancing

When working with the majority of machine learning algorithms (in particular, supervised ones), it's important to train the model with a dataset containing almost the same number of elements for each class. Remember that our goal is training models that can generalize in the best way for all of the possible classes and supposes that we have a binary dataset containing 1,000 samples with a proportion (0.95, 0.05). There are many scenarios where this proportion is very common. For example, a spam detector can collect lots of spam emails, but it's much more difficult to have access to personally accepted emails. However, we can suppose that some users (a very small percentage) decided to share anonymous regular messages so that our dataset consists of 5% non-spam entries.

Important Elements in Machine Learning

Now, let's consider a static algorithm that always outputs the label 0 (for example, due to a bug). Which is the final accuracy? In a similar situation, we would expect a behavior similar to a random oracle based on a coin flip, but, not surprisingly, we discover that this algorithm reaches an accuracy of 95%! How is this possible? Simply because the dataset is unbalanced and the number of samples with label 1 is negligible with respect to the total dataset size. For this reason, before any training process, it's important to perform a basic descriptive analysis of the data, trying to correct the balancing problems.

Resampling with replacement

The most common way to address this issue is based on a resampling procedure. This approach is extremely simple, but, unfortunately, it has many drawbacks. Considering the previous example, we could decide to upsample the class 1, so as to match the number of samples belonging to class 0. However, we can only use the existing data and, after every sampling step, we restart from the original dataset (replacement). To better understand the procedure, let's suppose that we generate the dataset by employing the scikit-learn `make_classification` function (we are going to use it lots of times in the upcoming chapters):

```
from sklearn.datasets import make_classification

nb_samples = 1000
weights = (0.95, 0.05)

X, Y = make_classification(n_samples=nb_samples, n_features=2, n_redundant=0, weights=weights, random_state=1000)
```

We can check the shape of the two subarrays like so:

```
print(X[Y==0].shape)
print(X[Y==1].shape)

(946, 2)
(54, 2)
```

As expected (we have imposed a class weighting), the first class is dominant. In upsampling with replacement, we proceed by sampling from the dataset that's limited to the minor class (1), until we reach the desired number of elements. As we perform the operation with replacement, it can be iterated any number of times, but the resultant dataset will always contain points sampled from 54 possible values. In scikit-learn, it's possible to perform this operation by using the built-in `resample` function:

```
import numpy as np

from sklearn.utils import resample

X_1_resampled = resample(X[Y==1], n_samples=X[Y==0].shape[0], random_state=1000)

Xu = np.concatenate((X[Y==0], X_1_resampled))
Yu = np.concatenate((Y[Y==0], np.ones(shape=(X[Y==0].shape[0], ), dtype=np.int32)))
```

The function samples from the subarray `X[Y==1]`, generating the number of samples selected through the `n_samples` parameters (in our case, we have chosen to create two classes with the same number of elements). In the end, it's necessary to concatenate the subarray containing the samples with label `0` to the upsampled one (the same is also done with the labels). If we check the new shapes, we obtain the following:

```
print(Xu[Yu==0].shape)
print(Xu[Yu==1].shape)

(946, 2)
(946, 2)
```

As expected, the classes are now balanced. Clearly, the same procedure can be done by downsampling the major class, but this choice should be carefully analyzed because, in this case, there is an information loss. Whenever the dataset contains many redundant samples, this operation is less dangerous, but, as can often happen, removing valid samples can negatively impact the final accuracy because some feature values could never be seeded during the training phase. Even if resampling with replacement is not extremely powerful (as it cannot generate new samples), I normally suggest upsampling as a default choice. Downsampling the major class is only justified when the variance of the samples is very small (there are many samples around the mean), and it's almost always an unacceptable choice for uniform distributions.

SMOTE resampling

A much stronger approach has been proposed by Chawla et al. (in *SMOTE: Synthetic Minority Over-sampling Technique, Chawla N. V., Bowyer K. W., Hall L. O., Kegelmeyer W. P., Journal of Artificial Intelligence Research, 16/2002*). The algorithm is called **Synthetic Minority Over-sampling Technique** (**SMOTE**) and, contrary to the previous one, has been designed to generate new samples that are coherent with the minor class distribution. A full description of the algorithm is beyond the scope of this book (it can be found in the aforementioned paper), however, the main idea is to consider the relationships that exist between samples and create new synthetic points along the segments connecting a group of neighbors. Let's consider the following diagram:

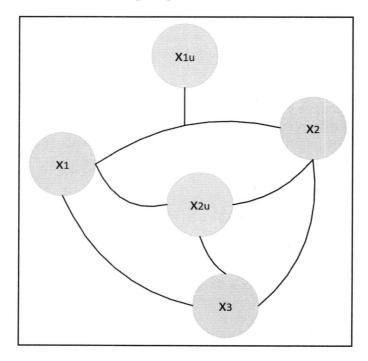

Example of a neighborhood containing three points (x_1, x_2, x_3) and two synthetic ones

The three points (x_1, x_2, x_3) belong to a minor class and are members of the same neighborhood (if the reader is not familiar with this concept, he/she can think of a group of points whose mutual distances are below a fixed threshold). SMOTE can upsample the class by generating the sample x_{1u} and x_{2u} and placing them on the segments, connecting the original samples. This procedure can be better understood by assuming that the properties of the samples are not changing below a certain neighborhood radius, hence it's possible to create synthetic variants that belong to the same original distribution. However, contrary to resampling with replacement, the new dataset has a higher variance, and a generic classifier can better find a suitable separation hypersurface.

In order to show how SMOTE works, we are going to employ a scikit-learn extension called imbalanced-learn (see the box at the end of this section), which implements many algorithms to manage this kind of problem. The balanced dataset (based on the one we previously generated) can be obtained by using an instance of the SMOTE class:

```
from imblearn.over_sampling import SMOTE

smote = SMOTE(random_state=1000)
X_resampled, Y_resampled = smote.fit_sample(X, Y)
```

The `fit_sample` method analyzes the original dataset and generates the synthetic samples from the minor class automatically. The most important parameters are as follows:

- `ratio` (default is `'auto'`): It determines which class must be resampled (acceptable values are `'minority'`, `'majority'`, `'all'`, and `'not minority'`). The meaning of each alternative is intuitive, but in general, we work by upsampling the `minority` class or, more seldom, by resampling (and balancing) the whole dataset.
- `k_neighbors` (default is 5): The number of neighbors to consider. Larger values yield more dense resamplings, and therefore I invite the reader to repeat this process by using `k_neighbors` equal to 2, 10, and 20, and compare the results. Remember that the underlying geometric structure is normally based on Euclidean distances, hence blobs are generally preferable to wireframe datasets. The value 5 is often a good trade-off between this condition and the *freedom* according to SMOTE in the generation process.

We can better understand this behavior by observing the following graph (we have upsampled the `minority` class with 5 neighbors):

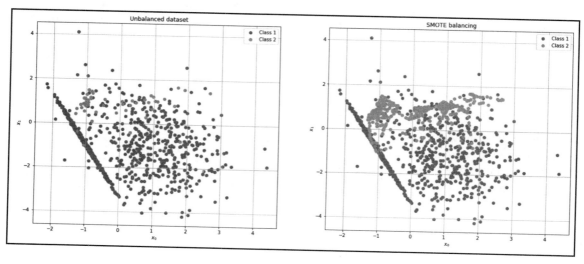

Unbalanced dataset (left) and the SMOTE balanced dataset (right)

As it's possible to see, the original dataset only has a few points belonging to the class **2** and they are all in the upper part of the graph. A resampling with replacement is able to increase the number of samples, but the resultant graph would be exactly the same, since the values are always taken from the existing set. On the other hand, SMOTE has generated the same number of samples by considering the neighborhoods (in this case, there's also an overlap in the original dataset). The final result is clearly acceptable and consistent with the data generating process. Moreover, it can help a classifier in finding out the optimal separating curve, which will probably be more centered (it could be a horizontal line passing through $x_1=0$) than the one associated with the unbalanced dataset.

 Imbalanced-learn (`http://contrib.scikit-learn.org/imbalanced-learn/stable/index.html`) can be installed by using the `pip install -U imbalanced-learn` command. It requires scikit-learn 0.19 or higher and is a growing project, so I suggest that the reader checks the website to discover all of its features and bug fixes.

Elements of information theory

A machine learning problem can also be analyzed in terms of information transfer or exchange. Our dataset is composed of n features, which are considered independent (for simplicity, even if it's often a realistic assumption) and drawn from n different statistical distributions. Therefore, there are n probability density functions $p_i(x)$ which must be approximated through other n $q_i(x)$ functions. In any machine learning task, it's very important to understand how two corresponding distributions diverge and what the amount of information we lose is when approximating the original dataset.

Entropy

The most useful measure in information theory (as well as in machine learning) is called **entropy**:

$$H(X) = -\sum_{x \in X} p(x) \, log \, p(x)$$

This value is proportional to the uncertainty of X and is measured in **bits** (if the logarithm has another base, this unit can change too). For many purposes, a high entropy is preferable, because it means that a certain feature contains more information. For example, in tossing a coin (two possible outcomes), H(X) = 1 bit, but if the number of outcomes grows, even with the same probability, H(X) also does because of a higher number of different values, and therefore has increased variability. It's possible to prove that for a Gaussian distribution (using natural logarithm):

$$H(X) = \frac{1}{2}\left(1 + log\left(2\pi\sigma^2\right)\right)$$

So, the entropy is proportional to the variance, which is a measure of the amount of information carried by a single feature. In the next chapter, *Chapter 3, Feature Selection and Feature Engineering,* we're going to discuss a method for feature selection based on the variance threshold. Gaussian distributions are very common, so this example can just be considered as a general approach to feature filtering: low variance implies low information level and a model could often discard all of those features.

In the following graph, there's a plot of H(X) for a Gaussian distribution expressed in **nats** (which is the corresponding unit measure when using natural logarithms):

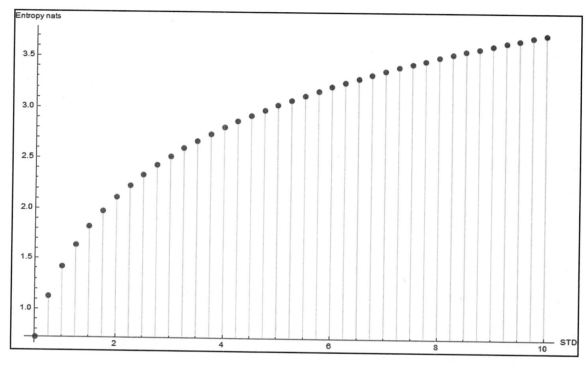

The entropy of a normal distribution as a function of the standard deviation

For example, if a dataset is made up of some features whose variance (here, it's more convenient to talk about standard deviation) is bounded between **8** and **10** and a few with **STD < 1.5**, the latter could be discarded with a limited loss in terms of information. These concepts are very important in real-life problems when large datasets must be cleaned and processed in an efficient way. Another important measure associated with the entropy is called **perplexity**:

$$Perplexity(p) = 2^{H(p)}$$

If the entropy is computed by using log_2, this measure is immediately associable with the total number of possible outcomes with the same probability. Whenever natural logarithm is used, the measure is still proportional to this quantity. Perplexity is very useful for assessing the amount of uncertainty in a distribution (it can be either a data distribution or the output of a model). For example, if we have a fair coin ($p_{head} = p_{tail} = 0.5$), the entropy is equal to the following:

$$H(p_{coin}) = -\frac{1}{2} log_2 \left(\frac{1}{2}\right) - \frac{1}{2} log_2 \left(\frac{1}{2}\right) = 1 \ bit$$

The perplexity is equal to 2, meaning that we expect 2 equally probable outcomes. As the total number of classes is 2, this situation is unacceptable, because we have no way to decide the right outcome (which is the entire goal of the experiment). On the other hand, if we have performed a previous analysis (imagine that the model has done it) and $p_{head} = 0.8$, the entropy becomes the following:

$$H(p_{coin}) = -0.1 \ log_2 (0.1) - 0.9 \ log_2 (0.9) \approx 0.47 \ bit$$

The perplexity drops to about *1.38* because, obviously, one outcome is not more likely than the other one.

Cross-entropy and mutual information

If we have a target probability distribution *p(x)*, which is approximated by another distribution *q(x)*, a useful measure is **cross-entropy** between *p* and *q* (we are using the discrete definition as our problems must be solved by using numerical computations):

$$H(p, q) = -\sum_{x \in X} p(x) \ log \ q(x)$$

If the logarithm base is 2, it measures the number of bits requested to decode an event drawn from *P* when using a code optimized for *Q*. In many machine learning problems, we have a source distribution and we need to train an estimator to be able to correctly identify the class of a sample. If the error is null, *P = Q* and the cross-entropy is minimum (corresponding to the entropy *H(P)*). However, as a null error is almost impossible when working with *Q*, we need to *pay* a price of *H(P, Q)* bits to determine the right class starting from a prediction. Our goal is often to minimize it so as to reduce this *price* under a threshold that cannot alter the predicted output if not paid. In other words, think about a binary output and a sigmoid function: we have a threshold of *0.5* (this is the maximum *price* we can pay) to identify the correct class using a step function (*0.6 -> 1, 0.1 -> 0, 0.4999 -> 0*, and so on). As we're not able to pay this *price*, since our classifier doesn't know the original distribution, it's necessary to reduce the cross-entropy under a tolerable noise-robustness threshold (which is always the smallest achievable one).

In order to understand how a machine learning approach is performing, it's also useful to introduce a **conditional** entropy or the uncertainty of *X* given the knowledge of *Y*:

$$H(X|Y) = - \sum_{x \in X, y \in Y} p(x,y) \, log \frac{p(x,y)}{p(y)}$$

Through this concept, it's possible to introduce the idea of **mutual information**, which is the amount of information shared by both variables and therefore, the reduction of uncertainty about *X* provided by the knowledge of *Y*:

$$I(X;Y) = H(X) - H(X|Y)$$

Intuitively, when *X* and *Y* are independent, they don't share any information. However, in machine learning tasks, there's a very tight dependence between an original feature and its prediction, so we want to maximize the information shared by both distributions. If the conditional entropy is small enough (so *Y* is able to describe *X* quite well), the mutual information gets close to the marginal entropy *H(X)*, which measures the amount of information we want to learn.

An interesting learning approach based on the information theory, called **minimum description length** (**MDL**), is discussed in *The Minimum Description Length Principle in Coding and Modeling, Barron A., Rissanen J., Yu B., IEEE Transaction on Information Theory, Vol. 44/6, 10/1998*.

Divergence measures between two probability distributions

Let's suppose you have a discrete data generating process $p_{data}(x)$ and a model that outputs a probability mass function $q(x)$. In many machine learning tasks, the goal is to tune up the parameter so that $q(x)$ becomes as similar to p_{data} as possible. A very useful measure is the **Kullback-Leibler divergence**:

$$D_{KL}(p_{data}||q) = \sum_x p_{data}(x) \, log \, \frac{p_{data}(x)}{q(x)}$$

This quantity (also known as **information gain**) expresses the gain obtained by using the approximation $q(x)$ instead of the original data generating process. It's immediate to see that if $q(x) = p_{data}(x) \Rightarrow D_{KL}(p_{data}||q) = 0$, while it's greater than 0 (unbounded) when there's a mismatch. Manipulating the previous expression, it's possible to gain a deeper understanding:

$$D_{KL}(p_{data}||q) = \sum_x p_{data}(x) \, log \, \frac{p_{data}(x)}{q(x)} =$$
$$= \sum_x p_{data}(x) \, log \, p_{data} - \sum_x p_{data}(x) \, log \, q(x) =$$
$$= -H(p_{data}) + H(p_{data}, q)$$

Important Elements in Machine Learning

The first term is the negative entropy of the data generating process, which is a constant. The second one, instead, is the cross-entropy between the two distributions. Hence, if we minimize it, we also minimize the Kullback-Leibler divergence. In the following chapters, we are going to analyze some models based on this loss function, which is extremely useful in multilabel classifications. Therefore, I invite the reader who is not familiar with these concepts to fully understand their rationale before proceeding.

In some cases, it's preferable to work with a symmetric and bounded measure. The **Jensen-Shannon divergence** is defined as follows:

$$D_{JS}(p_{data}||q) = \frac{1}{2}D_{KL}\left(p_{data}||\frac{p_{data}+q}{2}\right) + \frac{1}{2}D_{KL}\left(q||\frac{p_{data}+q}{2}\right)$$

Even if it seems more complex, its behavior is equivalent to the Kullback-Leibler divergence, with the main difference that the two distributions can now be swapped and, above all, $0 \leq D_{JS}(p_{data}||q) \leq log(2)$. As it's expressed as a function of $D_{KL}(p_{data}||q)$, it's easy to prove that its minimization is proportional to a cross-entropy reduction. The Jensen-Shannon divergence is employed in advanced models (such as **Generative Adversarial Networks (GANs)**, but it's helpful to know it because it can be useful in some tasks where the Kullback-Leibler divergence can lead to an overflow (that is, when the two distributions have no overlaps and $q(x) \to 0$).

Summary

In this chapter, we have introduced some main concepts about machine learning. We started with some basic mathematical definitions so that we have a clear view of data formats, standards, and certain kinds of functions. This notation will be adopted in the rest of the chapters in this book, and it's also the most diffused in technical publications. We also discussed how scikit-learn seamlessly works with multi-class problems, and when a strategy is preferable to another.

The next step was the introduction of some fundamental theoretical concepts regarding learnability. The main questions we tried to answer were: how can we decide if a problem can be learned by an algorithm and what is the maximum precision we can achieve? PAC learning is a generic but powerful definition that can be adopted when defining the boundaries of an algorithm. A PAC learnable problem, in fact, is not only manageable by a suitable algorithm, but is also fast enough to be computed in polynomial time. Then, we introduced some common statistical learning concepts, in particular, the MAP and ML learning approaches. The former tries to pick the hypothesis that maximizes the A Posteriori probability, while the latter optimizes the likelihood, looking for the hypothesis that best fits the data. This strategy is one of the most diffused in many machine learning problems because it's not affected by Apriori probabilities and it's very easy to implement in many different contexts. We also gave a physical interpretation of a loss function as an energy function. The goal of a training algorithm is to always try to find the global minimum point, which corresponds to the deepest valley in the error surface. At the end of this chapter, there was a brief introduction to information theory and how we can reinterpret our problems in terms of information gain and entropy. Every machine learning approach should work to minimize the amount of information needed to start from prediction and recover original (desired) outcomes.

In the next chapter, `Chapter 3`, *Feature Selection and Feature Engineering,* we're going to discuss the fundamental concepts of feature engineering, which is the first step in almost every machine learning pipeline. We're going to show you how to manage different kinds of data (numerical and categorical) and how it's possible to reduce dimensionality without a dramatic loss of information.

3
Feature Selection and Feature Engineering

Feature engineering is the first step in a machine learning pipeline and involves all the techniques adopted to clean existing datasets, increase their signal-noise ratio, and reduce their dimensionality. Most algorithms have strong assumptions about the input data, and their performances can be negatively affected when raw datasets are used. Moreover, the data is seldom isotropic; there are often features that determine the general behavior of a sample, while others that are correlated don't provide any additional pieces of information. So, it's important to have a clear view of a dataset and know the most common algorithms used to reduce the number of features or select only the best ones.

In particular, we are going to discuss the following topics:

- How to work with scikit-learn built-in datasets and split them into training and test sets
- How to manage missing and categorical features
- How to filter and select the features according to different criteria
- How to normalize, scale, and whiten a dataset
- How to reduce the dimensionality of a dataset using the **Principal Component Analysis (PCA)**
- How to perform a PCA on non-linear datasets
- How to extract independent components and create dictionaries of atoms
- How to visualize high-dimensional datasets using the t-SNE algorithm

scikit-learn toy datasets

scikit-learn provides some built-in datasets that can be used for prototyping purposes because they don't require very long training processes and offer different levels of complexity. They're all available in the `sklearn.datasets` package and have a common structure: the data instance variable contains the whole input set X while the target contains the labels for classification or target values for regression. For example, considering the Boston house pricing dataset (used for regression), we have the following:

```
from sklearn.datasets import load_boston

boston = load_boston()
X = boston.data
Y = boston.target

print(X.shape)
(506, 13)

print(Y.shape)
(506,)
```

In this case, we have 506 samples with 13 features and a single target value. In this book, we're going to use it for regressions and the MNIST handwritten digit dataset (`load_digits()`) for classification tasks. scikit-learn also provides functions for creating dummy datasets from scratch: `make_classification()`, `make_regression()`, and `make_blobs()` (which are particularly useful for testing cluster algorithms). They're very easy to use and, in many cases, it's the best choice to test a model without loading more complex datasets.

Visit http://scikit-learn.org/stable/datasets/ for further information.
The MNIST dataset provided by scikit-learn is limited for obvious reasons. If you want to experiment with the original one, refer to the website managed by Y. LeCun, C. Cortes, and C. Burges: http://yann.lecun.com/exdb/mnist/. Here, you can download a full version made up of 70,000 handwritten digits that are already split into training and test sets.

Creating training and test sets

When a dataset is large enough, it's a good practice to split it into training and test sets, the former to be used for training the model and the latter to test its performances. In the following diagram, there's a schematic representation of this process:

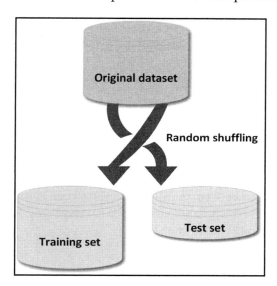

Training/test set split process schema

There are two main rules in performing such an operation:

- Both datasets must reflect the original distribution
- The original dataset must be randomly shuffled before the split phase in order to avoid a correlation between consequent elements

With scikit-learn, this can be achieved by using the `train_test_split()` function:

```
from sklearn.model_selection import train_test_split

X_train, X_test, Y_train, Y_test = train_test_split(X, Y, test_size=0.25, random_state=1000)
```

Feature Selection and Feature Engineering

The `test_size` parameter (as well as `training_size`) allows you to specify the percentage of elements to put into the test/training set. In this case, the ratio is 75 percent for training and 25 percent for the test phase. In classic machine learning tasks, this is a common ratio, however, in deep learning, it can be useful to extend the training set to 98% of the total data. The right split percentage depends on the specific scenario. In general, the rule of thumb is that the training set (as well as the test set) must represent the whole data generating process. Sometimes, it's necessary to rebalance the classes, as shown in the previous chapter, Chapter 2, *Important Elements in Machine Learning*, but it's extremely important not to exclude potential samples from the training phase – otherwise, the model won't ever be able to generalize correctly.

Another important parameter is `random_state`, which can accept a NumPy `RandomState` generator or an integer seed. In many cases, it's important to provide reproducibility for the experiments, so it's also necessary to avoid using different seeds and, consequently, different random splits:

In machine learning projects, I always suggest using the same random seed (it can also be 0 or completely omitted) or define a global `RandomState` which can be passed to all requiring functions. In this way, the reproducibility is guaranteed.

```
from sklearn.utils import check_random_state

rs = check_random_state(1000)

X_train, X_test, Y_train, Y_test = train_test_split(X, Y, test_size=0.25, random_state=rs)
```

In this way, if the seed is kept equal, all experiments have to lead to the same results and can be easily reproduced in different environments by other scientists.

For further information about NumPy random number generation, visit https://docs.scipy.org/doc/numpy/reference/generated/numpy.random.RandomState.html.

Managing categorical data

In many classification problems, the target dataset is made up of categorical labels that cannot immediately be processed by every algorithm. An encoding is needed, and scikit-learn offers at least two valid options. Let's consider a very small dataset made of 10 categorical samples with 2 features each:

```
import numpy as np

X = np.random.uniform(0.0, 1.0, size=(10, 2))
Y = np.random.choice(('Male', 'Female'), size=(10))

print(X[0])
array([ 0.8236887 ,  0.11975305])

print(Y[0])
'Male'
```

The first option is to use the `LabelEncoder` class, which adopts a dictionary-oriented approach, associating to each category label a progressive integer number, that is, an index of an instance array called `classes_`:

```
from sklearn.preprocessing import LabelEncoder

le = LabelEncoder()
yt = le.fit_transform(Y)

print(yt)
[0 0 0 1 0 1 1 0 0 1]

le.classes_array(['Female', 'Male'], dtype='|S6')
```

The inverse transformation can be obtained in this simple way:

```
output = [1, 0, 1, 1, 0, 0]
decoded_output = [le.classes_[int(i)] for i in output]
print(decoded_output)
['Male', 'Female', 'Male', 'Male', 'Female', 'Female']
```

Feature Selection and Feature Engineering

This approach is simple and works well in many cases, but it has a drawback: all labels are turned into sequential numbers. A classifier that works with real values will then consider similar numbers according to their distance, without any concern for semantics. For this reason, it's often preferable to use so-called one-hot encoding, which binarizes the data. For labels, this can be achieved by using the `LabelBinarizer` class:

```
from sklearn.preprocessing import LabelBinarizer

lb = LabelBinarizer()
Yb = lb.fit_transform(Y)
array([[1],
       [0],
       [1],
       [1],
       [1],
       [1],
       [0],
       [1],
       [1],
       [1]])

lb.inverse_transform(Yb)
array(['Male', 'Female', 'Male', 'Male', 'Male', 'Male', 'Female', 'Male',
       'Male', 'Male'], dtype='|S6')
```

In this case, each categorical label is first turned into a positive integer and then transformed into a vector where only one feature is 1 while all the others are 0. This means, for example, that using a softmax distribution with a peak corresponding to the main class can be easily turned into a discrete vector where the only non-null element corresponds to the right class. For example, consider the following code:

```
import numpy as np

Y = lb.fit_transform(Y)
array([[0, 1, 0, 0, 0],
       [0, 0, 0, 1, 0],
       [1, 0, 0, 0, 0]])

Yp = model.predict(X[0])
array([[0.002, 0.991, 0.001, 0.005, 0.001]])

Ypr = np.round(Yp)
array([[ 0.,  1.,  0.,  0.,  0.]])

lb.inverse_transform(Ypr)
array(['Female'], dtype='|S6')
```

Another approach to categorical features can be adopted when they're structured like a list of dictionaries (not necessarily dense; they can have values, but only for a few features). For example:

```
data = [
    { 'feature_1': 10.0, 'feature_2': 15.0 },
    { 'feature_1': -5.0, 'feature_3': 22.0 },
    { 'feature_3': -2.0, 'feature_4': 10.0 }
]
```

In this case, scikit-learn offers the `DictVectorizer` and `FeatureHasher` classes; they both produce sparse matrices of real numbers that can be fed into any machine learning model. The latter has a limited memory consumption and adopts **MurmurHash3** (refer to https://en.wikipedia.org/wiki/MurmurHash for further information), which is general-purpose (non-cryptographic, hence has a non-collision-resistant hash function with a 32-bit output). The code for these two methods is as follows:

```
from sklearn.feature_extraction import DictVectorizer, FeatureHasher

dv = DictVectorizer()
Y_dict = dv.fit_transform(data)

Y_dict.todense()
matrix([[ 10.,   15.,    0.,    0.],
        [ -5.,    0.,   22.,    0.],
        [  0.,    0.,   -2.,   10.]])

dv.vocabulary_
{'feature_1': 0, 'feature_2': 1, 'feature_3': 2, 'feature_4': 3}

fh = FeatureHasher()
Y_hashed = fh.fit_transform(data)

Y_hashed.todense()
matrix([[ 0.,  0.,  0., ...,  0.,  0.,  0.],
        [ 0.,  0.,  0., ...,  0.,  0.,  0.],
        [ 0.,  0.,  0., ...,  0.,  0.,  0.]])
```

In both cases, I suggest you read the original scikit-learn documentation so that you know all of the possible options and parameters.

Feature Selection and Feature Engineering

When working with categorical features (normally converted into positive integers through `LabelEncoder`), it's also possible to filter the dataset in order to apply one-hot encoding by using the `OneHotEncoder` class. In the following example, the first feature is a binary index that indicates `'Male'` or `'Female'`:

```
from sklearn.preprocessing import OneHotEncoder

data = [
    [0, 10],
    [1, 11],
    [1, 8],
    [0, 12],
    [0, 15]
]

oh = OneHotEncoder(categorical_features=[0])
Y_oh = oh.fit_transform(data)

>>> Y_oh.todense()
matrix([[ 1.,    0.,   10.],
        [ 0.,    1.,   11.],
        [ 0.,    1.,    8.],
        [ 1.,    0.,   12.],
        [ 1.,    0.,   15.]])
```

Considering that these approaches increase the number of values (also exponentially with binary versions), all the classes adopt sparse matrices based on the SciPy implementation. See https://docs.scipy.org/doc/scipy-0.18.1/reference/sparse.html for further information.

Managing missing features

Sometimes, a dataset can contain missing features, so there are a few options that can be taken into account:

- Removing the whole line
- Creating a submodel to predict those features
- Using an automatic strategy to input them according to the other known values

The first option is the most drastic one and should only be considered when the dataset is quite large, the number of missing features is high, and any prediction could be risky. The second option is much more difficult because it's necessary to determine a supervised strategy to train a model for each feature and, finally, to predict their value. Considering all pros and cons, the third option is likely to be the best choice. scikit-learn offers the `Imputer` class, which is responsible for filling the holes using a strategy based on the mean (default choice), median, or frequency (the most frequent entry will be used for all the missing ones).

The following snippet shows an example that's using the three approaches (the default value for a missing feature entry is `NaN`, however, it's possible to use a different placeholder through the `missing_values` parameter):

```
from sklearn.preprocessing import Imputer

data = np.array([[1, np.nan, 2], [2, 3, np.nan], [-1, 4, 2]])

imp = Imputer(strategy='mean')
imp.fit_transform(data)
array([[ 1. ,  3.5,  2. ],
       [ 2. ,  3. ,  2. ],
       [-1. ,  4. ,  2. ]])

imp = Imputer(strategy='median')
imp.fit_transform(data)
array([[ 1. ,  3.5,  2. ],
       [ 2. ,  3. ,  2. ],
       [-1. ,  4. ,  2. ]])

imp = Imputer(strategy='most_frequent')
imp.fit_transform(data)
array([[ 1.,  3.,  2.],
       [ 2.,  3.,  2.],
       [-1.,  4.,  2.]])
```

Data scaling and normalization

A generic dataset (we assume here that it is always numerical) is made up of different values that can be drawn from different distributions, having different scales and, sometimes, there are also outliers. A machine learning algorithm isn't naturally able to distinguish among these various situations, and therefore, it's always preferable to standardize datasets before processing them. A very common problem derives from having a non zero mean and a variance greater than 1. In the following graph, there's a comparison between a raw dataset and the same dataset scaled and centered:

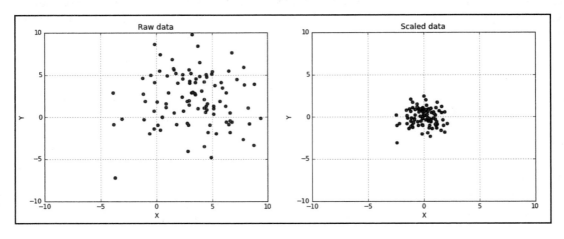

Original dataset (left) and the scaled one (right)

This result can be achieved by using the StandardScaler class (which implements feature-wise scaling):

```
from sklearn.preprocessing import StandardScaler

ss = StandardScaler()
scaled_data = ss.fit_transform(data)
```

It's possible to specify if the scaling process must include both mean and standard deviation by using the with_mean=True/False and with_std=True/False parameters (by default, they're both active). If you need a more powerful scaling feature, with a superior control on outliers and the possibility to select a quantile range, there's also the RobustScaler class. Here are some examples with different quantiles:

```
from sklearn.preprocessing import RobustScaler

rb1 = RobustScaler(quantile_range=(15, 85))
```

```
scaled_data1 = rb1.fit_transform(data)

rb1 = RobustScaler(quantile_range=(25, 75))
scaled_data1 = rb1.fit_transform(data)

rb2 = RobustScaler(quantile_range=(30, 60))
scaled_data2 = rb2.fit_transform(data)
```

The results are shown in the following graphs:

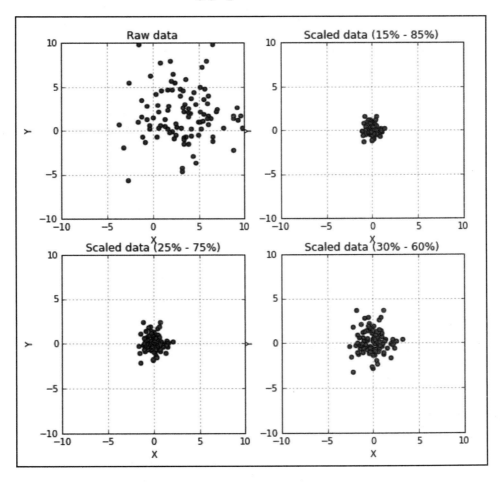

Original dataset (upper-left), together with different quantile scaling plots

Other options include `MinMaxScaler` and `MaxAbsScaler`, which scale data by removing elements that don't belong to a given range (the former) or by considering a maximum absolute value (the latter).

The scikit-learn library also provides a class for per-sample normalization, `Normalizer`. It can apply *Max*, *L1*, and *L2* norms to each element of a dataset. In a Euclidean space, these transformations are defined in the following way:

$$\begin{cases} Max\ norm: & \|X\|_{max} = \dfrac{X}{|max_i X|} \\ L1\ norm: & \|X\|_{L1} = \dfrac{X}{\sum_i |x_i|} \\ L2\ norm: & \|X\|_{L2} = \dfrac{X}{\sqrt{\sum_i |x_i|^2}} \end{cases}$$

An example of every normalization is shown in the following snippet:

```
from sklearn.preprocessing import Normalizer

data = np.array([1.0, 2.0])

n_max = Normalizer(norm='max')
n_max.fit_transform(data.reshape(1, -1))
[[ 0.5, 1. ]]

n_l1 = Normalizer(norm='l1')
n_l1.fit_transform(data.reshape(1, -1))
[[ 0.33333333,  0.66666667]]

n_l2 = Normalizer(norm='l2')
n_l2.fit_transform(data.reshape(1, -1))
[[ 0.4472136 ,  0.89442719]]
```

Whitening

The `StandardScaler` class operates in a feature-wise fashion, however, sometimes, it's useful to transform the whole dataset so as to force it to have an identity covariance matrix (to improve the performances of many algorithms that are sensitive to the number of independent components):

$$\frac{1}{N} X^T X \Rightarrow I$$

The goal is to find a transformation matrix A (called the **whitening matrix**) so that the new dataset $X' = XA^T$ has an identity covariance C' (we are assuming that X is zero-centered or, alternatively, it has zero mean). The procedure is quite simple (it can be found in *Mastering Machine Learning Algorithms, Bonaccorso G., Packt Publishing, 2018*), but it requires some linear algebra manipulations. In this context, we directly provide the final result. It's possible to prove that the **singular value decomposition** (**SVD**) (see the section on PCA) of $C' \propto X'^T X' = AX^T X A^T$ is:

$$AX^T X A^T = V\Omega V^T = I \Rightarrow AA^T = V\Omega^{-1}V^T$$

Ω is a diagonal matrix containing the eigenvalues of C', while the columns of V are the eigenvectors (the reader who is not familiar with this concept can simply consider this technique as a way to factorize a matrix). Considering the last equation, we obtain the whitening matrix:

$$AA^T = \left(V\Omega^{-\frac{1}{2}}\right)\left(\Omega^{-\frac{1}{2}}V^T\right) \Rightarrow A = V\Omega^{-\frac{1}{2}}$$

This is the whitening matrix for the dataset X. As Ω is diagonal, the square root of the inverse is as follows:

$$\Omega^{-\frac{1}{2}} = \begin{pmatrix} \sqrt{\frac{1}{\omega_{11}}} & \cdots & 0 \\ \vdots & \ddots & \vdots \\ 0 & \cdots & \sqrt{\frac{1}{\omega_{nn}}} \end{pmatrix}$$

The Python code to perform this operation (based on the SVD provided by NumPy) is as follows:

```
import numpy as np

def zero_center(X):
    return X - np.mean(X, axis=0)

def whiten(X, correct=True):
    Xc = zero_center(X)
    _, L, V = np.linalg.svd(Xc)
    W = np.dot(V.T, np.diag(1.0 / L))
    return np.dot(Xc, W) * np.sqrt(X.shape[0]) if correct else 1.0
```

In the following graph, there's an example based on the original dataset with a non-diagonal covariance matrix:

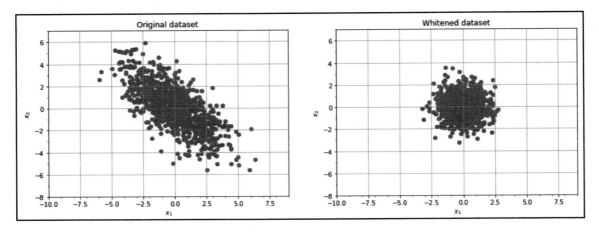

Original dataset (left), and the whitened one (right)

As it's possible to see, the whitened dataset is symmetric and its covariance matrix is approximately an identity (Xw is the output of the transformation):

```
import numpy as np

print(np.cov(Xw.T))

[[1.00100100e+00 5.26327952e-16]
 [5.26327952e-16 1.00100100e+00]]
```

To better understand the impact, I invite the reader to test this function with other datasets, comparing the performances of the same algorithms. It's important to remember that a whitening procedure works with the whole dataset, hence it could be unacceptable whenever the training process is performed online. However, in the majority of cases, it can be employed without restrictions and can provide a concrete benefit for both the training speed and accuracy.

Feature selection and filtering

An unnormalized dataset with many features contains information proportional to the independence of all features and their variance. Let's consider a small dataset with three features, generated with random Gaussian distributions:

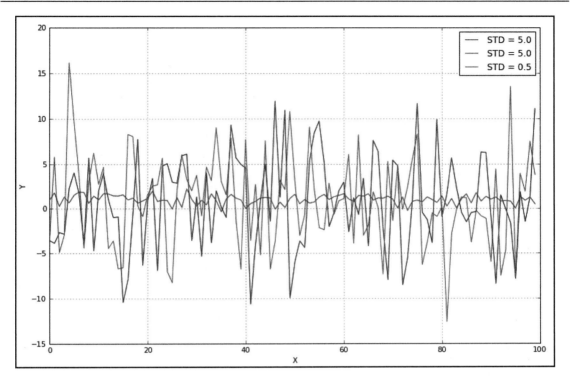

Sample dataset containing three Gaussian features with different standard deviations

Even without further analysis, it's obvious that the central line (with the lowest variance) is almost constant and doesn't provide any useful information. Recall from Chapter 2, *Important Elements in Machine Learning*, that the entropy $H(X)$ is quite small, while the other two variables carry more information. A variance threshold is, therefore, a useful approach to remove all those elements whose contribution (in terms of variability and so, information) is under a predefined level. The scikit-learn library provides the VarianceThreshold class which can easily solve this problem. By applying it to the previous dataset, we get the following result:

```
from sklearn.feature_selection import VarianceThreshold

X[0:3, :]
array([[-3.5077778 , -3.45267063,  0.9681903 ],
       [-3.82581314,  5.77984656,  1.78926338],
       [-2.62090281, -4.90597966,  0.27943565]])

vt = VarianceThreshold(threshold=1.5)
X_t = vt.fit_transform(X)
```

Feature Selection and Feature Engineering

```
X_t[0:3, :]
array([[-0.53478521, -2.69189452],
       [-5.33054034, -1.91730367],
       [-1.17004376,  6.32836981]])
```

The third feature has been completely removed because its variance is under the selected threshold (1.5, in this case).

There are also many univariate methods that can be used in order to select the best features according to specific criteria based on F-tests and p-values, such as chi-square or **Analysis of Variance** (**ANOVA**). However, their discussion is beyond the scope of this book and the reader can find further information in *Statistics for Machine Learning, Dangeti P., Packt Publishing, 2017*.

Two examples of feature selection that use the SelectKBest class (which selects the best *K* high-score features) and the SelectPercentile class (which selects only a subset of features belonging to a certain percentile) are shown next. It's possible to apply them both to regression and classification datasets, being careful to select appropriate score functions:

```
from sklearn.datasets import load_boston, load_iris
from sklearn.feature_selection import SelectKBest, SelectPercentile, chi2, f_regression

regr_data = load_boston()
print(regr_data.data.shape)
(506L, 13L)

kb_regr = SelectKBest(f_regression)
X_b = kb_regr.fit_transform(regr_data.data, regr_data.target)

print(X_b.shape)
(506L, 10L)

print(kb_regr.scores_)
array([  88.15124178,   75.2576423 ,  153.95488314,   15.97151242,
        112.59148028,  471.84673988,   83.47745922,   33.57957033,
         85.91427767,  141.76135658,  175.10554288,   63.05422911,
        601.61787111])

class_data = load_iris()
print(class_data.data.shape)
(150L, 4L)

perc_class = SelectPercentile(chi2, percentile=15)
X_p = perc_class.fit_transform(class_data.data, class_data.target)

print(X_p.shape)
```

```
(150L, 1L)

print(perc_class.scores_)
array([  10.81782088,    3.59449902,  116.16984746,   67.24482759])
```

 For further details about all scikit-learn score functions and their usage, visit http://scikit-learn.org/stable/modules/feature_selection.html#univariate-feature-selection.

Principal Component Analysis

In many cases, the dimensionality of the input dataset X is high and so is the complexity of every related machine learning algorithm. Moreover, the information is seldom spread uniformly across all the features and, as discussed in the previous chapter, Chapter 2, *Important Elements in Machine Learning*, there will be high-entropy features together with low-entropy ones, which, of course, don't contribute dramatically to the final outcome. This concept can also be expressed by considering a fundamental assumption of semi-supervised learning, called the **manifold assumption**. It states (without a formal proof, as it's an empirical hypothesis) that data with high dimensionality normally lies on lower-dimensional manifolds. If the reader is not familiar with the concept of a manifold, it's not necessary for our purpose to provide a complete rigorous definition. It's enough to say that a manifold is a non-Euclidean space that locally behaves like a Euclidean one. The simplest example is a spherical surface. Picking a point, it's possible to find a circle whose curvature is negligible and thus it resembles a circle on a bidimensional Euclidean space. In many machine learning and data science tasks, n-dimensional vectors often spread uniformly over all $\Re^n$. This means that there are dimensions whose associated information is almost null. However, this is not an error or a bad condition, because, in many cases, it's simpler to add dimensions to better summarize the data. On the other hand, an algorithm is agnostic to this kind of descriptive choice and, in many cases, it performs better when the dimensionality is reduced to its minimum value (cfr. Hughes phenomenon).

In general, if we consider a Euclidean space, we get the following:

$$X = \{\bar{x}_1, \bar{x}_2, \ldots, \bar{x}_n\} \quad where \quad \bar{x}_i = \sum_{j=1}^{m} x_i^{(j)} \bar{e}_j$$

Feature Selection and Feature Engineering

So, each point is expressed by using an orthonormal basis made of m linearly independent vectors (in general, this is the canonical base associated to $\Re^m$). Now, considering a dataset X, a natural question arises: how is it possible to reduce m without a drastic loss of information?

Let's consider the following graph (without any particular interpretation):

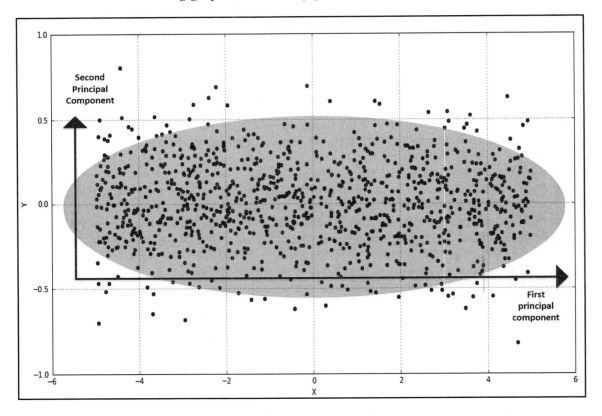

Sample bidimensional dataset with a larger horizontal variance

It doesn't matter which distributions generated $X=(x, y)$, however, the variance of the horizontal component is clearly higher than that of the vertical one. As discussed previously, this means that the amount of information provided by the first component is higher than the second and, for example, the x axis is stretched horizontally, keeping the vertical one fixed, and so the distribution becomes similar to a segment where the depth has lower and lower importance.

In order to assess how much information is carried by each component, and the correlation among them, a useful tool is the covariance matrix (if the dataset has zero mean, we can use the correlation matrix):

$$C = \frac{1}{n}X^T X = \frac{1}{n}\begin{pmatrix} \sum_j x_j^{(1)} x_j^{(1)} & \cdots & \sum_j x_j^{(1)} x_j^{(m)} \\ \vdots & \ddots & \vdots \\ \sum_j x_j^{(m)} x_j^{(1)} & \cdots & \sum_j x_j^{(m)} x_j^{(m)} \end{pmatrix} = \frac{1}{n}\begin{pmatrix} \sigma_{11}^2 & \cdots & \sigma_1 \sigma_m \\ \vdots & \ddots & \vdots \\ \sigma_m \sigma_1 & \cdots & \sigma_m^2 \end{pmatrix}$$

C is symmetric and a positive semidefinite, so all the eigenvalues are non-negative, but what's the meaning of each value? On the diagonal, there are the variances of each component σ_i^2 which encodes the deviation around the mean. Larger variances imply more spread components, while $\sigma_i^2 = 0$ is equivalent to a situation where all the values are equal to the mean. The elements $\sigma_i \sigma_j$ are the cross-variances (or cross-correlations) between components and their values are determined by the linear dependencies. Through an affine transform, it's possible to find a new basis where all these elements are null, as shown in the following graph:

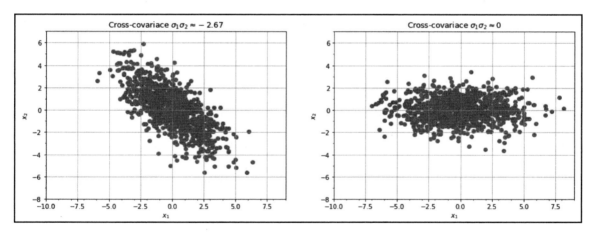

Dataset with non-null cross-covariances (left), and the rotated (with null-covariance) version (right)

In the first case, the dataset is rotated, therefore the components x_1 and x_2 become dependent on each other with respect to the canonical basis. Applying the affine transform and obtaining a new basis, the components become uncorrelated. Considering the previous example (without rotations), the covariance matrix is as follows:

$$C = \begin{pmatrix} 8.31 & -0.02 \\ -0.02 & 0.06 \end{pmatrix}$$

As expected, the horizontal variance is quite a bit higher than the vertical one. Moreover, the other values are close to zero. If you remember the definition and, for simplicity, remove the mean term, they represent the cross-correlation between couples of components. It's obvious that, in our example, X and Y are uncorrelated (they're orthogonal), but in real-life examples, there could be features that present a residual cross-correlation. In terms of information theory, it means that knowing Y gives us some information about X (which we already know), so they share information that is indeed doubled. So, our goal is to also decorrelate X while trying to reduce its dimensionality. The magnitude of the eigenvalues of C is proportional to the amount of information carried by the referenced component, and the associated eigenvector provides us with the direction of such a component. Hence, let's suppose that we compute the sorted set of eigenvalues Λ:

$$\Lambda = \{\lambda_1 \geqslant \lambda_2 \geqslant \ldots \geqslant \lambda_m\}$$

If we select $g < m$ values, the following relation holds for the relative subspace Λ_g:

$$\Lambda_g \subseteq \Lambda \ \ with \ \ dim(\Lambda_g) \leqslant dim(\Lambda)$$

At this point, we can build a transformation matrix based on the first g eigenvectors:

$$W = (\bar{w}_{\lambda_1}, \bar{w}_{\lambda_2}, \ldots, \bar{w}_{\lambda_g}) \ \ where \ \ W \in \mathbb{R}^{m \times g}$$

As $X \in \mathfrak{R}^{n \times m}$, we obtain the projections with the matrix multiplication:

$$X_g = XW \ \ where \ \ X_g \in \mathbb{R}^{n \times g}$$

So, it's possible to project the original feature vectors into this new (sub)space, where each component carries a portion of total variance and where the new covariance matrix is decorrelated to reduce useless information sharing (in terms of correlation) among different features. In other words, as we are considering the eigenvectors associated with the top g eigenvalues, the transformed dataset will be rotated so that W is now the canonical basis and all non-diagonal components are forced to zero. Before showing a practical example, it's useful to briefly discuss how to perform the eigendecomposition of C. Clearly, this operation can be done directly, but it's simpler to employ the SVD. For a generic matrix $A \in \Re^{m \times n}$ (the same result is valid for complex matrices, but in our case, we are going to use only real-valued ones), the SVD is a factorization that expresses A in the following way:

$$A = U\Sigma V^T \text{ where } U \in \mathbb{R}^{m \times m}, V \in \mathbb{R}^{n \times n} \text{ and } \Sigma \in diag_{\mathbb{R}}(m \times n)$$

We are going to employ this transformation in other contexts, but for now it's enough to say that the elements of Σ (called singular values) are the square root of the eigenvalues of both AA^T and A^TA, and that the columns of V (called right singular vectors) are the eigenvectors of A^TA. Therefore, we don't need to compute the covariance matrix. If we apply the SVD to dataset X, we can immediately obtain the transformation matrix. The scikit-learn library provides an implementation called `TruncatedSVD`, which performs the SVD, limited to the first top eigenvalues, and this is the most efficient way to perform a PCA. However, for simplicity, there's also the `PCA` class, which can do all of this in a very smooth way (without the need of any further operations):

```
from sklearn.datasets import load_digits
from sklearn.decomposition import PCA

digits = load_digits()
```

A screenshot with a few random MNIST handwritten digits follows:

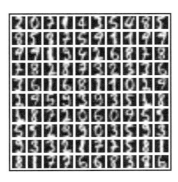

Examples of MNIST digits

Feature Selection and Feature Engineering

Each image is a vector of 64 unsigned int (8 bit) numbers (0, 255), so the initial number of components is indeed 64. However, the total amount of black pixels is often predominant and the basic signs needed to write 10 digits are similar, so it's reasonable to assume both high cross-correlation and a low variance on several components. Trying with 36 principal components, we get the following:

```
pca = PCA(n_components=36, whiten=True)
X_pca = pca.fit_transform(digits.data / 255)
```

In order to improve performance, all integer values are normalized into the range *[0, 1]* and, through the `whiten=True` parameter, the variance of each diagonal component is scaled to 1 (the covariance matrix is already decorrelated). As the official scikit-learn documentation says, this process is particularly useful when an isotropic distribution is needed for many algorithms to perform efficiently. It's possible to access the explained variance ratio through the `explained_variance_ratio_` instance variable, which shows which part of the total variance is carried by every single component:

```
print(pca.explained_variance_ratio_)
array([ 0.14890594,  0.13618771,  0.11794594,  0.08409979,  0.05782415,
        0.0491691 ,  0.04315987,  0.03661373,  0.03353248,  0.03078806,
        0.02372341,  0.02272697,  0.01821863,  0.01773855,  0.01467101,
        0.01409716,  0.01318589,  0.01248138,  0.01017718,  0.00905617,
        0.00889538,  0.00797123,  0.00767493,  0.00722904,  0.00695889,
        0.00596081,  0.00575615,  0.00515158,  0.00489539,  0.00428887,
        0.00373606,  0.00353274,  0.00336684,  0.00328029,  0.0030832 ,
        0.00293778])
```

A plot for the example of MNIST digits is shown next. The left-hand graph represents the **variance ratio**, while the right-hand one is the **cumulative variance**. It can be immediately seen how the first components are normally the most important ones in terms of information, while the following ones provide details that a classifier could also discard:

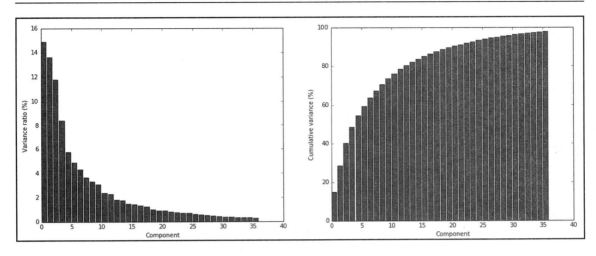

Explained variance per component (left), and the cumulative variance (right)

As expected, the contribution to the total variance decreases dramatically starting from the fifth component, so it's possible to reduce the original dimensionality without an unacceptable loss of information, which could drive an algorithm to learn incorrect classes. In the preceding graphs, you can see the same handwritten digits but rebuilt by using the first 36 components with whitening and normalization between 0 and 1. To obtain the original images, we need to inverse-transform all the new vectors and project them into the original space:

```
X_rebuilt = pca.inverse_transform(X_pca)
```

The result is shown in the following screenshot:

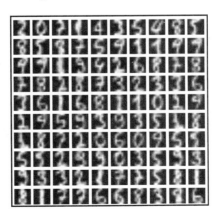

MNIST digits obtained by rebuilding them from the principal components

Feature Selection and Feature Engineering

This process can also partially denoise the original images by removing residual variance, which is often associated with noise or unwanted contributions (almost every calligraphy distorts some of the structural elements that are used for recognition).

I suggest that the reader try different numbers of components (using the explained variance data) and also `n_components='mle'`, which implements an automatic selection of the best dimensionality using a technique based on a Bayesian model selection (for further information, please refer to *Automatic Choice of Dimensionality for PCA, Minka T.P, NIPS 2000: 598-604, 2000*).

As explained, scikit-learn solves the PCA problem with SVD. It's possible to control the algorithm through the `svd_solver` parameter, whose values are `'auto'`, `'full'`, `'arpack'`, `'randomized'`. **ARnoldi PACKage (ARPACK)** implements a truncated SVD. `randomized` is based on an approximate algorithm that drops many singular vectors and can achieve very good performances with high-dimensional datasets where the actual number of components is sensibly smaller.

Non-Negative Matrix Factorization

When the dataset X is made up of non-negative elements, it's possible to use **Non-Negative Matrix Factorization** (**NNMF**) instead of standard PCA. This algorithm optimizes a loss function (alternatively on W and H) based on the Frobenius norm:

$$L = \|X - WH\|_{Frob} \quad where \quad \|A\|_{Frob} = \sqrt{\sum_i \sum_j |a_{ij}|^2}$$

If *dim(X)* = *n* × *m*, then *dim(W)* = *n* × *p* and *dim(H)* = *p* × *m* with *p* equal to the number of requested components (the `n_components` parameter), which is normally smaller than the original dimensions *n* and *m*. In some implementations, the loss function is squared and L1/L2 penalties are applied to both the W and H matrices:

$$L = \frac{1}{2}\|X - WH\|_{Frob}^2 + \alpha_1\|W\|_1 + \alpha_2\|H\|_1 + \alpha_3\|W\|_{Frob}^2 + \alpha_3\|H\|_{Frob}^2$$

As we are going to discuss in the next chapter, Chapter 4, *Regression Algorithms*, the regularization terms can improve the accuracy and increase the sparseness of *W* and *H* (*L1* penalty) by selecting all those elements that are rather greater than zero and forcing all the other ones to become null. The final reconstruction is purely additive and it has been shown that it's particularly efficient for images or text where there are normally no non-negative elements. The algorithms start by setting random values for both *W* and *H*, and then it alternates between two optimization steps:

1. *W* is kept constant and *L* is minimized with respect to *H*
2. *H* is kept constant and *L* is minimized with respect to *W*

The procedure is repeated until both *W* and *H* become stable. Clearly, the goal is to find a basis, *H*, so that $X \approx WH$. The main difference with PCA is that we don't have any constraint on the explained variance, but the matrices must be non-negative.

In the following snippet, there's an example of using the Iris dataset. The `init` parameter can assume different values (see the documentation) that determine how the data matrix is initially processed. A random choice is for non-negative matrices that are only scaled (no SVD is performed):

```
from sklearn.datasets import load_iris
from sklearn.decomposition import NMF

iris = load_iris()
print(iris.data.shape)
(150L, 4L)

nmf = NMF(n_components=3, init='random', l1_ratio=0.1)
Xt = nmf.fit_transform(iris.data)

print(nmf.reconstruction_err_)
1.8819327624141866

print(iris.data[0])
array([ 5.1,  3.5,  1.4,  0.2])

print(Xt[0])
array([ 0.20668461,  1.09973772,  0.0098996 ])

print(nmf.inverse_transform(Xt[0]))
array([ 5.10401653,  3.49666967,  1.3965409 ,  0.20610779])
```

Feature Selection and Feature Engineering

NNMF, together with other factorization methods, will be very useful for more advanced techniques, such as recommendation systems and topic modeling, hence I invite the reader to understand the logic of these methods. The factorization of a matrix allows you to express it as a product of two entities that *share* one dimension. For example, if X is made up of user vectors where each component refers to a product, W will associate users to some unknown (called latent) factors, and H will associate the latent factors with the products. As we are going to discuss in Chapter 12, *Introduction to Recommendation Systems*, this assumption becomes a fundamental concept. The latent factor behave as a basis in linear algebra. Hence it's a reference system that allows finding the components of any point in a specific subspace.

 NNMF is very sensitive to its parameters (in particular, initialization and regularization), so I suggest reading the original documentation for further information: http://scikit-learn.org/stable/modules/generated/sklearn.decomposition.NMF.html.

Sparse PCA

The scikit-learn library provides different PCA variants that can solve particular problems. I do suggest reading the original documentation. However, I'd like to mention SparsePCA, which allows exploiting the natural sparsity of data while extracting principal components. If you think about the handwritten digits or other images that must be classified, their initial dimensionality can be quite high (a 10 x 10 image has 100 features). However, applying a standard PCA selects only the average most important features, assuming that every sample can be rebuilt using the same components. Simplifying this, this is equivalent to the following:

$$y_R = c_1 y_{R_1} + c_2 y_{R_2} + \ldots + c_g y_{R_g} = \sum_i c_i y_{R_i}$$

On the other hand, we can always use a limited number of components, but without the limitation given by a dense projection matrix. This can be achieved by using sparse matrices (or vectors), where the number of non zero elements is quite low. In this way, each element can be rebuilt using its specific components (in most cases, they will always be the most important), which can include elements normally discarded by a dense PCA. The previous expression now becomes the following:

$$y_R = \left(c_1 y_{R_1} + c_2 y_{R_2} + \ldots + c_g y_{R_g}\right) + \left(0 \cdot y_{R_{g+1}} + 0 \cdot y_{R_{g+2}} + \ldots + 0 \cdot y_{R_m}\right) = \sum_{i=1}^{g} c_i y_{R_i}$$

Here, the non-null components have been put into the first block (they don't have the same order as the previous expression), while all the other zero terms have been separated. In terms of linear algebra, the vectorial space now has the original dimensions. However, using the power of sparse matrices (provided by scipy.sparse), scikit-learn can solve this problem much more efficiently than a classical PCA.

The following snippet shows a SparsePCA with 60 components. In this context, they're usually called *atoms*, and the amount of sparsity can be controlled via *L1*-norm regularization (higher alpha parameter values lead to more sparse results). This approach is very common in classification algorithms and will be discussed in the upcoming *Atom extraction and dictionary learning* section and also in the following chapters:

```
from sklearn.decomposition import SparsePCA

spca = SparsePCA(n_components=60, alpha=0.1)
X_spca = spca.fit_transform(digits.data / 255)

print(spca.components_.shape)
(60L, 64L)
```

As we are going to discuss, the extraction of sparse components is very helpful whenever it's necessary to rebuild each sample, starting from a finite subset of features. In this particular case, we are not considering the explained variance anymore, but we are focusing on finding out all those elements that can be used as distinctive atoms. For example, we could employ a SparsePCA (which is equivalent to a dictionary learning scikit-learn) to the MNIST dataset in order to find the geometrical base components (such as vertical/horizontal lines) without caring about the actual dimensionality reduction (which becomes a secondary goal, in this case).

For further information about SciPy sparse matrices, visit https://docs.scipy.org/doc/scipy-0.18.1/reference/sparse.html.

Kernel PCA

We're going to discuss kernel methods in Chapter 7, *Support Vector Machines*, however, it's useful to mention the KernelPCA class, which performs a PCA with non-linearly separable datasets. This approach is analogous to a standard PCA with a particular preprocessing step. Contrary to what many people can expect, a non-linear low-dimensional dataset can often become linearly separable when projected onto special higher-dimensional spaces. On the other hand, we prefer not to introduce a major complexity that could even result in an unsolvable problem. The **kernel trick** can help us in achieving this goal without the burden of hard, non-linear operations. The complete mathematical proofs are beyond the scope of this book, however, we can define a **kernel** as a real-valued vectorial function that has the following property:

$$K(\bar{x}_i, \bar{x}_j) = \Psi(\bar{x}_i)^T \Psi(\bar{x}_j)$$

In particular, if $x_i \in \Re^n$, the vectorial transformation $\Psi(x_i)$ projects the original vector onto a larger subspace. In fact, its codomain is normally $\Re^p$ with $p > n$. The advantage of the kernel $K(x_i, x_j)$ derives from the fact that computing it with two values, x_i and x_j, its output corresponds to the dot product of the transformed vectors, therefore no additional computations are required. Now, let's consider the standard PCA transformation for a single sample ($x_i \in \Re^{n \times 1}$):

$$\bar{x}_i^g = \bar{x}_i^T W$$

When using a Kernel PCA, we need to re-express the basis matrix W using the projection $W^g = \Psi(x)v^g$, where v^g are the eigenvectors computed considering the kernel transformation K instead of the covariance matrix C (remember that the dot product of a scalar is equivalent to a standard multiplication) and $\Psi(x)$ is a matrix whose rows are $\Psi(x_i)$ for all x_i. Therefore, the Kernel PCA transformation becomes the following:

$$\Psi(\bar{x}_i^g) = \Psi(\bar{x}_i)^T \Psi(\bar{x}) W^g$$

However, the dot products can be efficiently computed using the kernel trick, and so the final computation corresponds to the extraction of the components of the dataset projected on the higher-dimensional subspace (where it's likely to be linearly separable). In other words, we are assuming that in a different (high-dimensional) space, there are components whose explained variance is negligible with respect to other ones. To understand this concept, let's consider a dataset made up of a circle with a blob inside it:

```
from sklearn.datasets import make_circles

Xb, Yb = make_circles(n_samples=500, factor=0.1, noise=0.05)
```

The graphical representation is shown in the following graph. In this case, a classic PCA approach isn't able to capture the non-linear dependency of existing components (the reader can verify that the projection is equivalent to the original dataset). However, looking at the samples and using polar coordinates, it's easy to separate the two sets, only considering the radius (in this case, the explained variance is no more a function of both components because the radius is almost constant):

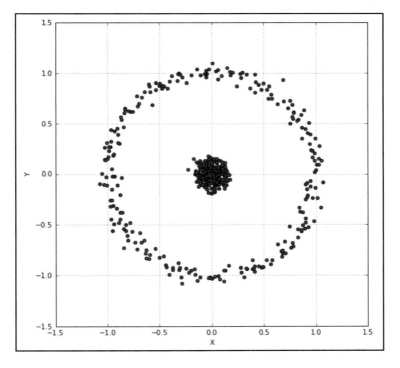

A non-linearly separable dataset containing a dense blob surrounded by a circular set

Feature Selection and Feature Engineering

Considering the structure of the dataset, it's possible to investigate the behavior of a PCA with a radial basis function kernel (which is sensitive to the distance from the origin):

$$K(\bar{x}_i, \bar{x}_j) = e^{-\frac{\|\bar{x}_i - \bar{x}_j\|^2}{2\sigma^2}}$$

As the default value for gamma is 1.0/number of features (for now, consider that this parameter as inversely proportional to the variance of a Gaussian σ^2), we need to increase it to capture the external circle. A value of 1.0 is enough:

```
from sklearn.decomposition import KernelPCA

kpca = KernelPCA(n_components=2, kernel='rbf', fit_inverse_transform=True, gamma=1.0)
X_kpca = kpca.fit_transform(Xb)
```

The X_transformed_fit_ instance variable will contain the projection of our dataset into the new space. Plotting it, we get the following:

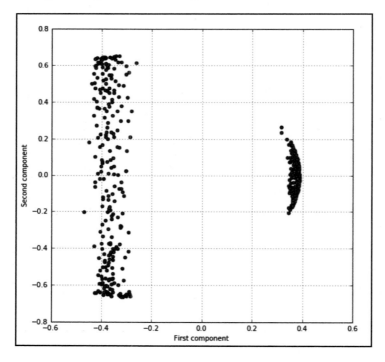

Kernel PCA graph where the left-hand strip represents the outer circle, while the smaller half-moon represents the inner blob

The plot shows a separation just like expected, and it's also possible to see that the points belonging to the central blob have a curve distribution because they are more sensitive to the distance from the center.

The Kernel PCA is a powerful instrument when we think of our dataset as being made up of elements that can be a function of components (in particular, radial-basis or polynomials), but we aren't able to determine a linear relationship among them.

 For more information about the different kernels supported by scikit-learn, visit `http://scikit-learn.org/stable/modules/metrics.html#linear-kernel`.

Independent Component Analysis

Sometimes, it's useful to process the data in order to extract components that are uncorrelated and independent. To better understand this scenario, let's suppose that we record two people while they sing different songs. The result is clearly very noisy, but we know that the stochastic signal could be decomposed into the following:

$$s(t) = s_1(t) + s_2(t) + n(t)$$

The first two terms are the single music sources (modeled as stochastic processes), while $n(t)$ is additive Gaussian noise. Our goal is to find $s_1(t) + n_1(t)$ and $s_2(t) + n_1(t)$ in order to remove one of the two sources (with a part of the additive noise that cannot be filtered out). Performing this task using a standard PCA is very difficult because there are no constraints on the independence of the components. This problem has been widely studied by Hyvärinen and Oja (please refer to *Independent Component Analysis: Algorithms and Applications, Hyvarinen A., Oja E., Neural Networks 13/2000*), who considered the conditions that must be enforced in order to extract independent components. The most important thing is that we cannot assume Gaussian distributions, but instead, we need to look for more peaked ones (with heavy tails). The reason is based on the statistical property of the Gaussian distribution and, in particular, we know that if some random variables $s_i \sim N(\mu, \sigma)$ are independent, they are also jointly normally distributed (that is, $(s_1, s_2, ..., s_i) \sim N(\mu, \Sigma)$).

Feature Selection and Feature Engineering

Considering the previous example, this result implies that if s_1 and s_2 are assumed to be independent and normal, the additive signal will be represented by a multivariate normal distribution. Unfortunately, this result makes the research of independent components impossible. As this book is more introductory, I omit the proofs (which can be rather complex, but can be found in the aforementioned paper and in *Mastering Machine Learning Algorithms, Bonaccorso G., Packt Publishing, 2018*). However, in order to understand the logic, it's helpful to consider a statistical index called **Kurtosis** (which corresponds to the fourth moment):

$$Kurt(X) = E_{x \sim X}\left[\left(\frac{x - \mu_x}{\sigma_x}\right)^4\right]$$

This measure is normally referred to as a normal distribution, whose value is $Kurt(N) = 3$. All the distributions with a Kurtosis greater than 3 are called super-gaussian and they are very peaked around the mean. This implies that the probability is high, but only by a very small region, and it's close to zero elsewhere. An example of this is shown in the following graph:

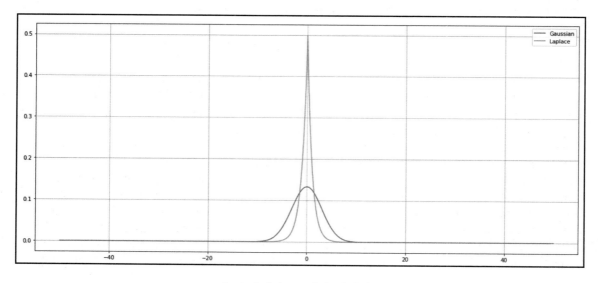

Gaussian distribution versus Laplace distribution

The Laplace distribution is an example of super-gaussian distribution with *Kurtosis* = 6. As it's possible to see, it's very peaked around the mean and the tails are heavy. The probability density function is as follows:

$$p_{Laplace}(x) = \frac{1}{2\beta} e^{-\frac{|x-\mu|}{\beta}}$$

Hence, the exponent now has degree 1 (while Gaussians are squared), and the joint probability of different components can never be represented as a Gaussian distribution. Assuming that the sources are modeled with such kinds of distributions (there are many other alternatives) means that we are implicitly saying that they are separable and independent (this also forces sparsity). Hence, if the final samples are obtained with super-gaussian additive contributions, such an approach can help us in identifying and isolating the sources because they are no more jointly normally distributed and, moreover, the probability distributions of the single component have a potential smaller overlap.

However, even if this measure is very powerful, it's extremely sensitive to outliers (due to the fourth power), and therefore the authors proposed an alternative algorithm that they called **Fast Independent Component Analysis (FastICA)**. A full description of the mathematical background is beyond the scope of this book, but I want to show how to extract 256 independent components from the original MNIST dataset (which can be downloaded by using the scikit-learn built-in `fetch_mldata('MNIST original')` function).

The first step is loading and zero-centering the dataset (this algorithm is very sensitive to symmetric data):

```
import numpy as np

from sklearn.datasets import fetch_mldata

def zero_center(Xd):
    return Xd - np.mean(Xd, axis=0)

digits = fetch_mldata('MNIST original')
X = zero_center(digits['data'].astype(np.float64))
np.random.shuffle(X)
```

Feature Selection and Feature Engineering

At this point, we can instantiate the `FastICA` class, selecting `n_components=256` and `max_iter=5000` (sometimes, it's necessary to raise this value to guarantee the convergence), and train the model:

```
from sklearn.decomposition import FastICA

fastica = FastICA(n_components=256, max_iter=5000, random_state=1000)
fastica.fit(X)
```

At the end of the process, the `components_` instance variable contains all the extracted values. In the following screenshot, we can see the representation of the first 64 independent components:

64 independent components extracted from the MNIST dataset

Contrary to PCA, we can now detect many small *building blocks* that represent portions of the digits that can be shared among many of them. For example, it's possible to detect small horizontal strokes, as well as several rounded angles in different positions. These components are almost independent, and a sample can be rebuilt as a weighted sum of them. In particular, when the number of components is very large, the resultant weight vector will be sparse because many components won't be used in all of the samples. I invite the reader to repeat this exercise with different n_component values, trying to understand when the components contain overlaps (so they are made up of a group of independent parts) and when they begin to become *basic elements*. From a neuroscientific viewpoint, the fundamental mechanism employed by the brain to decode images is based on neurons that are only receptive to specific patterns. Therefore, using this algorithm, it is possible to implement a filtering mechanism that splits a complex source (for example, an image) into blocks that can elicit a selective response. For example, it's possible to analyze a dataset and exclude all of those samples where a particular component has a large weight (such as background noises or echoes) or, conversely, it's possible to select only those elements that are helpful for a specific task (for example, containing strong, vertical components).

Atom extraction and dictionary learning

Dictionary learning is a technique that allows you to rebuild a sample starting from a sparse dictionary of atoms (similar to principal components, but without constraints about the independence). Conventionally, when the dictionary contains a number of elements less than the dimensionality of the samples m, it is called *under-complete*, and on the other hand it's called *over-complete* when the number of atoms is larger (sometimes much larger) than m.

In *Online Dictionary Learning for Sparse Coding, Mairal J., Bach F., Ponce J., Sapiro G., Proceedings of the 29th International Conference on Machine Learning, 2009*, there's a description of the same online strategy adopted by scikit-learn, which can be summarized as a double optimization problem.

Let's suppose that we have a dataset, X:

$$X = \{\bar{x}_1, \bar{x}_2, \ldots, \bar{x}_n\} \quad \text{where} \quad \bar{x}_i \in \mathbb{R}^m$$

Our goal is to find both a dictionary D (that is a matrix containing m k-dimensional vectors) and a set of vectorial weights for each sample:

$$D \in \mathbb{R}^{m \times k} \quad \text{and} \quad A = \{\bar{\alpha}_1, \bar{\alpha}_2, \ldots, \bar{\alpha}_m\} \quad \text{where} \quad \bar{\alpha}_i \in \mathbb{R}^k$$

After the training process, an input vector x_i can be computed as follows:

$$\bar{x}_i = D\bar{\alpha}_i$$

The reader can recognize an approach which is very similar to what we discussed for PCA. In fact, each sample x_i is re-expressed using a transformation matrix and a projection α_i.

The optimization problem (which involves both D and alpha vectors) can be expressed as the minimization of the following loss function:

$$L(D, A) = \frac{1}{2} \sum_i \|\bar{x}_i - D\bar{\alpha}_i\|_2^2 + c\|\bar{\alpha}_i\|_1$$

Here, the parameter c controls the level of sparsity (which is proportional to the strength of *L1* normalization). This problem, analogously to NNMF, can be solved by alternating the least square variable until a stable point is reached. Dictionary learning can be used to perform a Sparse PCA, but, in some cases, it can be helpful to consider over-complete dictionaries with more complex datasets. For example, a set of pictures representing cars can be decomposed into all possible components, considering that a single sample must be rebuilt using a large number of atoms (even $k \gg m$). In fact, when $k \ll n$, the components are forced to represent more complex parts, while $k \gg n$ allows extracting all the basic constituent elements.

A very important element to consider is the L1-normalization. Unfortunately, it's rather difficult to prove why this term forces sparseness (we are going to show some other examples in the next chapter), however, a very simple explanation can be obtained considering the L0-norm, which (rather informally) is the number of non-null elements of a vector. Minimizing the L0-norm means forcing all the components to reduce their magnitude until they become as close as possible to zero. Unfortunately, this norm is not differentiable and it cannot be used in an optimization process. The L1-norm is hence the best candidate because, using Lp-norms, the *amount of sparseness* decreases while p increases. This is not a rigorous proof, but it can be useful to understand the logic of this approach. In many other algorithms, the presence of an *L1* penalty indicates that a specific quantity is desired to be sparse.

In scikit-learn, we can implement such an algorithm with the
DictionaryLearning class (using the usual MNIST datasets), where n_components, as usual, determines the number of atoms:

```
from sklearn.decomposition import DictionaryLearning

dl = DictionaryLearning(n_components=36, fit_algorithm='lars',
transform_algorithm='lasso_lars')
X_dict = dl.fit_transform(digits.data)
```

A plot of each atom (component) is shown in the following screenshot:

A set of atoms extracted from the MNIST dataset

In this example, the number of components is smaller than the dimensionality, and the atoms contain more complex shapes. I invite the reader to repeat this exercise with different values, observing the resultant plot and comparing the granularity of the features.

This process can be very long on low-end machines. In such a case, I suggest limiting the number of samples to 20 or 30 or that you work with smaller datasets.

Visualizing high-dimensional datasets using t-SNE

Before ending this chapter, I want to introduce the reader to a very powerful algorithm called **t-Distributed Stochastic Neighbor Embedding** (**t-SNE**), which can be employed to visualize high-dimensional dataset also in 2D plots. In fact, one the hardest problems that every data scientist has to face is to understand the structure of a complex dataset without the support of graphs. This algorithm has been proposed by Van der Maaten and Hinton (in *Visualizing High-Dimensional Data Using t-SNE, Van der Maaten L.J.P., Hinton G.E., Journal of Machine Learning Research 9 (Nov), 2008*), and can be used to reduce the dimensionality trying to preserve the internal relationships. A complete discussion is beyond the scope of this book (but the reader can check out the aforementioned paper and *Mastering Machine Learning Algorithms, Bonaccorso G., Packt Publishing, 2018*), however, the key concept is to find a low-dimensional distribution so as to minimize the Kullback-Leibler divergence between it and the data generating process. Clearly, a few mathematical tricks are needed to carry out this task efficiently, but the only fundamental concept that is useful for a beginner is the *perplexity*, which is defined as follows:

$$Perplexity(p) = 2^{H(p)}$$

In other words, the perplexity is directly proportional to the entropy of distribution (when computed using $log_2(x)$, it becomes a sort of inverse operation) and by minimizing it, we reduce the uncertainty about the target distribution, given the original data generating process. Therefore, all t-SNE implementations require that you specify the target perplexity (normally in the range 10 ÷ 30), and the computational time grows inversely proportionally.

To show the power of this algorithm, let's suppose that we want to visualize the digits of the MNIST dataset in a bidimensional plot to check whether similar digits are closer than different ones. Let's start by loading and normalizing the dataset:

```
import numpy as np

from sklearn.datasets import load_digits

digits = load_digits()
X = digits['data'] / np.max(digits['data'])
```

At this point, we can instantiate the scikit-learn `TSNE` class with `n_components=2` and `perplexity=20`, and fit it with the original dataset (which has 64 dimensions):

```
from sklearn.manifold import TSNE
```

```
tsne = TSNE(n_components=2, perplexity=20, random_state=1000)
X_tsne = tsne.fit_transform(X)
```

The final result is shown in the following screenshot (where 400 samples are shown):

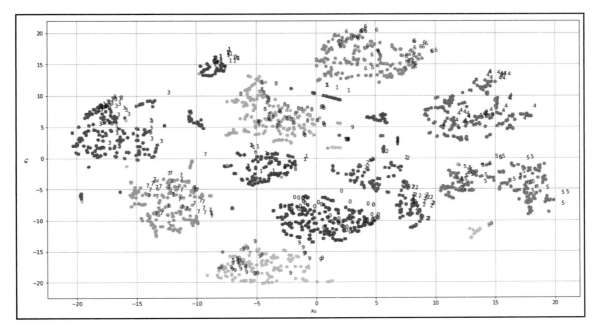

t-SNE plot of the MNIST dataset (limited to 400 digits)

As it's possible to see, the digits are grouped (clustered) coherently with their original distribution with only a few errors, which is probably due to the excessive deformations (different `random_state` values can yield slightly different final configurations). However, t-SNE is extremely powerful, and most of the current implementations are fast enough to process very large high-dimensional datasets in a short amount of time. Therefore, I suggest that you employ it whenever it's helpful to have a graphical representation of the data. This algorithm, which belongs to the family of manifold learning ones, implicitly confirms the manifold assumption. In fact, even if we have 64 original dimensions, the samples are grouped into dense chunks that can be represented by using a smaller number of dimensions. In this particular case, we have also obtained an implicit clustering because many blocks are not only very dense, but they are separated from the other ones and, considering the central point (centroid), it's possible to predict if a digit belongs to the group by computing the distance from the center. In general, this is not always so simple, but we have a further confirmation that the dimensionality reduction has a strong mathematical rationale and it mainly works thanks to the internal structure of real datasets.

Summary

Feature selection is the first (and sometimes the most important) step in a machine learning pipeline. Not all of these features are useful for our purposes, and some of them are expressed using different notations, so it's often necessary to preprocess our dataset before any further operations.

We saw how we can split the data into training and test sets using a random shuffle and how to manage missing elements. Another very important section covered the techniques used to manage categorical data or labels, which are very common when a certain feature only assumes a discrete set of values.

Then, we analyzed the problem of dimensionality. Some datasets contain many features that are correlated with each other, so they don't provide any new information but increase the computational complexity and reduce the overall performances. The PCA is a method to select only a subset of features that contain the largest amount of total variance. This approach, together with its variants, allows you to decorrelate the features and reduce the dimensionality without a drastic loss in terms of accuracy. Dictionary learning is another technique that's used to extract a limited number of building blocks from a dataset, together with the information needed to rebuild each sample. This approach is particularly useful when the dataset is made up of different versions of similar elements (such as images, letters, or digits).

In the next chapter, `Chapter 4`, *Regression Algorithms*, we're going to discuss linear regression, which is the most diffused and simplest supervised approach to predicting continuous values. We'll also analyze how to overcome some limitations and how to solve non-linear problems using the same algorithms.

4
Regression Algorithms

Linear models are the simplest parametric methods and always deserve the right attention, because many problems, even intrinsically non-linear ones, can be easily solved with these models. As discussed previously, a *regression* is a prediction where the target is continuous and it has several applications, so it's important to understand how a linear model can fit the data, what its strengths and weaknesses are, and when it's preferable to pick an alternative. In the last part of the chapter, we're going to discuss an interesting method to work efficiently with non-linear data using the same models.

In particular, we are going to discuss the following:

- Standard linear regression
- Regularized regressions (Ridge regression, Lasso, and ElasticNet)
- Robust regression (**Random Sample Consensus** (**RANSAC**) and Huber regression)
- Polynomial regression
- Bayesian regression (this topic is more advanced and can be skipped at the beginning)
- Isotonic regression
- Regression evaluation metrics

Linear models for regression

Let's consider a dataset of real-value vectors drawn from a data generating process p_{data}:

$$X = \{\bar{x}_1, \bar{x}_2, \ldots, \bar{x}_n\} \quad where \quad \bar{x}_i \in \mathbb{R}^m$$

Each input vector is associated with a real value y_i:

$$Y = \{y_1, y_2, \ldots, y_n\} \quad where \quad y_i \in \mathbb{R}$$

Regression Algorithms

A linear model is based on the assumption that it's possible to approximate the output values through a regression process based on this rule:

$$\tilde{y}_k = \alpha_0 + \sum_{i=1}^{m} \alpha_i \bar{x}_k^{(i)} + \eta_k \text{ where } \alpha_i \in \mathbb{R} \text{ and } \eta_i \sim N(0, \Sigma)$$

In other words, the strong assumption is that our dataset and all other unknown points lie in the volume defined by a hyperplane and random normal noise that depends on the single point. In many cases, the covariance matrix is $\Sigma = \sigma^2 I_m$ (that is, *homoscedastic* noise); hence, the noise has the same impact on all the features. Whenever this doesn't happen (that is, when the noise is *heteroscedastic*), it's not possible to simplify the expression of Σ. It's helpful to understand that this situation is more common than expected, and it means that the uncertainty is higher for some features and the model can fail to explain them with enough accuracy. In general, the maximum error is proportional to both the training quality and the adaptability of the original dataset, which is proportional to the variance of the random noise. In the following graph, there's an example of two possible scenarios:

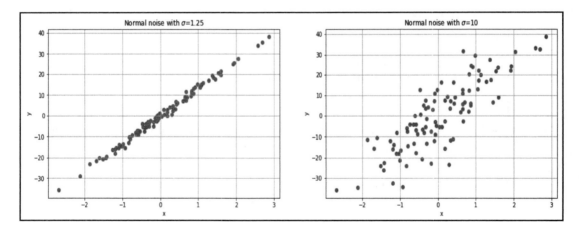

Time series with low random noise (left). Time series with high random noise (right).

A linear regression approach is based on *flat* structures (lines, planes, hyperplanes); therefore, it's not able to adapt to datasets with high dispersion. One of the most common problems arises when the dataset is clearly non-linear and other models have to be considered (such as **polynomial regression, neural networks,** or **kernel support vector machines**). In this chapter, we are going to analyze different situations, showing how to measure the performance of an algorithm and how to make the most appropriate decision to solve specific problems.

A bidimensional example

Let's consider a small dataset built by adding some uniform noise to the points belonging to a segment bounded between **-6** and **6**. The original equation is $y = x + 2 + \eta$, where η is a noise term.

In the following graph, there's a plot with a *candidate* regression function:

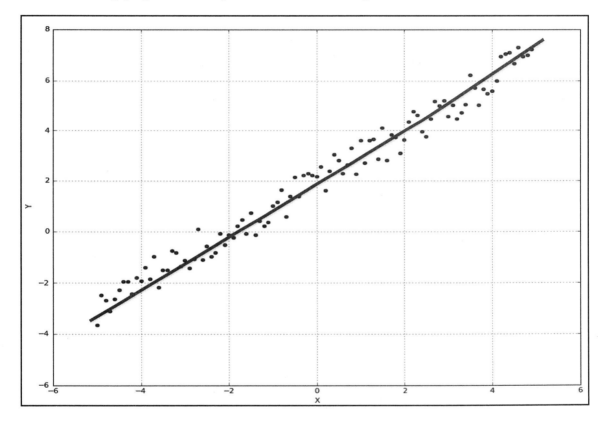

A simple bidimensional dataset with a candidate regression line

The dataset is defined as follows:

```
import numpy as np

nb_samples = 200

X = np.arange(-5, 5, 0.05)

Y = X + 2
Y += np.random.normal(0.0, 0.5, size=nb_samples)
```

As we're working on a plane, the regressor we're looking for is a function of only two parameters (the intercept and the only multiplicative coefficient) with an additive random normal noise term that is associated with every data point x_i (formally, all η_i are **independent and identically distributed (i.i.d)** variables):

$$\tilde{y} = \alpha + \beta x + \eta_i$$

To fit our model, we must find the best parameters and we start with an **Ordinary Least Squares** (**OLS**) approach based on the known data points (x_i, y_i). The cost function to minimize is as follows:

$$L = \frac{1}{2}\sum_{i=1}^{n} \|\tilde{y}_i - y_i\|_2^2 \quad \Rightarrow \quad L = \frac{1}{2}\sum_{i=1}^{m}(\alpha + \beta x_i - y_i)^2$$

With an analytic approach, to find the global minimum we must impose both derivatives to be equal to 0 (in the general case, the same must be done using the gradient $\nabla L = 0$):

$$\begin{cases} \frac{\partial L}{\partial \alpha} = \sum_{i=1}^{n}(\alpha + \beta x_i - y_i) = 0 \\ \frac{\partial L}{\partial \beta} = \sum_{i=1}^{n}(\alpha + \beta x_i - y_i)x_i = 0 \end{cases}$$

So, we can define the function in Python, using an input vector containing both α and β:

```
import numpy as np

def loss(v):
    e = 0.0
    for i in range(nb_samples):
        e += np.square(v[0] + v[1]*X[i] - Y[i])
    return 0.5 * e
```

And the gradient can be defined as follows:

```
import numpy as np

def gradient(v):
    g = np.zeros(shape=2)
    for i in range(nb_samples):
        g[0] += (v[0] + v[1]*X[i] - Y[i])
        g[1] += ((v[0] + v[1]*X[i] - Y[i]) * X[i])
    return g
```

The optimization problem can now be solved using SciPy:

```
from scipy.optimize import minimize

minimize(fun=loss, x0=[0.0, 0.0], jac=gradient, method='L-BFGS-B')

fun: 25.224432728145842
 hess_inv: <2x2 LbfgsInvHessProduct with dtype=float64>
      jac: array([ -8.03369622e-07, 3.60194360e-06])
  message: b'CONVERGENCE: NORM_OF_PROJECTED_GRADIENT_<=_PGTOL'
     nfev: 8
      nit: 7
   status: 0
  success: True
        x: array([ 1.96534464, 0.98451589])
```

As expected, the regression denoised our dataset, rebuilding the original equation: y = x + 2 (with a negligible approximation error). This procedure is absolutely acceptable; however, it's not difficult to understand that the problem can be solved in closed form in a single step. The first thing to do is get rid of the intercept by adding an extra feature equal to 1:

$$\bar{x} \Rightarrow (\bar{x}^T, 1)^T \text{ and } X = (\bar{x}_1^T, \bar{x}_2^T, \ldots, \bar{x}_n^T)^T \text{ so } X \in \mathbb{R}^{n \times m}$$

At this point, the generic problem can be expressed in vectorial notation using a coefficient vector θ:

$$\tilde{y}_i = \bar{x}_i^T \bar{\theta} + \eta_i$$

Regression Algorithms

In the bidimensional case $\theta = (\beta, \alpha)$ because the intercept always corresponds to the last value. If we assume that the noise is homoscedastic with a variance equal to σ^2 (that is, $\eta \sim N(0, \sigma^2 I)$), the cost function can be rewritten as follows:

$$L = (Y - X\bar{\theta})^T \cdot (Y - X\bar{\theta}) \Rightarrow \nabla L = -2X^T \cdot (Y - X\bar{\theta})$$

If the matrix $X^T X$ has *full rank* (that is, $det(X^T X) \neq 0$), it's easy to find the solution $\nabla L = 0$ using the Moore-Penrose pseudo-inverse (which is an extension of matrix inversion when the shape is not square):

$$\bar{\theta}_{opt} = (X^T X)^{-1} X^T Y$$

The previous example becomes this:

```
import numpy as np

Xs = np.expand_dims(X, axis=1)
Ys = np.expand_dims(Y, axis=1)
Xs = np.concatenate((Xs, np.ones_like(Xs)), axis=1)

result = np.linalg.inv(np.dot(Xs.T, Xs)).dot(Xs.T).dot(Y)

print('y = %.2fx + %2.f' % (result[0], result[1]))
y = 0.98x +  2
```

Clearly, this approach is much more efficient and exploits the vectorization features provided by NumPy (the computation on large datasets is extremely fast, and there's no more need for multiple gradient evaluations). According to the **Gauss-Markov theorem**, this is the **Best Linear Unbiased Estimator (BLUE)**, meaning that there are no other solutions with a smaller coefficient variance. We omit the proofs (which are easy to obtain by applying the standard formulas), but we have $E[\theta_{opt}] = 0$ and $Var[\theta_{opt}] = \sigma^2(X^T X)^{-1}$ that depends on σ^2 and on the inverse of $X^T X$. For our example, the coefficient covariance matrix is as follows:

```
import numpy as np

covariance = (0.5 ** 2) * np.linalg.inv(np.dot(Xs.T, Xs))

print(covariance)
[[ 1.50003750e-04  3.75009375e-06]
 [ 3.75009375e-06  1.25009375e-03]]
```

Chapter 4

The covariance matrix is decorrelated (all non-diagonal terms are close to *0*) and the variances for the coefficient and the intercept are extremely small (thanks also to the simplicity of the problem). It's interesting to notice that the variance of the intercept is ten times larger than the variance of the coefficient. This is due to the fact that we expect half of the points to be above the regression line and the other half below it. Therefore, small vertical shifts reduce the error for 50% of points and increase it for the remaining part. On the other side, a small change in the slope has an impact on all points, increasing always the mean squared error.

Considering *m* samples and *m* i.i.d normal noise terms, the probability of the output *Y* given the sample set *X*, the parameter vector θ_{opt}, and the noise variance σ^2 is Gaussian, with a density function equal to the following:

$$p(Y|X; \bar{\theta}, \sigma^2) \propto \frac{1}{\sigma^m} e^{-\frac{(Y - X\bar{\theta}_{opt}^T) \cdot (Y - X\bar{\theta}_{opt})}{2\sigma^2}}$$

This means that, once the model has been trained and the θ_{opt} has been found, we expect all samples to have a Gaussian distribution centered on the regression hyperplane. In the following diagram, there's a bidimensional example of this concept:

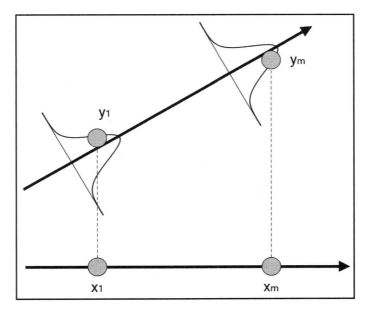

Expected distribution of the outputs *y* around a regression line

In the general case of heteroscedastic noise, we can express the noise covariance matrix as $\Sigma = \sigma^2 C$, where C is a generic square and invertible matrix whose values are normally bounded between 0 and 1. This is not a limitation, because, if we select σ^2 as the maximum variance, a generic element C_{ij} becomes a weight for the covariance between the parameter features θ_i and θ_j (obviously, on the diagonal, there are the variances for all the features).

The cost function is slightly different because we must take into account the different impact of the noise on the single features:

$$L = (Y - X\bar{\theta})^T \cdot C^{-1} \cdot (Y - X\bar{\theta})$$

It's easy to compute the gradient and derive an expression for the parameters, similar to what we have obtained in the previous case:

$$\bar{\theta}_{opt} = (X^T C^{-1} X)^{-1} X^T C^{-1} Y$$

For a general linear regression problem based on the minimization of the squared error, it's possible to prove that the optimal prediction (that is, with the minimum variance) of a new sample's x_j corresponds to this:

$$\tilde{y}_j = E[\tilde{y}|\bar{x} = \bar{x}_j] = \bar{x}_j^T \bar{\theta}_{opt} + E[\eta_j] = \bar{x}_j^T \bar{\theta}_{opt}$$

This result confirms our expectations: the optimal regressor will always predict the expected value of the dependent variable y conditioned to the input x_j. Hence, considering the previous diagram, the Gaussians are optimized to have their means as close as possible to each training sample (of course, with the constraint of a global linearity).

Linear regression with scikit-learn and higher dimensionality

The scikit-learn library offers the `LinearRegression` class, which works with n-dimensional spaces. For this purpose, we're going to use the Boston dataset:

```
from sklearn.datasets import load_boston

boston = load_boston()

print(boston.data.shape)
```

Chapter 4

```
(506L, 13L)

print(boston.target.shape)
(506L,)
```

It has `506` samples with `13` input features and one output. In the following graph, there's a collection of the plots of the first 12 features:

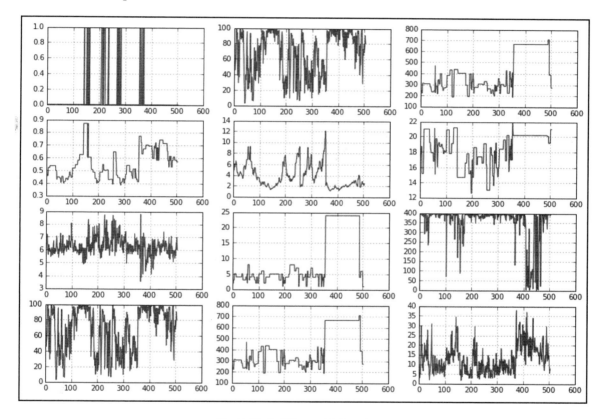

The plot of the first 12 features of the Boston dataset

When working with datasets, it's useful to have a tabular view to manipulate data. **Pandas** is a perfect framework for this task, and even though it's beyond the scope of this book, I suggest you create a data frame with the `pandas.DataFrame(boston.data, columns=boston.feature_names)` command and use Jupyter to visualize it. For further information, refer to *Learning pandas - Python Data Discovery and Analysis Made Easy*, Heydt M., Packt Publishing, 2017.

Regression Algorithms

There are different scales and outliers (which can be removed using the methods studied in the previous chapters), so it's better to ask the model to normalize the data before processing it (setting the `normalize=True` parameter). Moreover, for testing purposes, we split the original dataset into training (90%) and test (10%) sets:

```
from sklearn.linear_model import LinearRegression
from sklearn.model_selection import train_test_split

X_train, X_test, Y_train, Y_test = train_test_split(boston.data,
boston.target, test_size=0.1)

lr = LinearRegression(normalize=True)
lr.fit(X_train, Y_train)

LinearRegression(copy_X=True, fit_intercept=True, n_jobs=1, normalize=True)
```

When the original dataset isn't large enough, splitting it into training and test sets may reduce the number of samples that can be used for fitting the model. As we are assuming that the dataset represents an underlying data generating process, it's absolutely necessary that both the training and test sets obey this rule. Small datasets, can happen to have only a few samples representing a specific area of the data generating process, and it's important to include them in the training set, to avoid a lack of generalization ability.

K-fold cross-validation can help solve this problem with a different strategy. The whole dataset is split into *k* folds using always *k-1* folds for training and the remaining one to validate the model. *K* iterations will be performed, using always a different validation fold. In the following diagram, there's an example with three folds/iterations:

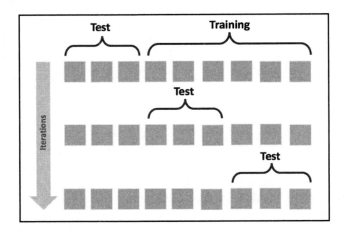

K-Fold cross-validation schema

In this way, the final score can be determined as the average of all values, and all samples are selected for training *k-1* times. Whenever the training time is not extremely long, this approach is the best way to assess the performance of a model, in particular the generalization ability, that can be compromised when some samples are not present in the training set.

To check the accuracy of a regression, scikit-learn provides the internal method `score(X, y)` method, which evaluates the model on test data using the R^2 score (see the next section):

```
print(lr.score(X_test, Y_test))
0.77371996006718879
```

So, the overall accuracy is about 77%, which is an acceptable result considering the non-linearity of the original dataset, but it can be also influenced by the subdivision made by `train_test_split` (as in our case). In fact, the test set can contain points that can also be easily predicted also when the overall accuracy is unacceptable. For this reason, it's preferable not to trust this measure immediately and default on a cross-validation (**CV**) evaluation.

To perform a k-fold cross-validation, we can use the `cross_val_score()` function, which works with all the classifiers. The `scoring` parameter is very important because it determines which metric will be adopted for tests. As `LinearRegression` works with ordinary least squares, we preferred the negative mean squared error, which is a cumulative measure that must be evaluated according to the actual values (it's not relative):

```
from sklearn.model_selection import cross_val_score

scores = cross_val_score(lr, boston.data, boston.target, cv=7,
scoring='neg_mean_squared_error')
array([ -11.32601065,  -10.96365388,  -32.12770594,  -33.62294354,
        -10.55957139, -146.42926647,  -12.98538412])

print(scores.mean())
-36.859219426420601

print(scores.std())
45.704973900600457
```

The high standard deviation confirms that this dataset is very sensitive to the split strategy. In some cases, the probability distribution of both training and test sets are rather similar, but in other situations (at least three with seven folds), they are different; hence, the algorithm cannot learn to predict correctly.

R² score

Another very important metric used in regressions is called the **coefficient of determination** or R^2. It measures the amount of variance of the prediction which is explained by the dataset. In other words, given the variance of the data generating process p_{data}, this metric is proportional to the probability of predicting new samples that actually belong to p_{data}. Intuitively, if the regression hyperplane approximated the majority of samples with an error below a fixed threshold, we can assume that future values will be correctly estimated. On the other side, if, for example, the slope allows to have a small error only for a part of the dataset, the probability of future wrong prediction increases because the model is not able to capture the complete dynamics.

To introduce the measure, let's define as **residual** the following quantity:

$$r_i = y_i - \tilde{y}_i \ \forall \, i \in (0, n)$$

In other words, it is the difference between the sample and the prediction. So, the R^2 is defined as follows:

$$R^2 = 1 - \frac{\sum_i r_i^2}{\sum_i (y_i - \bar{y})^2}$$

The term $\bar{y}$ represents the average computed over all samples. For our purposes, R^2 values close to 1 mean an almost-perfect regression, while values close to 0 (or negative) imply a bad model. It's very easy to use this metric together with CV:

```
print(cross_val_score(lr, X, Y, cv=10, scoring='r2').mean())
0.2
```

The result is low, meaning that the model can easily fail on future prediction (contrary to the result provided by the `score()` method). In fact, we can have a confirmation considering the standard deviation:

```
print(cross_val_score(lr, X, Y, cv=10, scoring='r2').std())
0.599
```

This is not surprising, considering the variance of the original dataset and the presence of several *anomalies*. The great advantage of a CV method is to minimize the risk of selecting only points that are close to the regression hyperplane.

The reader should be aware that such a model will be very likely to yield inaccurate predictions and shouldn't try to find a better train/test split. A reasonable solution is characterized by a high CV score mean and low standard deviation. On the other side, when the CV standard deviation is high, another solution should be employed because the algorithm is very sensitive to the structure training set. We are going to analyze some more powerful models at the end of this chapter; however, the choice must normally be restricted to non-linear solutions, which are able to capture complex dynamics (and, thus, to explain more variance).

Explained variance

In a linear regression problem (as well as in a **Principal Component Analysis (PCA)**), it's helpful to know how much original variance can be explained by the model. This concept is useful to understand the amount of information that we lose by approximating the dataset. When this value is small, it means that the data generating process has strong oscillations and a linear model fails to capture them. A very simple but effective measure (not very different from R^2) is defined as follows:

$$EV = 1 - \frac{Var[Y - \widetilde{Y}]}{Var[Y]}$$

When Y is well approximated, the numerator is close to 0 and $EV \rightarrow 1$, which is the optimal value. In all the other cases, the index represents the ratio between the variance of the errors and the variance of the original process. We can compute this score for our example using the same CV strategy employed previously:

```
print(cross_val_score(lr, X, Y, cv=10,
scoring='explained_variance').mean())
0.271
```

Regression Algorithms

The value, similarly to R^2, is not acceptable, even if it's a little bit higher. It's obvious that the dynamics of the dataset cannot be modeled using a linear system because only a few features show a behavior which can be represented as a generic line with additive Gaussian noise. In general, when R^2 is unacceptable, it doesn't make sense to compute other measures because the accuracy of the model will be normally low. Instead, it's preferable to analyze the data, to have a better understanding of the single time-series. In the Boston dataset, many values show an extremely non-linear behavior and the most regular features don't seem to be stationary. This means that there are (often unknown) events that alter dramatically the dynamics. If the time window is short, it's possible to imagine the presence of oscillations and seasonalities, but when the samples are collected over a sufficiently large period, it's more likely supposing the presence of factors that have not been included in the model. For example, the price of the houses can dramatically change after a natural catastrophe, and such a situation is clearly unpredictable. Whenever the number of irregular samples is small, it's possible to consider them as outliers and filter them out with particular algorithms, but when they are more frequent, it's better to employ a model that can learn non-linear dynamics. In this chapter, and in the rest of the book, we are going to discuss methods that can solve or mitigate these problems.

Regressor analytic expression

If we want to have an analytical expression of our model (a hyperplane), `LinearRegression` offers two instance variables, `intercept_`, and `coef_`:

```
print('y = ' + str(lr.intercept_) + ' ')
for i, c in enumerate(lr.coef_):
    print(str(c) + ' * x' + str(i))

y = 38.0974166342
-0.105375005552 * x0
0.0494815380304 * x1
0.0371643549528 * x2
3.37092201039 * x3
-18.9885299511 * x4
3.73331692311 * x5
0.00111437695492 * x6
-1.55681538908 * x7
0.325992743837 * x8
-0.01252057277 * x9
-0.978221746439 * x10
0.0101679515792 * x11
-0.550117114635 * x12
```

As for any other model, a prediction can be obtained through the `predict(X)` method. As an experiment, we can try to add some Gaussian noise to our training data and predict the value:

```
X = boston.data[0:10] + np.random.normal(0.0, 0.1)

lr.predict(X)
array([ 29.5588731 ,  24.49601998,  30.0981552 ,  28.01864586,
        27.28870704,  24.65881135,  22.46335968,  18.79690943,
        10.53493932,  18.18093544])

print(boston.target[0:10])
[ 24. ,  21.6,  34.7,  33.4,  36.2,  28.7,  22.9,  27.1,  16.5,  18.9]
```

It's obvious that the model is not performing in an ideal way and there are many possible reasons, the foremost being non-linearities and the presence of outliers. In general, a linear regression model is not a perfectly robust solution. However, in this context, a common threat is represented by collinearities that lead to low-rank X matrix. This determines an ill-conditioned matrix that is particularly sensitive to noise, causing the explosion of some parameters as well. The following methods have been studied, to mitigate this risk and to provide more robust solutions.

Ridge, Lasso, and ElasticNet

In this section, we are going to analyze the most common regularization methods and how they can impact the performance of a linear regressor. In real-life scenarios, it's very common to work with *dirty* datasets, containing outliers, inter-dependent features, and different sensitivity to noise. These methods can help the data scientist mitigate the problems, yielding more effective and accurate solutions.

Ridge

Ridge regression (also known as **Tikhonov regularization**) imposes an additional shrinkage penalty to the ordinary least squares cost function to limit its squared L_2 norm:

$$L = \|Y - X\bar{\theta}\|_2^2 + \alpha\|\bar{\theta}\|_2^2$$

X is a matrix containing all samples as rows and the term θ represents the weight vector. The additional term (through the alpha coefficient—if large it implies a stronger regularization and smaller values) forces the loss function to disallow an infinite growth of w, which can be caused by multicollinearity or ill-conditioning.

In the following diagram, there's a representation of what happens when a Ridge penalty is applied:

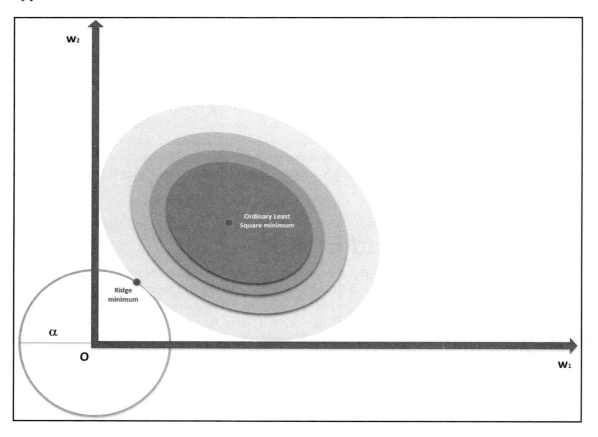

Bidimensional solution using a Ridge penalty (the gray surface represents the cost function)

The gray surface represents the cost function (here, for simplicity, we're working with only two weights), while the circle center **O** is the boundary imposed by the Ridge condition. The minimum will have smaller w values and potential explosions are avoided. It's very easy to verify that, in this case, the optimal coefficient vector is a slightly different version of the formula obtained for a standard linear regression:

$$\bar{\theta}_{opt} = (X^T X + \alpha I_m)^{-1} X^T Y$$

Therefore, with the right choice of α, $det(X^TX + \alpha I_m) \neq 0$ and the solution can always be found. In general, the shrinkage term plays a fundamental role when the dataset contains linearly dependent features (which, to some extent, is very often possible). In this case, the matrix X^TX can be ill-conditioned (for example, $det(X^TX) \approx 0$, and the inverse can be very sensitive to small changes in the values), and, hence, the coefficient variance can become extremely high. Imposing a constraint on their growth guarantees a more stable solution and an increased robustness to noise.

In the following snippet, we're going to compare `LinearRegression` and `Ridge` with a cross-validation:

```
from sklearn.datasets import load_boston
from sklearn.linear_model import LinearRegression, Ridge

boston = load_boston()

lr = LinearRegression(normalize=True)
rg = Ridge(0.05, normalize=True)

lr_scores = cross_val_score(lr, boston.data, boston.target, cv=10)
print(lr_scores.mean())
0.200138

rg_scores = cross_val_score(rg, boston.data, boston.target, cv=10)
print(rg_scores.mean())
0.287476
```

As the house pricing is based on 13 variables that are very likely to be partially dependent (for example, the number of non-occupied homes and the crime rate), we can observe a sensible increase in the CV score when imposing an L_2 penalty.

Sometimes, finding the right value for alpha (Ridge coefficient) is not so immediate. The scikit-learn library provides the `RidgeCV` class, which allows performing an automatic grid search among a set and returning the best estimation:

```
from sklearn.linear_model import RidgeCV

rg = RidgeCV(alphas=(1.0, 0.1, 0.01, 0.005, 0.0025, 0.001, 0.00025),
normalize=True)
rg.fit(boston.data, boston.target)

print(rg.alpha_)
0.010
```

Lasso

A **Lasso** regressor imposes a penalty on the L_1 norm of w to determine a potentially higher number of null coefficients:

$$L = \|Y - X\bar{\theta}\|_2^2 + \alpha\|\bar{\theta}\|_1$$

The sparsity is a consequence of the penalty term (the mathematical proof is non-trivial, but we have provided an intuitive explanation in Chapter 2, *Important Elements in Machine Learning*). A Lasso constraint yields a shrinkage too, but the dynamic is a little bit different from Ridge. When using an L_1 norm, the partial derivatives computed with respect to the coefficients can be only +1 or -1, according to the sign of θ_i. Hence, the constraint forces the smallest components to move towards zero with higher speed because the effect on the minimization is independent of the magnitude of the coefficients. Conversely, the derivative of an L_2 norm is proportional to the magnitude of the parameters and its effect decreases for small values. That's why Lasso is commonly used to induce sparsity with an implicit feature selection. When the number of features is large, Lasso selects a subset, discarding (that is, setting $\theta_i \approx 0$) the other features, which are not taken into account in future predictions. In the following diagram, there's a representation of the effect of a Lasso regularization:

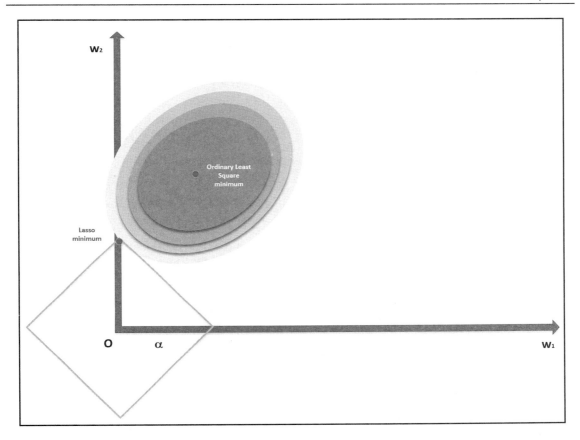

Bidimensional solution using a Lasso penalty (the gray surface represents the cost function)

In this case, there are vertices where a component is non-null while all the other weights are zero. The probability of an intersection with a vertex is proportional to the dimensionality of w, and, therefore, it's normal to discover a rather sparse model after training a Lasso regressor.

In the following snippet, the Boston dataset is used to fit a Lasso model:

```
from sklearn.linear_model import Lasso

ls = Lasso(alpha=0.01, normalize=True)
ls_scores = cross_val_score(ls, boston.data, boston.target, cv=10)

print(ls_scores.mean())
0.270937
```

Also, for Lasso, there's the possibility of running a grid search for the best `alpha` parameter. The class, in this case, is `LassoCV`, and its internal dynamics are similar to what was already seen for Ridge. Lasso can also perform efficiently on the sparse data generated through the `scipy.sparse` class, allowing for training bigger models without the need for partial fitting:

```
from scipy import sparse

ls = Lasso(alpha=0.001, normalize=True)
ls.fit(sparse.coo_matrix(boston.data), boston.target)

Lasso(alpha=0.001, copy_X=True, fit_intercept=True, max_iter=1000,
    normalize=True, positive=False, precompute=False, random_state=None,
    selection='cyclic', tol=0.0001, warm_start=False)
```

When working with a huge amount of data, some models cannot fit completely in memory, so it's impossible to train them. The scikit-learn library offers some models, such as **stochastic gradient descent (SGD)**, which work in a way quite similar to `LinearRegression` with Ridge/Lasso; however, they also implement the `partial_fit()` method, which also allows continuous training through Python generators. See http://scikit-learn.org/stable/modules/linear_model.html#stochastic-gradient-descent-sgd for further details.

ElasticNet

The last alternative is called **ElasticNet** and combines both Lasso and Ridge into a single model with two penalty factors: one proportional to L_1 norm and the other to L_2 norm. In this way, the resulting model will be sparse like a pure Lasso, but with the same regularization ability as provided by Ridge. The resulting loss function is as follows:

$$L = \|Y - X\bar{\theta}\|_2^2 + \alpha\beta\|\bar{\theta}\|_1 + \frac{\alpha(1-\beta)}{2}\|\theta\|_2^2$$

The `ElasticNet` class provides an implementation where the `alpha` parameter works in conjunction with `l1_ratio` (beta in the formula). The main peculiarity of `ElasticNet` is avoiding a selective exclusion of correlated features, thanks to the balanced action of the L_1 and L_2 norms.

In the following snippet, there's an example using both the `ElasticNet` and `ElasticNetCV` classes:

```
from sklearn.linear_model import ElasticNet, ElasticNetCV

en = ElasticNet(alpha=0.001, l1_ratio=0.8, normalize=True)
en_scores = cross_val_score(en, boston.data, boston.target, cv=10)

print(en_scores.mean())
0.324458

encv = ElasticNetCV(alphas=(0.1, 0.01, 0.005, 0.0025, 0.001),
   l1_ratio=(0.1, 0.25, 0.5, 0.75, 0.8), normalize=True)
encv.fit(boston.data, boston.target)

ElasticNetCV(alphas=(0.1, 0.01, 0.005, 0.0025, 0.001), copy_X=True,
cv=None,
       eps=0.001, fit_intercept=True, l1_ratio=(0.1, 0.25, 0.5, 0.75, 0.8),
       max_iter=1000, n_alphas=100, n_jobs=1, normalize=True,
       positive=False, precompute='auto', random_state=None,
       selection='cyclic', tol=0.0001, verbose=0)

print(encv.alpha_)
0.005

print(encv.l1_ratio_)
0.7500
```

As expected, the performance of `ElasticNet` is superior to both Ridge and Lasso because it combines the shrinkage effect of the former and the feature selection of the latter. However, as the interaction of the two penalties is more complex to predict, I always suggest performing a grid search with a large range of parameters. Whenever necessary, it's possible to repeat the process by *zooming* into the range containing the parameters previously selected, so to find out possibly the most performing combination.

Robust regression

In this section, we are going to consider two solutions that can be employed when the dataset contains outliers. Unfortunately, a linear regression is very sensitive to them because the coefficients are forced to minimize the squared error and hence, the hyperplane is forced to move closer to the outliers (which yield a higher error). However, in the majority of real-life applications, we expect a good ability to discriminate between points belonging to data-generating processes and outliers. The algorithms presented in this section have been designed to mitigate this very problem.

RANSAC

A common problem with linear regressions is caused by the presence of outliers. An ordinary least-square approach will take them into account and the result (in terms of coefficients) will be therefore biased. In the following graph there's an example of such a behavior:

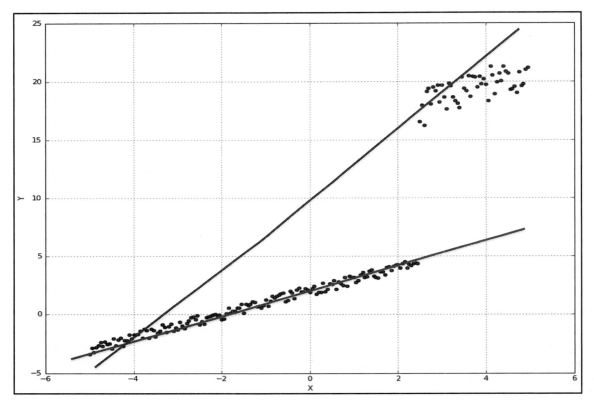

Example of a dataset containing outliers

The shallower-sloped line represents an acceptable regression that discards the outliers, while the other one is influenced by them. An interesting approach to avoid this problem is offered by RANSAC, which works with every regressor by subsequent iterations, after splitting the dataset into inliers and outliers. The model is trained only with valid samples (evaluated internally or through the callable `is_data_valid()`), and all samples are re-evaluated to verify whether they're still inliers or whether they have become outliers. The process ends after a fixed number of iterations, or when the desired score is achieved.

In the following snippet, there's an example of a simple linear regression applied to the dataset shown in the previous diagram:

```
from sklearn.linear_model import LinearRegression

lr = LinearRegression(normalize=True)
lr.fit(X.reshape((-1, 1)), Y.reshape((-1, 1)))

print(lr.intercept_)
5.500572

print(lr.coef_)
2.53688672
```

As imagined, the slope is high due to the presence of outliers. The resulting regressor is $y = 5.5 + 2.5x$ (slightly less sloped than what was shown in the graph). Now we're going to use RANSAC with the same linear regressor:

```
from sklearn.linear_model import RANSACRegressor

rs = RANSACRegressor(lr)
rs.fit(X.reshape((-1, 1)), Y.reshape((-1, 1)))

print(rs.estimator_.intercept_)
2.03602026

print(es.estimator_.coef_)
0.99545348
```

Regression Algorithms

In this case, the regressor is about y = 2 + x (which is the original clean dataset without outliers).

 If you want to have further information, I suggest visiting the page http://scikit-learn.org/stable/modules/generated/sklearn.linear_model.RANSACRegressor.html. For other robust regression techniques, visit http://scikit-learn.org/stable/modules/linear_model.html#robustness-regression-outliers-and-modeling-errors.

Huber regression

An alternative approach is based on a slightly modified loss function, called **Huber loss** (for a single sample):

$$L = \begin{cases} \frac{1}{2}\|y - \bar{x}^T\bar{\theta}\|_2^2 & if\ |y - \bar{x}^T\bar{\theta}| \leqslant t_H \\ t_H |y - \bar{x}^T\bar{\theta}| - \frac{t_H}{2} & otherwise \end{cases}$$

The parameter t_H (called epsilon in scikit-learn) defines a threshold (based on the distance between target and prediction) that makes the loss function switch from a squared error to an absolute one. In this way, the magnitude of the loss changes accordingly, passing from a quadratic behavior to a linear one when the points are supposed to be outliers. In this way, their contribution to the global cost function is reduced and the hyperplane will remain closer to the majority of points even in presence of outliers. Of course, the parameter t_H must be chosen properly, in fact, a very small value (as well as a very large one) will make the loss function prefer almost always one metric (either quadratic or linear), excluding the effect due to the copresence of both terms.

Let's consider a simple bidimensional example made up of 500 points with 50 outliers (not normally distributed):

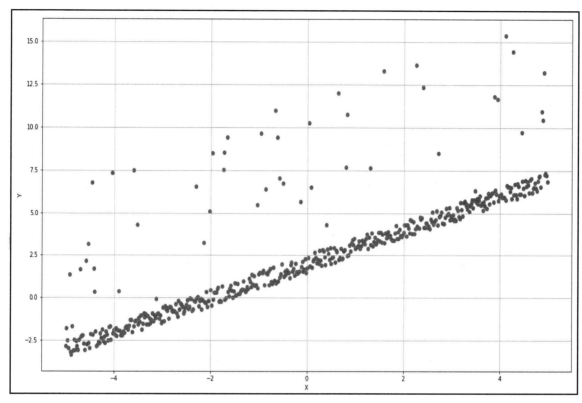

A dataset with a linear structure and some outliers

The linear part of the dataset can be successfully interpolated by the $y = x + 2$ function; however, we expect that the presence of the outliers will have an impact on both the slope and the intercept. Let's check our hypothesis with a standard linear regression:

```
from sklearn.linear_model import LinearRegression

lr = LinearRegression(normalize=True)
lr.fit(X.reshape((-1, 1)), Y.reshape((-1, 1)))

print(lr.intercept_)
2.550

print(lr.coef_)
0.978
```

The slope shows a minimum (negligible) impact because the outliers are uniformly distributed on the same side; however, the intercept is now higher than expected. This is clearly due to the large error determined by the outliers. To minimize it, the regressor has pushed the line towards the outliers, yielding a higher relative error for all other samples. In fact, remember that we expect them to be Gaussian distributed along the regression line? In this case, the majority of them are below the line and their probability is lower than the optimal one.

Let's now repeat the regression using the `HuberRegressor` class, with an `epsilon=1.25` parameter:

```
from sklearn.linear_model import HuberRegressor

hr = HuberRegressor(epsilon=1.25)
hr.fit(X.reshape(-1, 1), Y)

print(hr.intercept_)
2.033

print(hr.coef_)
1.001
```

In this case, the interpolation is almost perfect and the effect of the outliers has been filtered out. I always suggest using a grid search to find the optimal `epsilon` value with a strategy that can be based on the evaluation of the model without the outliers (which can be easily removed considering a variance threshold) and some evaluations (for example, with the R^2 score) based on a Huber regressor with different parameters. Real scenarios are generally more complex than the example discussed previously; however, strong outliers are normally easy to detect, and their effect can be minimized.

Bayesian regression

At the beginning of the chapter, we discussed how the samples are distributed after the linear regression model has been fitted:

$$p(Y|X; \bar{\theta}, \sigma^2) \propto \frac{1}{\sigma^m} e^{-\frac{(Y - X\bar{\theta}_{opt}^T) \cdot (Y - X\bar{\theta}_{opt})}{2\sigma^2}}$$

Clearly, the Gaussian itself is *agnostic* to the way the coefficients have been determined, and by employing a standard method such as OLS or the closed-form expression, we are implicitly relying only on the dataset. Our assumption is that we have enough samples to represent the underlying data generating process correctly and the coefficients must be chosen in a way that minimizes the squared error. However, we may have some prior beliefs about the distribution of all parameters (for example, we could imagine that θ_i is drawn from a Gaussian distribution) and we would like to include this piece of information in our model. As we are going to discuss in Chapter 6, *Naive Bayes and Discriminant Analysis* (for further details, please refer also to *Mastering Machine Learning Algorithms*, Bonaccorso G., Packt Publishing, 2018), the main concept behind the adoption of a Bayesian framework is that we can derive a posterior distribution that is based on the likelihood of the data and on prior beliefs.

In our specific case, we are always assuming that the dependent variables y_i are normally distributed:

$$p(y_i|X, \bar{\theta}, \sigma^2) \sim N(\bar{\mu}, \Sigma)$$

The mean vector and the covariance matrix are a function of dataset *X*, coefficient vector *θ*, and the noise variance. Our goal is to estimate the parameters, considering a prior distribution over them. Exploiting the rule for conditional probabilities, we can write the joint prior probability distribution as a factor of two distributions:

$$p(\bar{\theta}, \sigma^2) = p(\sigma^2)p(\sigma^2|\bar{\theta})$$

The choice of the right distribution is generally non-trivial and, in many cases, it's made so to make the problem tractable in closed form. As the topic is beyond the scope of this book, we are not going to define all the details, but it's useful to consider the concept of **conjugate prior**. In Chapter 6, *Naive Bayes and Discriminant Analysis*, everything will be clearer, but for readers without background knowledge, it's helpful to show the generic Bayes formula:

$$p(a|b) \propto p(b|a)p(a)$$

The posterior probability *p(a|b)* is proportional (up to a normalization factor) to a likelihood *p(b|a)* and a prior distribution *p(a)*. In our case, the term *a* represents the parameters, while *b* is the set of independent variables *X*. Given a likelihood *p(b|a)*, a prior is said to conjugate to it, if the posterior distribution has the same distribution as the prior. This *trick* allows us to simplify the computation and obtain closed-form expressions.

In our specific problem, let's consider a potential $p(\theta|\sigma^2)$. The likelihood $p(Y|X; \theta,\sigma^2)$ is Gaussian and also the prior $p(\theta|\sigma^2)$ can be chosen to be Gaussian (which is, in this case, conjugate to itself), as the likelihood is quadratic in θ. The main problem arises with $p(\sigma^2)$. In fact, this term appears in the denominator of the exponent and in the multiplication factor (always as reciprocal); hence, the prior cannot be Gaussian. The most appropriate distribution for the variance is an inverse-gamma, whose probability density function (with a support, $\sigma^2 > 0$, which is always true) is defined as follows:

$$p(\sigma^2; a, b) = \frac{b^a \sigma^{-2a-2}}{\Gamma(a)} e^{-\frac{b}{\sigma^2}}$$

With this choice, we can derive a posterior distribution that can be computed analytically. In fact, the σ variable appears both in the multiplication factor (as σ^{-k}) and in the denominator of the exponent. Alternatively, it's possible to consider the reciprocal of the variance, called *precision*:

$$\tau = \frac{1}{\sigma^2}$$

In this case, τ appears at the numerator of the exponent and as a multiplicative factor, and the conjugate prior is the gamma distribution. However, this choice is equivalent to the previous one and it has been included only for didactic purposes.

The general expression for the posterior (without substituting the single terms) is this:

$$p(\bar{\theta}, \sigma^2 | Y, X) \propto p(Y|X; \bar{\theta}, \sigma^2) p(\bar{\theta}|\sigma^2) p(\sigma^2)$$

At this point, how can we estimate the parameters? In the Bayesian framework, this can be done by maximizing the marginal likelihood $p(Y|model)$ with respect to all parameters (sometimes called **model evidence**). The model utility variable groups the likelihood and the priors into a single term. Hence, by maximizing the model evidence, we are going to look for parameters that allow us to find the model that best describes the specific problem, or whose $p(Y|model)$ is the maximum. In other words, given the likelihood and the priors that have been chosen for the parameters, we want to find the optimal distribution whose distance from the data generating process is the least. In the following diagram, there's a representation of the process that leads to a posterior distribution for a parameter given a prior:

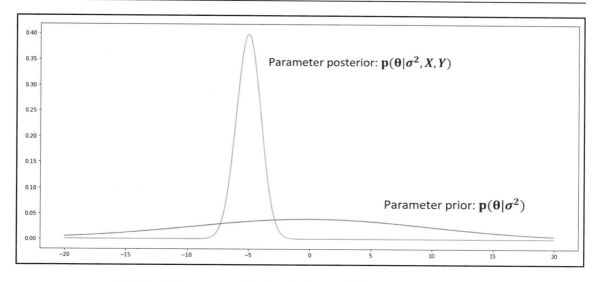

Prior distribution for a parameter (very high variance), with a posterior distribution (peaks around the mean value)

In general, a prior distribution is chosen to be non-informative or with a very high entropy (this condition can be compatible with a proper conjugate prior). In this way, no specific values have a higher chance to be selected (that is, the distribution is very flat). Conversely, we expect a very peaked posterior, thanks to the contribution of the likelihood. As pointed out by D. MacKay (in *Bayesian Interpolation, MacKay D. J. C., Neural Computation, 1991*), this process resembles the **Occam's razor** concept, as it excludes all those hypotheses that are more uncertain and focuses only on the simplest (and most likely) one.

This method is very effective in many situations (in particular, when we want to update the model with new samples), but it requires a more complex computation for the inference part. In fact, to compute the probability of an output, we need to integrate over all possible values of θ and σ^2, considering the current dataset of independent variables X:

$$p(y|model) = \iint p(y|X; \bar{\theta}, \sigma^2) p(\sigma^2) p(\bar{\theta}|\sigma^2) d\bar{\theta} d\sigma^2$$

With the choices previously made, this double integral can be solved in closed form, but its value depends on the gamma function $\Gamma(x)$ and requires some the computation of determinants. Therefore, it's clearly more computationally expensive than all the other solutions presented in this chapter.

The scikit-learn library implements a Bayesian regression based on a Ridge penalty through the BayesianRidge class. In the next example, we are going to employ it with the Boston dataset:

```
from sklearn.linear_model import BayesianRidge

br = BayesianRidge(n_iter=1000)
br.fit(X_train, Y_train)

print(br.score(X_test, Y_test))
0.702
```

The n_iter parameter allows us to specify the maximum number of iterations for the estimation of the parameters. The score for a fixed training/test split is comparable to the one achieved with a standard linear regression. Let's now compute the average $CV\ R^2$ score:

```
r2_scores = cross_val_score(br, boston.data, boston.target, cv=10, scoring='r2')

print(r2_scores.mean())
0.257
```

This value is slightly higher than the one obtained with a linear regression (but still quite small). The reason is due to both the presence of the Ridge penalty and the superior adaptability of the parameters. As an exercise, I invite readers to test this model (compared with a standard Ridge regression) with other datasets. Some possibilities are offered by scikit-learn, but an extremely useful resource is the UCI Machine-Learning Repository (https://archive.ics.uci.edu/ml/datasets.html), where it's possible to find free datasets with different levels of complexity for any kind of problem.

Polynomial regression

Polynomial regression is a technique based on a trick that allows the use of linear models even when the dataset has strong non-linearities. The idea is to add some extra variables computed from the existing ones and using (in this case) only polynomial combinations:

$$\tilde{y}_k = \alpha_0 + \sum_{i=1}^{m} \alpha_i \bar{x}_k^{(i)} + \sum_{j=m+1}^{k} \alpha_j f_{P_j}\left(\bar{x}_k^{(1)}, \bar{x}_k^{(2)}, \ldots, \bar{x}_k^{(m)}\right)$$

In the previous expression, every $f_{Pj}(\bullet)$ is a polynomial function of a single feature. For example, with two variables, it's possible to extend to a second-degree problem by transforming the initial vector (whose dimension is equal to m) into another one with higher dimensionality (whose dimension is $k > m$):

$$\bar{x} = (x_1, x_2) \quad \Rightarrow \quad \bar{x}_t = \left(x_1, x_2, x_1^2, x_2^2, x_1 x_2\right)$$

In this case, the model remains externally linear, but it can capture internal non-linearities. To show how scikit-learn implements this technique, let's consider the dataset shown in the following graph:

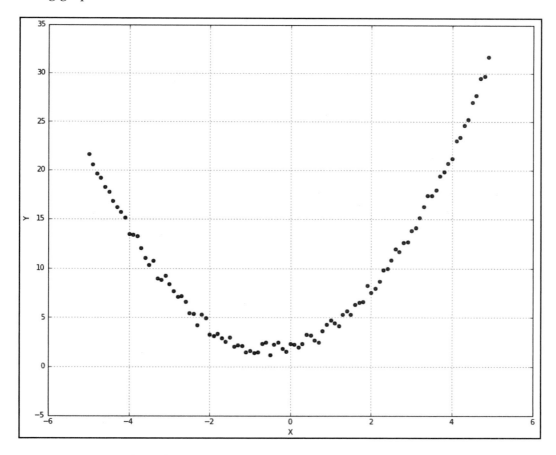

Example of a non-linear dataset that can be efficiently interpolated using a parabolic regression

This is clearly a non-linear dataset, and any linear regression based only on the original two-dimensional points cannot capture the dynamics. Just to try this, we can train a simple model (testing it on the same dataset):

```
from sklearn.linear_model import LinearRegression

lr = LinearRegression(normalize=True)
lr.fit(X.reshape((-1, 1)), Y.reshape((-1, 1)))

print(lr.score(X.reshape((-1, 1)), Y.reshape((-1, 1))))
0.10888218817034558
```

The accuracy is extremely poor, as expected. However, looking at the diagram, we might suppose that a quadratic regression could easily solve this problem. The scikit-learn library provides the `PolynomialFeatures` class, which transforms an original set into an expanded one according to the `degree` parameter:

```
from sklearn.preprocessing import PolynomialFeatures

pf = PolynomialFeatures(degree=2)
Xp = pf.fit_transform(X.reshape(-1, 1))

print(Xp.shape)
(100L, 3L)
```

As expected, the old x_1 coordinate has been replaced by a triplet, which also contains the quadratic and mixed terms. At this point, a linear regression model can be trained:

```
lr.fit(Xp, Y.reshape((-1, 1)))

print(lr.score(Xp, Y.reshape((-1, 1))))
0.99692778265941961
```

The score is a lot higher and the only price we have paid is an increase in terms of features. In general, this is feasible; however, if the number goes over an accepted threshold, it's useful to try dimensionality reduction or, as an extreme solution, to move to a non-linear model (such as **Support Vector Machine-Kernel** (**SVM-Kernel**)). Usually, a good approach is to use the `SelectFromModel` class to let scikit-learn select the best features, based on their importance. In fact, when the number of features increases, the probability that all of them have the same importance reduces. This is the result of mutual correlation or the co-presence of major and minor trends, which act like noise and don't have the strength to perceptibility alter the hyperplane slope. Moreover, when using a polynomial expansion, some weak features (which cannot be used for a linear separation) are substituted by their functions, and so the actual number of strong features decreases.

In the following snippet, there's an example with the previously employed Boston dataset. The `threshold` parameter is used to set a minimum importance level. If missing, the class will try to maximize the efficiency by removing the highest possible number of features:

```
from sklearn.feature_selection import SelectFromModel

boston = load_boston()

pf = PolynomialFeatures(degree=2)
Xp = pf.fit_transform(boston.data)

print(Xp.shape)
(506L, 105L)

lr = LinearRegression(normalize=True)
lr.fit(Xp, boston.target)

print(lr.score(Xp, boston.target))
0.91795268869997404

sm = SelectFromModel(lr, threshold=10)
Xt = sm.fit_transform(Xp, boston.target)

print(sm.estimator_.score(Xp, boston.target))
0.91795268869997404

print(Xt.shape)
(506L, 8L)
```

After selecting only the best features (with the threshold set to 10), the score remains approximately the same, with a consistent dimensionality reduction (only 8 features are considered important for the prediction). Moreover, the overall R^2 score (not the CV one) is now rather high. I invite readers to test the CV scores, to find out whether this solution is definitely acceptable or not. In the latter case, is it possible to obtain better results using a higher degree?

If, after any other processing steps, it's necessary to return to the original dataset, it's possible to use inverse transformation:

```
Xo = sm.inverse_transform(Xt)

print(Xo.shape)
(506L, 105L)
```

Isotonic regression

There are situations when we need to find a regressor for a dataset of non-decreasing points that can present low-level oscillations (such as noise). A linear regression can easily achieve a very high score (considering that the slope is about constant), but it works like a denoiser, producing a line that can't capture the internal dynamics we'd like to model. For these situations, scikit-learn offers the `IsotonicRegression` class, which produces a piecewise interpolating function, minimizing the following functional:

$$L = \sum_i \bar{\theta}_i (y_i - \tilde{y})^2 \quad where \quad y_0 \leqslant y_1 \leqslant \ldots \leqslant y_m$$

An example (with a toy dataset) is provided next:

```
import numpy as np

X = np.arange(-5, 5, 0.1)
Y = X + np.random.uniform(-0.5, 1, size=X.shape)
```

The following graph shows a plot of the dataset. As everyone can see, it can be easily modeled by a linear regressor, but without a high non-linear function, it is very difficult to capture the slight (and local) modifications in the slope:

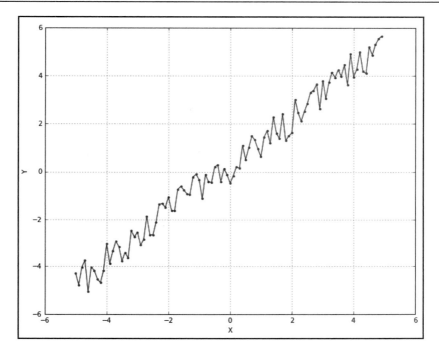

Example of an isotonic dataset

The IsotonicRegression class needs to know y_{min} and y_{max} (which correspond to the y_0 and y_n variables in the loss function). In this case, we have chosen to impose the range bounded by −6 and 10:

```
from sklearn.isotonic import IsotonicRegression

ir = IsotonicRegression(-6, 10)
Yi = ir.fit_transform(X, Y)
```

The result is provided through three instance variables:

```
print(ir.X_min_)
-5.0

print(ir.X_max_)
4.8999999999999648

print(ir.f_)
<scipy.interpolate.interpolate.interp1d at 0x126edef8>
```

Regression Algorithms

The last one, (ir.f_), is an interpolating function that can be evaluated in the domain $[x_{min}, x_{max}]$, for example:

```
print(ir.f_(2))
1.7294334618146134
```

A plot of this function (the solid green line), together with the original dataset, is shown in the following graph:

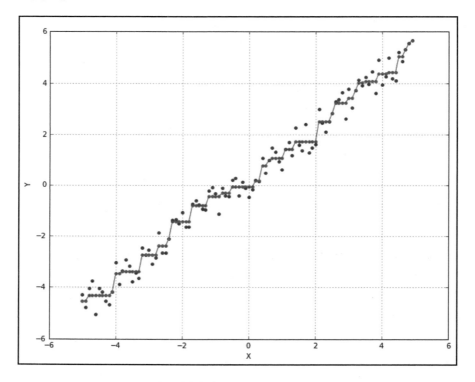

Interpolation performed using isotonic regression

 For further information about interpolation with SciPy, visit https://docs.scipy.org/doc/scipy-0.18.1/reference/interpolate.html.

Summary

In this chapter, we introduced the important concepts of linear models and described how linear regression works. In particular, we focused on the basic model and its main variants: Lasso, Ridge, and ElasticNet. They don't modify the internal dynamics, but work as normalizers for the weights to avoid common problems when the dataset contains unscaled samples. These penalties have specific peculiarities. While Lasso promotes sparsity, Ridge tries to find a minimum with the constraint that the weights must lie within a circle centered at the origin (whose radius is parametrized to increase/decrease the normalization strength). ElasticNet is a mix of both these techniques, and it tries to find a minimum where the weights are small enough and a certain degree of sparsity is achieved.

We also discussed advanced techniques such as RANSAC, which allows us to cope with outliers in a very robust way, and polynomial regression, which is a very smart way to include virtual non-linear features in our model and continue working with them with the same linear approach. In this way, it's possible to create another dataset, containing the original columns together with polynomial combinations of them. This new dataset can be used to train a linear regression model, and then it's possible to select only those features that contributed towards achieving good performance. The last method we saw was isotonic regression, which is particularly useful when the function to interpolate never decreases. Moreover, it can capture small oscillations that would be flattened by a generic linear regression.

In the next chapter, Chapter 5, *Linear Classification Algorithms*, we're going to discuss some linear models for classifications. In particular, we'll focus our attention on the logistic regression and SGD algorithms. Moreover, we're going to introduce some useful metrics to evaluate the accuracy of a classification system, and a powerful technique to automatically find the best hyperparameters.

5
Linear Classification Algorithms

This chapter begins by analyzing linear classification problems, with a particular focus on **logistic regression** (despite its name, it's a classification algorithm) and the **stochastic gradient descent (SGD)** approach. Even if these strategies appear too simple, they're still the main choices in many classification tasks.

Speaking of which, it's useful to remember a very important philosophical principle: **Occam's razor**.

In our context, it states that the first choice must always be the simplest and only if it doesn't fit, it's necessary to move on to more complex models. In the second part of the chapter, we're going to discuss some common metrics that are helpful when evaluating a classification task. They are not limited to linear models, so we use them when talking about different strategies as well.

In particular, we are going to discuss the following:

- The general structure of a linear classification problem
- Logistic regression (with and without regularization)
- SGD algorithms and perceptron
- Passive-aggressive algorithms
- Grid search of optimal hyperparameters
- The most important classification metrics
- The **Receiver Operating Characteristic** (**ROC**) curve

Linear classification

Let's consider a generic linear classification problem with two classes. In the following graph, there's an example:

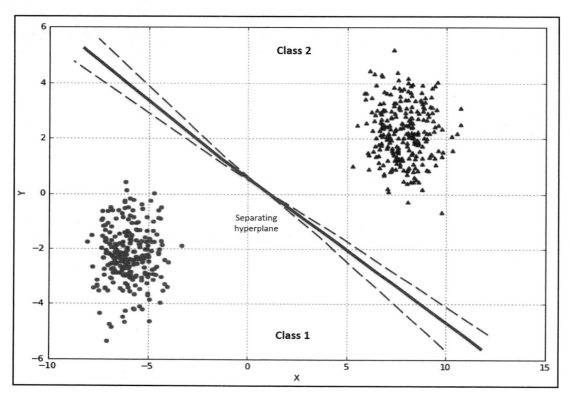

Bidimensional scenario for a linear classification problem

Our goal is to find an optimal hyperplane, that separates the two classes. In multi-class problems, the one-vs-all strategy is normally adopted, so the discussion can focus only on binary classifications. Suppose we have the following dataset made up of n m-dimensional samples:

$$X = \{\bar{x}_1, \bar{x}_2, \ldots, \bar{x}_n\} \quad \bar{x}_i \in \mathbb{R}^m$$

This dataset is associated with the following target set:

$$Y = \{y_1, y_2, \ldots, y_n\} \quad where \quad y_i \in \{0, 1\} \quad or \quad y_i \in \{-1, 1\}$$

Generally, there are two equivalent options; binary and bipolar outputs and different algorithms are based on the former or the latter without any substantial difference. Normally, the choice is made to simplify the computation and has no impact on the results.

We can now define a weight vector made of *m* continuous components:

$$W = (w_1, w_2, \ldots, w_m)^T \quad where \quad w_i \in \mathbb{R}$$

We can also define the quantity, *z*:

$$z = \bar{x} \cdot \bar{w} = \sum_i w_i x_i \quad \forall \bar{x} \in \mathbb{R}^m$$

If *x* is a variable, *z* is the value determined by the hyperplane equation. Therefore, in a bipolar scenario, if the set of coefficients *w* that has been determined is correct, the following happens:

$$sign(z) = \begin{cases} +1 & if \ \bar{x} \in Class\ 1 \\ -1 & if \ \bar{x} \in Class\ 2 \end{cases}$$

When working with binary outputs, the decision is normally made according to a threshold. For example, if the output *z* ∈ *(0, 1)*, the previous condition becomes the following:

$$z = \begin{cases} 1 & if \ z \geqslant 0.5 \\ 0 & if \ z < 0.5 \end{cases}$$

Now, we must find a way to optimize *w* to reduce the classification error. If such a combination exists (with a certain error threshold), we say that our problem is **linearly separable**. On the other hand, when it's impossible to find a linear classifier, the problem is defined as **non-linearly separable**.

A very simple but famous example belonging to the second class is given by the XOR logical operator:

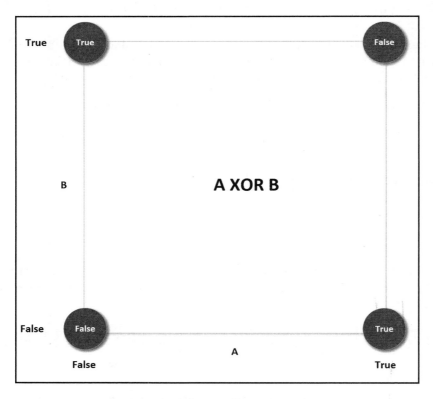

Schema representing the non-linearly-separable problem of binary XOR

As you can see, any line will always include a wrong sample. Hence, to solve this problem, it is necessary to involve non-linear techniques involving high-order curves (for example, two parabolas). However, in many real-life cases, it's possible to use linear techniques (which are often simpler and faster) for non-linear problems too, under the condition of accepting a tolerable misclassification error.

Logistic regression

Even if called regression, this is a classification method that is based on the probability of a sample belonging to a class. As our probabilities must be continuous in $\Re$ and bounded between $(0, 1)$, it's necessary to introduce a threshold function to filter the term z. As already done with linear regression, we can get rid of the extra parameter corresponding to the intercept by adding a *1* element at the end of each input vector:

$$\bar{x}_i \Rightarrow (\bar{x}_i^T, 1)^T$$

In this way, we can consider a single parameter vector θ, containing *m + 1* elements, and compute the *z*-value with a dot product:

$$z_i = \bar{\theta}^T \cdot \bar{x}_i$$

Now, let's suppose we introduce the probability $p(x_i)$ that an element belongs to class 1. Clearly, the same element belongs to class 0 with a probability $1 - p(x_i)$. Logistic regression is mainly based on the idea of modeling the odds of belonging to class 1 using an exponential function:

$$odds = \frac{p(\bar{x}_i)}{1 - p(\bar{x}_i)} = e^{\bar{\theta}^T \cdot \bar{x}_i} = e^{z_i}$$

This function is continuous and differentiable on $\Re$, always positive, and tends to infinite when the argument $x \to \infty$. These conditions are necessary to correctly represent the odds, because when $p \to 0$, $odds \to 0$, but when $p \to 1$, $odds \to \infty$. If we take the logit (which is the natural logarithm of the odds), we obtain an expression for the structure of z_i:

$$z_i = log\left(\frac{p(\bar{x}_i)}{1 - p(\bar{x}_i)}\right)$$

Linear Classification Algorithms

A function with a support on $\Re$, bounded between *0* and *1*, and which satisfies the previous expression is the sigmoid (or logistic) function:

$$p(\bar{x}_i) = \sigma(\bar{x}_i) = \frac{1}{1+e^{-z_i}}$$

A partial plot of this function is shown in the following graph:

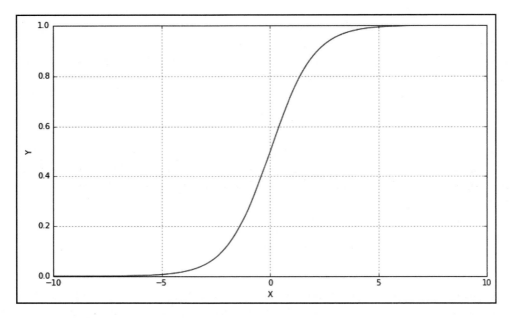

Plot of the sigmoid function; σ(x) → 0 when x → -∞ and σ(x) → 1 when x → ∞

As it's possible to see, the function intersects *x=0* in the ordinate *0.5*, and *y<0.5* for *x<0* and *y>0.5* for *x>0*. Moreover, it's possible to compute it for every *x* and it's almost linear around *0* (this element is very helpful when using shrinkage because the function is much more sensitive when *x* is close to *0*). Hence, if we model the probability distribution using a sigmoid, we obtain the following:

$$z_i = \log\left(\frac{p(\bar{x}_i)}{1-p(\bar{x}_i)}\right) = \log\left(\frac{\frac{1}{1+e^{-z_i}}}{1-\frac{1}{1+e^{-z_i}}}\right) = \log\left(\frac{1}{e^{-z_i}}\right) = z_i$$

So, we can define the probability of a sample belonging to a class (from now on, we'll call them *0* and *1*) as follows:

$$p(y|\bar{x}_i) = \sigma(\bar{x}_i, \bar{\theta})$$

At this point, finding the optimal parameters is equivalent to maximizing the log-likelihood given the target output class:

$$L(\bar{\theta}; Y, X) = \log \prod_i p(y_i|\bar{x}_i; \bar{\theta}) = \sum_i \log p(y_i|\bar{x}_i; \bar{\theta})$$

Each event is based on a Bernoulli distribution; therefore, the optimization problem can be expressed, using the indicator notation, as the minimizing of the loss function:

$$L = -\sum_i \log p(y_i|\bar{x}_i, \bar{\theta}) = -\sum_i \left(y_i \log \sigma(z_i) + (1 - y_i) \log (1 - \sigma(z_i)) \right)$$

If *y=0*, the first term becomes null and the second one becomes *log(1-x)*, which is the log-probability of class *0*. On the other hand, if *y=1*, the second term is *0* and the first one represents the log-probability of *x*. In this way, both cases are embedded in a single expression. The optimization can be achieved by computing the gradient ∇L with respect to the parameters and setting it equal to *0*. The full derivation is not complex but requires several manipulations; hence, it's left as an exercise to the reader.

In terms of information theory, it is equivalent to minimizing the cross-entropy between a target distribution *p(x)* and an approximated one *q(x)*:

$$H(p, q) = -\sum_{x \in X} p(x) \log q(x)$$

In particular, if log_2 is adopted, the function expresses the number of extra bits requested to encode the original distribution with the predicted one. It's obvious that when *L = 0*, the two distributions are equal. Therefore, minimizing the cross-entropy is an elegant way to optimize the prediction error when the target distributions are categorical.

Implementation and optimizations

scikit-learn implements the `LogisticRegression` class, which can solve this problem using optimized algorithms. Let's consider a toy dataset made of 500 samples:

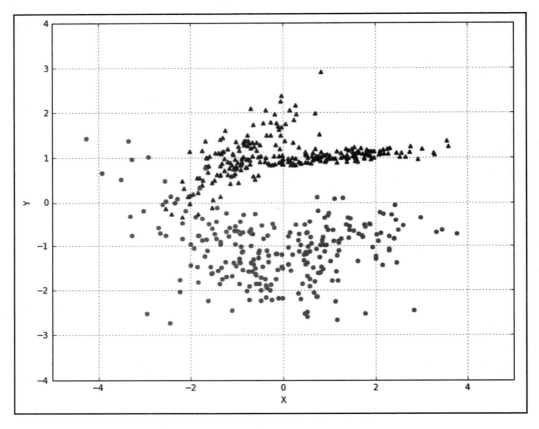

Dataset employed to test the logistic regression

The dots belong to class **0**, while the triangles belong to class **1**. To immediately test the accuracy of our classification, it's useful to split the dataset into training and test sets:

```
from sklearn.model_selection import train_test_split

X_train, X_test, Y_train, Y_test = train_test_split(X, Y, test_size=0.25)
```

Now, we can train the model using these default parameters:

```
from sklearn.linear_model import LogisticRegression

lr = LogisticRegression()
lr.fit(X_train, Y_train)

LogisticRegression(C=1.0, class_weight=None, dual=False, fit_intercept=True,
          intercept_scaling=1, max_iter=100, multi_class='ovr', n_jobs=1,
          penalty='l2', random_state=None, solver='liblinear', tol=0.0001,
          verbose=0, warm_start=False)

print(lr.score(X_test, Y_test))
0.95199999999999996
```

It's also possible to check the quality through a cross-validation (like for linear regression):

```
from sklearn.model_selection import cross_val_score

print(cross_val_score(lr, X, Y, scoring='accuracy', cv=10))
array([ 0.96078431,  0.92156863,  0.96      ,  0.98      ,  0.96      ,
        0.98      ,  0.96      ,  0.96      ,  0.91836735,  0.97959184])
```

The classification task was successful without any further action (also confirmed by cross-validation) and it's also possible to check the resulting hyperplane parameters:

```
print(lr.intercept_)
-0.64154943

print(lr.coef_)
[ 0.34417875,  3.89362924]
```

In the following diagram, there's a representation of this hyperplane (a line), where it's possible to see how the classification works and what samples are misclassified. Considering the local density of the two blocks, it's easy to see that the misclassifications happened for outliers and for some borderline samples. The latter can be controlled by adjusting the hyperparameters, even if a trade-off is often necessary. For example, if we want to include the four right dots on the separation line, this could exclude some elements in the right part. Later on, we're going to see how to find the optimal solution. However, when a linear classifier can easily find a separating hyperplane (even with a few outliers), we can say that the problem is linearly modelable; otherwise, more sophisticated non-linear techniques must be taken into account:

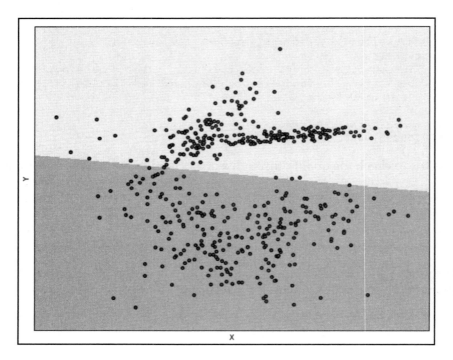

Separating line and classification areas

Just like for linear regression, it's possible to impose norm conditions on the weights. In particular, the actual function becomes this:

$$L_{reg} = \begin{cases} L + \alpha \|\bar{\theta}\|_1 \\ L + \alpha \|\bar{\theta}\|_2^2 \end{cases}$$

The behavior is the same as explained in Chapter 4, *Regression Algorithms*. Both produce a shrinkage, but L_1 forces sparsity. This can be controlled using the parameter penalty (whose values can be L_1 or L_2) and C, which is the inverse regularization factor (α^{-1}), so bigger values reduce the strength, while smaller ones (in particular, those less than *1*) force the weights to move closer to the origin. Moreover, L_1 will prefer vertexes (where all but one component are null), so it's a good idea to apply SelectFromModel to optimize the actual features after shrinkage.

Stochastic gradient descent algorithms

After discussing the basics of logistic regression, it's useful to introduce the SGDClassifier class, which implements a very common algorithm that can be applied to several different loss functions. The idea behind SGD is to minimize a cost function by iterating a weight update based on the gradient:

$$\bar{\theta}(k+1) = \bar{\theta}(k) - \gamma \nabla L$$

However, instead of considering the whole dataset, the update procedure is applied on batches randomly extracted from it (for this reason, it is often also called **mini-batch gradient descent**). In the preceding formula, *L* is the cost function we want to minimize with respect to the parameters (as discussed in Chapter 2, *Important Elements in Machine Learning*) and γ (eta0 in scikit-learn) is the learning rate, a parameter that can be constant or decayed while the learning process proceeds. The learning_rate hyperparameter can also be left with its default value (optimal), which is computed internally according to the regularization factor. In the following diagram, there's a schematic representation of the process:

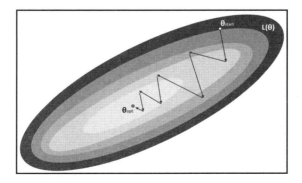

Example of gradient descent; the step size is proportional to the learning rate γ

The process should end when the weights stop modifying or their variation keeps itself under a selected threshold. This means that an optimum has been reached. In the simplest scenarios, we can be sure that θ_{opt} is the global minimum; however, in more complex models (for example, deep neural networks), the risk of sub-optimal equilibrium points is very high. In these case, more complex strategies must be implemented to reach maximum accuracy (for further details about the optimization algorithm, please read *Chapter 9* of *Mastering Machine Learning Algorithms, Bonaccorso G., Packt Publishing, 2018*). The scikit-learn implementation uses the `n_iter` parameter to define the maximum number of desired iterations and a `tol` tolerance that allows specifying a threshold for the parameter modification ratio.

There are many possible loss functions, but in this chapter we consider only log (which corresponds to cross-entropy) and perceptron. Other alternatives (such as the hinge loss) will be discussed in the next chapters. The former implements a logistic regression, while the latter (which is also available as the `Perceptron` autonomous class) is the simplest neural network, composed of a single layer of weights θ, a fixed constant called *bias* (*b*), and a binary output function:

$$z = \bar{\theta}^T \cdot \bar{x} + b$$

The output function (which classifies in two classes) is as follows:

$$y = \begin{cases} 1 & if \ z > 0 \\ 0 & if \ z \leqslant 0 \end{cases}$$

The differences between a Perceptron and a logistic regression are the output function (sign versus sigmoid) and the training model (with the loss function). A Perceptron, in fact, is normally trained by minimizing the mean square distance between the actual value and prediction:

$$L = \frac{1}{n} \sum_i \|y_i - \tilde{y}_i\|_2^2$$

Just like any other linear classifier, a Perceptron is not able to solve complex non-linear problems. Hence, our example will be generated using the built-in `make_classification` function:

```
from sklearn.datasets import make_classification

nb_samples = 500

X, Y = make_classification(n_samples=nb_samples, n_features=2,
n_informative=2, n_redundant=0, n_clusters_per_class=1)
```

In this way, we can generate 500 samples split into two classes:

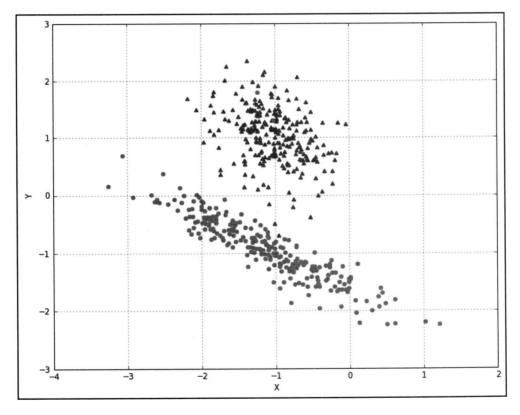

Bidimensional dataset used to test the Perceptron

Linear Classification Algorithms

This problem, under a determined precision threshold (that is, assuming that a small percentage of samples will be always misclassified), can be linearly solved, so our expectations are equivalent for both the Perceptron and logistic regression. In the latter case, the training strategy is focused on maximizing the likelihood of a probability distribution. Considering the dataset, the probability of a red sample belonging to class 0 must be greater than 0.5 (it's equal to 0.5 when $z = 0$, so when the point lies on the separating hyperplane) and vice versa. On the other hand, a Perceptron will adjust the hyperplane so that the dot product between a sample and the weights would be positive or negative, according to the class. In the following diagram, there's a geometrical representation of a Perceptron (where the bias is 0):

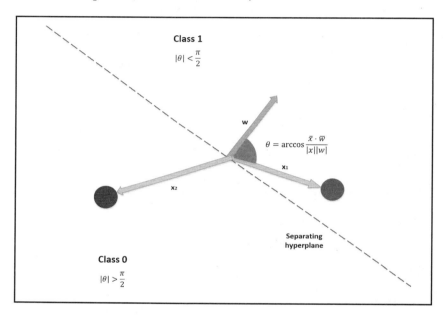

Geometrical representation of adjusting the separating hyperplane during the learning phase

The weight vector is orthogonal to the **separating hyperplane** so that the discrimination can happen only considering the sign of the dot product. An example of SGD with `perceptron` loss (without L_1/L_2 constraints) is shown as follows:

```
from sklearn.linear_model import SGDClassifier

sgd = SGDClassifier(loss='perceptron', learning_rate='optimal', n_iter=10)

cross_val_score(sgd, X, Y, scoring='accuracy', cv=10).mean()
0.98595918367346935
```

The same result can be obtained by directly using the `Perceptron` class:

```
from sklearn.linear_model import Perceptron

perc = Perceptron(n_iter=10)

cross_val_score(perc, X, Y, scoring='accuracy', cv=10).mean()
0.98195918367346935
```

The Perceptron was the first neural network, proposed by F. Rosenblatt in 1957. However, it was criticized by many researchers (in particular, by Marvin Minsky) because of its inability to solve the XOR problem. I invite the reader to test this model on this simple dataset (made up of four bidimensional values), to better understand the limits of linear separability.

Passive-aggressive algorithms

In this section, we are going to analyze an algorithm that can be efficiently employed for online linear classifications. In fact, one of the problems with other methods is that when new samples are collected, the whole model must be retrained. The main idea proposed by Crammer et al. (in *Online Passive-Aggressive Algorithms, Crammer K., Dekel O., Keshet J., Shalev-Shwartz S., Singer Y., Journal of Machine Learning Research 7 (2006) 551–585*) is to train a model incrementally, allowing modifications of the parameters only when needed, while discarding all the updates that don't alter the equilibrium. In the original paper, three variants were proposed. In this description, we are considering the one called *PA-II* (which is the most flexible).

For simplicity, in this description we are assuming bipolar outputs *(-1, +1)*; however, there are no particular restrictions on the labeling (in the aforementioned paper, some more complex approaches are discussed; however, the general logic doesn't change). This algorithm is based on the minimization of a particular loss function (in this case, we are considering only a single value; therefore, we're adopting the naming convention *loss* instead of *cost*) called hinge (we're going to reuse it when discussing SVM):

$$L = max(0, 1 - y_i z_i) \ where \ z_i = \bar{\theta}^T \cdot \bar{x}_i$$

Linear Classification Algorithms

The value z_i is the dot product between the parameter vector θ and the sample x_i so, it can be positive, null, or negative. The actual output of the classifier is obtained as follows:

$$\tilde{y}_i = sign(z_i) = sign(\bar{\theta}^T \cdot \bar{x}_i)$$

Let's suppose that the expected class is $y_i = +1$; therefore, a correct classification occurs whenever $z_i \geq 0$ (conventionally, zero is considered as part of the +1 class). In this case, when $y_i z_i \geq 1$, $L = 0$ (which corresponds to a match with high confidence), while $L = 1 - y_i z_i$ (which is $0 < L < 1$) when the classification is correct but with a low confidence (a slight modification in θ could alter the result). On the other side, if $y_i = -1$ and $z_i \geq 0$, when $y_i z_i \geq 0$, $L = 1 - y_i z_i$ (which is always positive). Hence, our goal is always to minimize the loss function in order to obtain correct predictions with high confidence. As we are going to see in Chapter 7, *Support Vector Machines*, this strategy is equivalent to searching for a separating hyperplane so that the distance between the two closest points belonging to different classes is higher than a minimum threshold (fulfilled by the condition $y_i z_i = 1$).

The algorithm works with a sequence of couples (x_i, y_i) that are assumed not to be present at the same time. Therefore, starting with a random or null parameter vector θ_0, the update rule for a single step is as follows:

$$\begin{cases} \bar{\theta}_{i+1} = argmin_{\bar{\theta}} \frac{1}{2} \|\bar{\theta} - \bar{\theta}_i\|^2 + \alpha \zeta^2 \\ subject\ to\ L(\bar{\theta}, \bar{x}_i, y_i) \leq \zeta\ and\ \zeta \geq 0 \end{cases}$$

Instead of imposing $L = 0$, which can lead to a too restrictive classification, a slack variable ζ is employed (so that $L \leq \zeta$). With this trick, the model becomes more flexible and can converge even if some samples belonging to a class are located in the space of another class. Let's now analyze how the rule works (setting $\zeta = 0$ for simplicity). If $(\theta_i^T \cdot x_i) = y_i$, the classification is correct (with high confidence), the loss function is zero, and the optimal new parameter vector is θ_i. This means that no changes are required and the algorithm remains passive (when the classification is correct and $L > 0$, the behavior is similar to the one employed for a misclassification, with a moderate effect).

Instead, if $(\theta_i^T \cdot x_i) \neq y_i$ (let's suppose, for example, that $y_i = +1$), a misclassification has occurred and the angle γ between x_i and θ_i is greater than 90° (remember that the dot product is $|x_i||\theta_i| \cos(\gamma)$, so it's negative). This condition is represented in the following diagram:

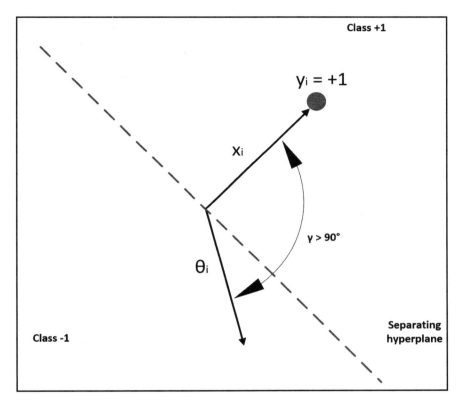

Misclassification of a sample ($y_i = +1$ and $\gamma > 90°$)

Linear Classification Algorithms

In the previous case, the solution to the problem is a new parameter vector that is as close as possible to θ_i, subject to the condition of minimizing the loss function. The algorithm is defined as aggressive for this specific reason. The research of the next parameter vector θ_{i+1} is constrained by two conditions: the first one is its proximity to the previous value (clearly we don't want to completely discard the knowledge already acquired), while the second is the minimization of L. The introduction of slack variable ζ has a mitigating effect that can be modulated by choosing the value of parameter α (C in scikit-learn). Small values ($\alpha \ll 1$) correspond to a situation very similar to the case without the slack variable. Conversely, large positive values increase the aggressiveness by reducing the magnitude of the constraint imposed on the loss function and increasing the length of the parameter vector. Hence, the problem can be solved more easily without a large separation between the classes (corresponding to L = 0 and, consequently, a tolerance for partial class overlaps) and an abrupt change in the magnitude of θ_{i+1}. The update process is shown in the following diagram:

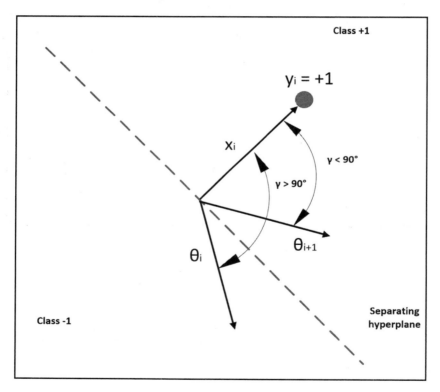

Computation of θ_{i+1} after a misclassification ($y_i = +1$)

Geometrically, the update rule for the previous case corresponds to a rotation of θ, until the angle γ becomes smaller than 90° (and so the dot product becomes positive). With the slack variable and a large α, the new parameter vector $\bar{\theta}_{i+1}$ will be forced to get closer to x_i and its magnitude is amplified (however, the overall dynamic remains unchanged). This can result in partial overlaps of the classes (which is a very common condition), but also in a reduced robustness to noise. In fact, a strong outlier can destabilize the model, but at the same time its effect can be quickly mitigated by using a few normal samples.

With a few mathematical manipulations, it's possible to solve the optimization problem in closed form:

$$\bar{\theta}_{i+1} = \bar{\theta}_i + \frac{max(0, 1 - y_i(\bar{\theta}_i^T \cdot \bar{x}_i))}{\|\bar{x}_i\|^2 + \frac{1}{2\alpha}} y_i \bar{x}_i$$

This formula simplifies the computation and speeds up the algorithm. If $x_i \in \Re^m$, the algorithm operates approximately in time $O(m^2)$ because of the dot product and the norm. However, considering the vectorization features of modern frameworks (such as NumPy), the actual computational complexity is almost constant.

To test this algorithm, we are going to employ the `PassiveAggressiveClassifier` class with the Iris dataset. The idea is to start with an initial fit using 66% of the training samples (startup set) and continuing with the remaining ones, evaluating the accuracy of the test set at each incremental step:

```
from sklearn.datasets import load_iris
from sklearn.preprocessing import StandardScaler
from sklearn.model_selection import train_test_split

iris = load_iris()

ss = StandardScaler()

X = ss.fit_transform(iris['data'])
Y = iris['target']

X_train, X_test, Y_train, Y_test = train_test_split(X, Y, test_size=0.3, random_state=1000)

nb_initial_samples = int(X_train.shape[0] / 1.5)
```

Linear Classification Algorithms

We have scaled the dataset because this kind of algorithm is very sensitive to magnitude differences; hence, it's better to have a unit variance and a zero mean. At this point, we can instantiate the model and start the initial training:

```
from sklearn.linear_model import PassiveAggressiveClassifier

pac = PassiveAggressiveClassifier(C=0.05, loss='squared_hinge',
max_iter=2000, random_state=1000)
pac.fit(X_train[0:nb_initial_samples], Y_train[0:nb_initial_samples])
```

We have chosen to reduce the aggressiveness by setting a small `C` (the default value is `1.0`). Moreover, we have specified `loss='squared_hinge'`, which corresponds to the `loss` function previously defined (PA-II) and the maximum number of iterations `max_iter=2000` when working with an SGD optimization approach. Once the model has been trained with an initial number of samples, we can proceed incrementally using the `partial_fit()` method that also requires (at least for the first call) the array of class labels (through the `classes` parameter):

```
import numpy as np

validation_accuracies = []

for (x, y) in zip(X_train[nb_initial_samples:],
Y_train[nb_initial_samples:]):
    pac.partial_fit(x.reshape(1, -1), y.ravel(),
classes=np.unique(iris['target']))
    validation_accuracies.append(pac.score(X_test, Y_test))
```

At every iteration, we compute the validation accuracy using the test set. The final plot is shown in the following graph:

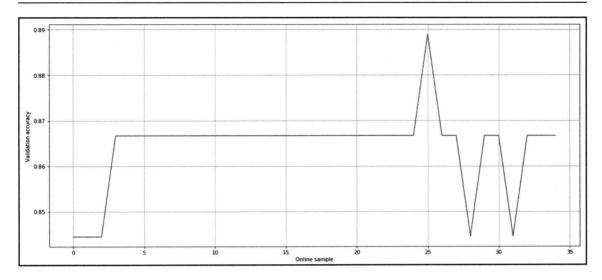

Validation accuracy after the partial fit with incremental samples

As it's possible to see, the new samples don't alter the existing equilibrium and the oscillations are always extremely small, with an average accuracy of about 87%. I invite the reader to test the algorithm with different datasets, higher C values, and smaller startup sets.

Passive-aggressive regression

We discussed the most common regression algorithms in Chapter 4, *Regression Algorithms*; however, the passive-aggressive strategy can also be employed to obtain efficient step-wise regression algorithms. We are not going to discuss all theoretical details (which can be found in the aforementioned paper), but it's helpful to introduce the *ε-insensitive* loss:

$$L = \begin{cases} 0 \ if \ |y_i - \tilde{y}_i| - \epsilon \leqslant 0 \\ |y_i - \tilde{y}_i| - \epsilon \ otherwise \end{cases}$$

This loss function is analogous to a standard hinge loss, but it has been designed to work with continuous data. The role of parameter ε is to allow a low tolerance of prediction errors. In fact the following conditions hold:

$$\begin{cases} |y_i - \tilde{y}_i| - \epsilon > 0 \Rightarrow L = |y_i - \tilde{y}_i| - \epsilon \\ |y_i - \tilde{y}_i| - \epsilon \leq 0 \Rightarrow L = 0 \end{cases}$$

In the first case, the prediction error is too high and the loss function is positive to force an adjustment (aggressive behavior). In the second case the error is tolerable and the loss function is forced to be zero (passive behavior) instead. The update rule for the parameters is analogous to the classification one:

$$\begin{cases} \bar{\theta}_{i+1} = argmin_{\bar{\theta}} \frac{1}{2} \|\bar{\theta} - \bar{\theta}_i\|^2 \\ subject\ to\ L(\bar{\theta}, \epsilon, \bar{x}_i, y_i) = 0\ and\ \epsilon \geq 0 \end{cases}$$

Considering the general logic of linear regression and passive-aggressive classification, it's not difficult to understand that when a new sample is presented, the goal of the algorithm is to find the new parameter vector θ_{i+1}; only then $L > 0$ (when $L = 0$, $\theta_{i+1} = \theta_i$). In this case, θ_{i+1} will be chosen to be as close as possible to θ_i with the condition of reducing the prediction error below threshold ε.

To test this method, we are going to create a synthetic regression dataset with 5 features and 300 samples:

```
from sklearn.datasets import make_regression

nb_samples = 300

X, Y = make_regression(n_samples=nb_samples, n_features=5,
random_state=1000)
```

Now, we can instantiate the `PassiveAggressiveRegressor` class, setting C = 0.01 (we want to avoid excessive regularization as the task is very easy); epsilon = 0.001 (the threshold for an update); loss='squared_epsilon_insensitive', which corresponds to a ε-insensitive hinge loss with a behavior analogous to a classification PA-II variant; and max_iter = 2000. In this case, however, we fit the model directly using `partial_fit()` and measure the error after each update:

```
from sklearn.linear_model import PassiveAggressiveRegressor

par = PassiveAggressiveRegressor(C=0.01,
loss='squared_epsilon_insensitive', epsilon=0.001, max_iter=2000,
```

```
random_state=1000)

squared_errors = []

for (x, y) in zip(X, Y):
    par.partial_fit(x.reshape(1, -1), y.ravel())
    y_pred = par.predict(x.reshape(1, -1))
    squared_errors.append(np.power(y_pred - y, 2))
```

The plot with the squared errors is shown in the following graph:

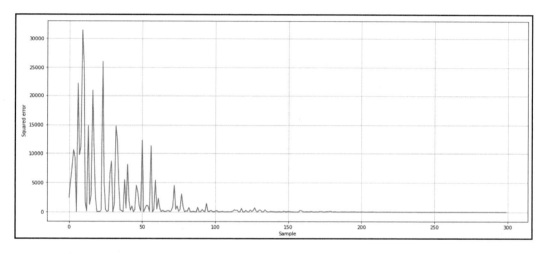

Squared errors after each passive-aggressive update

At the beginning, the algorithm is uncertain, and some corrections are canceled by subsequent ones; however, after about **100** samples, the error drops to about zero and remains constant until the last sample. It's interesting to note that, contrary to a standard linear regression, in this case the algorithm is able to change the interpolating function (without the knowledge already acquired) when new data points alter the existing trend. Unfortunately, this flexibility must be paid for with a delay between the oscillating phase and the second steady state, which is noisier than the first one because of the previous trend. For example, we can consider this dataset:

```
import numpy as np

from sklearn.datasets import make_regression

nb_samples = 500

X1, Y1 = make_regression(n_samples=nb_samples, n_features=5,
```

```
random_state=1000)
X2, Y2 = make_regression(n_samples=nb_samples, n_features=5,
random_state=1000)

X2 += np.max(X1)
Y2 += 0.5

X = np.concatenate((X1, X2))
Y = np.concatenate((Y1, Y2))
```

The first 500 samples are generated as a basic regression set. The following ones instead are shifted vertically by adding 0.5 to each of them. If we repeat the previous exercise, we get the following squared error plot:

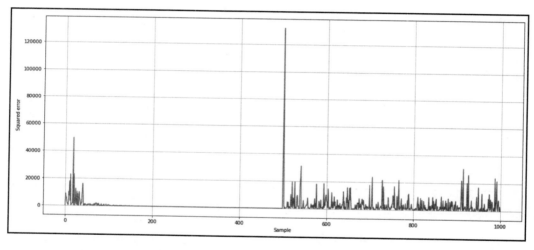

Squared errors after each passive-aggressive update in a regression with an abrupt change occurring after 500 samples

The first part is exactly as expected. However, when the 501st sample is presented, the previous model fails to predict it correctly. All the subsequent samples force the model to readapt its parameters; however, as a passive-aggressive strategy is stateful, the impact of the first samples continues to affect the result. On the other side, this kind of problem can't be easily solved using a linear regression because the step in the second dataset needs an adjustment of both coefficients and intercept. Clearly, this modification has a negative impact on performance. Moreover, passive-aggressive algorithms tend to be *conservative* with respect to previous knowledge and once they learn a long-term dependency, they are almost unable to forget it completely. I invite the reader to repeat the exercise, comparing this result with the one obtained by linear regression trained on the whole dataset (X, Y). I anticipate that the R^2 score evaluated on (X, Y) will be well below 0.5 for the reasons explained previously.

Finding the optimal hyperparameters through a grid search

Finding the best hyperparameters (they are called this because they influence the parameters learned during the training phase) is not always easy, and there are seldom good methods to start from. Personal experience (a fundamental element) must be aided by an efficient tool, such as `GridSearchCV`, which automates the training process of different models and provides the user with optimal values using cross-validation.

As an example, we show how to use grid search to find the best penalty and strength factors for logistic regression based on the Iris dataset:

```
import multiprocessing

from sklearn.datasets import load_iris
from sklearn.model_selection import GridSearchCV

iris = load_iris()

param_grid = [
    {
        'penalty': [ 'l1', 'l2' ],
        'C': [ 0.5, 1.0, 1.5, 1.8, 2.0, 2.5]
    }
]

gs = GridSearchCV(estimator=LogisticRegression(), param_grid=param_grid,
    scoring='accuracy', cv=10, n_jobs=multiprocessing.cpu_count())

gs.fit(iris.data, iris.target)

GridSearchCV(cv=10, error_score='raise',
       estimator=LogisticRegression(C=1.0, class_weight=None, dual=False,
fit_intercept=True,
          intercept_scaling=1, max_iter=100, multi_class='ovr', n_jobs=1,
          penalty='l2', random_state=None, solver='liblinear', tol=0.0001,
          verbose=0, warm_start=False),
       fit_params={}, iid=True, n_jobs=8,
       param_grid=[{'penalty': ['l1', 'l2'], 'C': [0.1, 0.2, 0.4, 0.5, 1.0,
1.5, 1.8, 2.0, 2.5]}],
       pre_dispatch='2*n_jobs', refit=True, return_train_score=True,
       scoring='accuracy', verbose=0)

print(gs.best_estimator_)
LogisticRegression(C=1.5, class_weight=None, dual=False,
```

```
fit_intercept=True,
        intercept_scaling=1, max_iter=100, multi_class='ovr', n_jobs=1,
        penalty='l1', random_state=None, solver='liblinear', tol=0.0001,
        verbose=0, warm_start=False)
print(cross_val_score(gs.best_estimator_, iris.data, iris.target,
scoring='accuracy', cv=10).mean())
0.96666666666666679
```

It's possible to insert any parameter supported by the model into the `param` dictionary with a list of values. `GridSearchCV` will process in parallel and return the best estimator (through the `best_estimator_` instance variable, which is an instance of the same classifier specified through the `estimator` parameter).

When working with parallel algorithms, scikit-learn provides the `n_jobs` parameter, which allows us to specify how many threads must be used. Setting `n_jobs=multiprocessing.cpu_count()` is useful to exploit all CPU cores available on the current machine.

In the next example, we're going to find the best parameters of an `SGDClassifier` trained with `perceptron` loss. The dataset is plotted in the following graph:

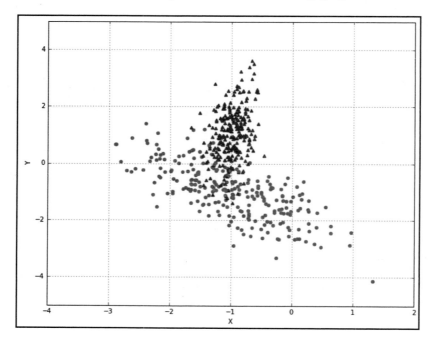

A sample dataset for grid search testing

At this point, we can create our parameter grid and perform the research:

```
import multiprocessing

from sklearn.model_selection import GridSearchCV

param_grid = [
    {
        'penalty': [ 'l1', 'l2', 'elasticnet' ],
        'alpha': [ 1e-5, 1e-4, 5e-4, 1e-3, 2.3e-3, 5e-3, 1e-2],
        'l1_ratio': [0.01, 0.05, 0.1, 0.15, 0.25, 0.35, 0.5, 0.75, 0.8]
    }
]

sgd = SGDClassifier(loss='perceptron', learning_rate='optimal')
gs = GridSearchCV(estimator=sgd, param_grid=param_grid, scoring='accuracy',
cv=10, n_jobs=multiprocessing.cpu_count())

gs.fit(X, Y)

GridSearchCV(cv=10, error_score='raise',
       estimator=SGDClassifier(alpha=0.0001, average=False,
class_weight=None, epsilon=0.1,
       eta0=0.0, fit_intercept=True, l1_ratio=0.15,
       learning_rate='optimal', loss='perceptron', n_iter=5, n_jobs=1,
       penalty='l2', power_t=0.5, random_state=None, shuffle=True,
       verbose=0, warm_start=False),
       fit_params={}, iid=True, n_jobs=8,
       param_grid=[{'penalty': ['l1', 'l2', 'elasticnet'], 'alpha': [1e-05,
0.0001, 0.0005, 0.001, 0.0023, 0.005, 0.01], 'l1_ratio': [0.01, 0.05, 0.1,
0.15, 0.25, 0.35, 0.5, 0.75, 0.8]}],
       pre_dispatch='2*n_jobs', refit=True, return_train_score=True,
       scoring='accuracy', verbose=0)

print(gs.best_score_)
0.89400000000000002

print(gs.best_estimator_)
SGDClassifier(alpha=0.001, average=False, class_weight=None, epsilon=0.1,
       eta0=0.0, fit_intercept=True, l1_ratio=0.1, learning_rate='optimal',
       loss='perceptron', n_iter=5, n_jobs=1, penalty='elasticnet',
       power_t=0.5, random_state=None, shuffle=True, verbose=0,
       warm_start=False)
```

Classification metrics

A classification task can be evaluated in many different ways to achieve specific objectives. Of course, the most important metric is the accuracy, often expressed as follows:

$$Accuracy = 1 - \frac{Number\ of\ misclassified\ samples}{Total\ number\ of\ samples}$$

For this example, we are going to use a binary test dataset obtained as follows:

```
from sklearn.datasets import make_classification
from sklearn.model_selection import train_test_split

X, Y = make_classification(n_samples=nb_samples, n_features=2,
n_informative=2, n_redundant=0,
 n_clusters_per_class=1, random_state=1000)

X_train, X_test, Y_train, Y_test = train_test_split(X, Y, test_size=0.25,
random_state=1000)
```

In scikit-learn, the accuracy can be assessed using the built-in `accuracy_score()` function:

```
from sklearn.metrics import accuracy_score

print(accuracy_score(Y_test, lr.predict(X_test)))
0.968
```

Another very common approach is based on the zero-one loss function, which we saw in `Chapter 2`, *Important Elements in Machine Learning*. It is defined as the normalized average of $L_{0/1}$ (where *1* is assigned to misclassifications) over all samples. In the following example, we show a normalized score (if it's close to zero, it's better) and then the same unnormalized value (which is the actual number of misclassifications):

```
from sklearn.metrics import zero_one_loss

print(zero_one_loss(Y_test, lr.predict(X_test)))
0.032

print(zero_one_loss(Y_test, lr.predict(X_test), normalize=False))
4.000
```

A similar but opposite metric is the **Jaccard similarity coefficient**, defined as follows:

$$J(A, B) = \frac{|A \cap B|}{|A \cup B|}$$

$$where \begin{cases} A = \{y_i : y_i \text{ is a true label}\} \\ B = \{\tilde{y}_i : \tilde{y}_i \text{ is a predicted label}\} \end{cases}$$

This index measures the similarity between two sets, A and B (the $|\cdot|$ operator is the cardinality of the set) and is bounded between 0 (worst performance) and 1 (best performance). In the former case, the intersection is null, while in the latter, the intersection and union are equal because there are no misclassifications. In scikit-learn, the implementation is as follows:

```
from sklearn.metrics import jaccard_similarity_score

print(jaccard_similarity_score(Y_test, lr.predict(X_test)))
0.968
```

These measures provide a good insight into our classification algorithms. However, in many cases, it's necessary to be able to differentiate between different kinds of misclassifications (we're considering the binary case with the conventional notation: 0-negative, 1-positive), because the relative weight is quite different. For this reason, given the ground-truth sets P and N, containing respectively the true and false samples, we introduce the following definitions:

- **True positive (TP)**: A positive sample correctly classified.
- **False positive (FP)**: A negative sample classified as positive. In statistical terms, this is called a **type I** error because the hypothesis that is checked (in this case, the negativity) is true, but it's rejected. In many scenarios, it represents a *false alert* that can always be canceled without particular implications.
- **True negative (TN)**: A negative sample correctly classified.
- **False negative (FN)**: A positive sample classified as negative. Contrary to the case of *FP*, this is called a **type II** *error*, because the hypothesis that is checked (negativity) is false, but it's not correctly rejected. In this case, the problem is generally harder, because it's equivalent to a miss that cannot be easily recovered (for example, a sick patient can be declared healthy and no countermeasures are employed immediately).

Clearly, $P = TP + FN$, while $N = TN + FP$. With this convention, the previously defined accuracy measure becomes this:

$$Accuracy = \frac{TP + TN}{P + N}$$

Confusion matrix

At first glance, a false positive and a false negative can be considered as similar errors, but think about a medical prediction: while a false positive can be easily discovered with further tests, a false negative is often neglected, with repercussions as a result of this. For this reason, it's useful to introduce the concept of a confusion matrix:

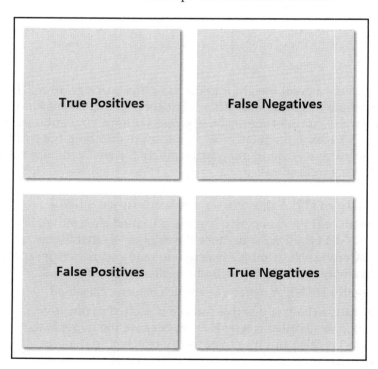

The standard structure of a binary confusion matrix

In scikit-learn, it's possible to build a confusion matrix using the `confusion_matrix` built-in function. Let's consider a generic logistic regression on dataset X with Y labels:

```
X_train, X_test, Y_train, Y_test = train_test_split(X, Y, test_size=0.25, random_state=1000)

lr = LogisticRegression()
lr.fit(X_train, Y_train)

LogisticRegression(C=1.0, class_weight=None, dual=False, fit_intercept=True,
        intercept_scaling=1, max_iter=100, multi_class='ovr', n_jobs=1,
        penalty='l2', random_state=None, solver='liblinear', tol=0.0001,
        verbose=0, warm_start=False)
```

Now we can compute our confusion matrix and immediately see how the classifier is working:

```
from sklearn.metrics import confusion_matrix

cm = confusion_matrix(y_true=Y_test, y_pred=lr.predict(X_test))

print(cm[::-1, ::-1])
[[61  1]
 [ 3 60]]
```

The last operation (`cm[::-1, ::-1]`) is needed because scikit-learn adopts inverse axes (this is a consequence of labeling the classes as 0 and 1: negatives are placed in the first row and the positives in the second one). However, in many books the confusion matrix has true values on the main diagonal, so I prefer to invert the axes.

To avoid mistakes, I suggest the reader visits http://scikit-learn.org/stable/modules/generated/sklearn.metrics.confusion_matrix.html and checks for true/false positive/negative positions.

So, we have five false negatives and two false positives. If needed, further analysis can detect the misclassifications to decide how to treat them (for example, if their variance goes over a predefined threshold, it's possible to consider them as outliers and remove them). Let's now compute a multiclass confusion matrix using the Wine dataset and logistic regression:

```
from sklearn.datasets import load_wine
from sklearn.linear_model import LogisticRegression
from sklearn.model_selection import train_test_split
```

Linear Classification Algorithms

```
wine = load_wine()

X_train, X_test, Y_train, Y_test = train_test_split(wine['data'],
wine['target'], test_size=0.25)

lr = LogisticRegression()
lr.fit(X_train, Y_train)
```

The dataset is made up of 178 samples with 13 features describing the chemical composition of each wine, and there are three target classes. In this particular example, we don't care about the accuracy; however, a standard logistic regression with an L_2 penalty achieves about 88% with a fixed split. Let's now compute the confusion matrix (I've provided the matplotlib code):

```
import matplotlib.cm as cm
import matplotlib.pyplot as plt

from sklearn.metrics import confusion_matrix

def plot_confusion_matrix(Y_test, Y_pred, targets):
    cmatrix = confusion_matrix(y_true=Y_test, y_pred=Y_pred)
    cm_fig, cm_ax = plt.subplots(figsize=(8.0, 8.0))
    cm_ax.matshow(cmatrix, cmap=cm.GnBu)

    cm_ax.set_xticklabels([''] + targets)
    cm_ax.set_yticklabels([''] + targets)

    for i in range(len(targets)):
        for j in range(len(targets)):
            cm_ax.text(x=j, y=i, s=cmatrix[i, j], va='center', ha='center', size='x-large')

    plt.title('Confusion matrix')
    plt.show()

plot_confusion_matrix(lr.predict(X_test), Y_test,
list(wine['target_names']))
```

The result is shown in the following screenshot:

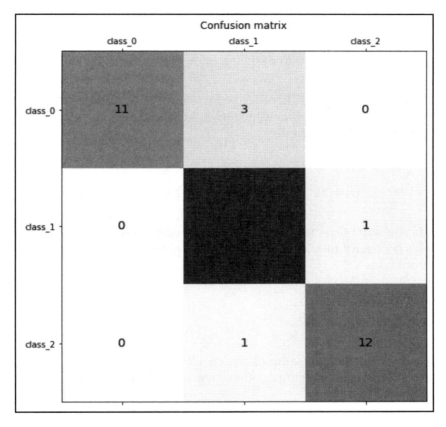

Confusion matrix for the Wine dataset classified using logistic regression

The diagonal contains the number of samples assigned to the right class, while all the other elements represent misclassifications (in this case, it doesn't make sense to consider true/false positives/negatives). For example, **class_0** is the hardest to separate and there are three samples that are wrongly assigned to **class_1**. Moreover, we can observe that the misclassifications are always in a single class. This is probably due to the differences in the chemical compositions, which are proportionally different (that is, **class_0** is more similar to **class_1** than **class_1** is to **class_2**, and so on). In general, a confusion matrix is a very powerful tool that can help the data scientist identify possible critical areas to check whether the *confusion* is due to colinearities, strong similarities, or outliers. I invite the reader to use the function shown previously with other datasets (for example, MNIST) to try to understand which sample classes are more difficult to separate and whether other models can solve this problem.

Precision

Another useful direct measure is the *precision* (or **positive predictive value**):

$$Precision = \frac{TP}{TP+FP}$$

This is directly connected to the ability to capture features that determine the positiveness of a sample, to avoid misclassification as negative. In scikit-learn, the implementation is based on the previous binary example:

```
from sklearn.metrics import precision_score
print(precision_score(Y_test, lr.predict(X_test)))
0.953
```

 If you don't flip the confusion matrix, but want to get the same measures, it's necessary to add the `pos_label=0` parameter to all metric score functions.

Recall

The ability to detect true positive samples among all the potential positives (including the false negatives) can be assessed using another measure called *recall* (sometimes defined as **sensitivity**):

$$Recall = \frac{TP}{TP+FN} = \frac{TP}{P}$$

The scikit-learn implementation is as follows:

```
from sklearn.metrics import recall_score
print(recall_score(Y_test, lr.predict(X_test)))
0.984
```

It's not surprising that we have a 98% recall with 95% precision because the number of false negatives (which directly impact the recall) is proportionally lower (1) than the number of false positives (3), which impact the precision.

The reader should understand the subtle difference between these two measures because they can dramatically affect the validity of an algorithm. In a generic scenario, *FP* and *FN* are equally weighted and the goal is therefore to maximize both precision and recall. However, there are many situations where the importance of *FN* is extremely higher. Let's suppose we are working on a fraud detection system: it's obvious that a false alarm (*FP*) is preferable than a miss (*FN*). In the former case, a further check can solve the problem without consequences, while if a fraudulent transaction is authorized, in many cases there's nothing more to do. The same happens in medical diagnostics, where an *FN* can cause the death of a patient (think about a heart attack that is misclassified as indigestion). Clearly, in these scenarios, having a high recall is a higher priority than having a system with a high precision (for example, 98%) but *FN* > *FP* (for example, *recall* = 85%).

F-Beta

A weighted harmonic mean between *precision* and *recall* is provided as follows:

$$F_{beta} = (\beta^2 + 1)\frac{Precision \cdot Recall}{\beta^2 Precision + Recall} = (\beta^2 + 1)\frac{TP}{\beta^2(FN + FP) + (1 - \beta^2)TP}$$

A beta value equal to *1* determines the so-called F_1 score, which is a perfect balance between the two measures. A beta less than *1* gives more importance to *precision* and a value greater than *1* gives more importance to *recall*. The following snippet shows how to implement this with scikit-learn:

```
from sklearn.metrics import fbeta_score

print(fbeta_score(Y_test, lr.predict(X_test), beta=1))
0.968

print(fbeta_score(Y_test, lr.predict(X_test), beta=0.75))
0.964

print(fbeta_score(Y_test, lr.predict(X_test), beta=1.25))
0.972
```

For an F_1 score, scikit-learn provides the `f1_score()` function, which is equivalent to `fbeta_score()` with `beta=1`.

The highest score is achieved by giving more importance to precision (which is higher), while the least one corresponds to a recall predominance. F_{Beta} is therefore useful to have a compact picture of the accuracy as a trade-off between high precision and a limited number of false negatives.

Cohen's Kappa

This metric is different than the other ones discussed in this section because its goal is to measure the agreement between two raters (for example, ground truth, human labeling, and estimators) considering the possibility that the raters agree without full awareness (in general, by chance). It is computed as follows:

$$\kappa = 1 - \frac{1 - p_{observed}}{1 - p_{chance}}$$

The two values represent, respectively, the observed agreement between the raters and the probability of a chance agreement. Coefficient κ is bounded between 0 (no agreement) and 1 (total agreement). In fact, if $p_{observed} = 1$ and $p_{chance} = 0$, $\kappa = 1$, while $p_{observed} = 0$ and $p_{chance} = 0$, $k = 0$. All intermediate values indicate a disagreement that can be caused either by specific choices or chance. Hence, this metric is useful when it's necessary to evaluate the possible impact of a random selection in the computation of standard accuracy. For example, a classifier could show 90% accuracy, but we know that 60% of predictions are randomized. The term p_{chance} takes into account the possibility of a correct chance prediction and corrects the estimation proportionally. In the following snippet, we show how to compute the Cohen's Kappa coefficient using the ground truth and logistic regression (as in the previous metrics):

```
from sklearn.metrics import cohen_kappa_score

print(cohen_kappa_score(Y_test, lr.predict(X_test)))
0.936
```

This value indicates that the probability of a chance agreement is almost negligible; therefore, we can fully trust the estimator. In general, when $\kappa > 0.8$, the agreement is extremely high, while a value between 0.4 and 0.8 indicates a discrete agreement with some uncertainty. Lower results show an almost complete disagreement and the estimator cannot be trusted.

There's a strong relationship between this metric and a binary confusion matrix. In fact, in this case we have obtained this result:

```
[[61  1]
 [ 3 60]]
```

We have TP = 61, FN = 1, FP = 3, and TN = 60. The parameters of Cohen's Kappa are computed as follows:

$$p_{observed} = \frac{TP + TN}{TP + TN + FP + FN} = \frac{61 + 60}{60 + 61 + 3 + 1} = 0.968$$

The probability of a chance agreement must be split into $p_{positive}$ and $p_{negative}$. Observing the confusion matrix and applying some basic probability rules, it's easy to understand that in both cases, we need to consider the product of the probabilities that both raters output a random positive/negative label.

In the first case, we get this:

$$p_{positive} = \frac{TP + FN}{TP + TN + FP + FN} \cdot \frac{TP + FP}{TP + TN + FP + FN} = \frac{62}{125} \cdot \frac{64}{125} = 0.254$$

For readers without detailed knowledge of probability theory, $p_{positive} = p^1_{positive} \cdot p^2_{positive}$ (the superscript indicates the rater). Therefore, considering the ground truth against an estimator, the first term represents the probability of total true positives (TP + FN), while the second is the probability of total predicted positives (TP + FP).

In the same way, we obtain $p_{negative}$:

$$p_{negative} = \frac{FP + TN}{TP + TN + FP + FN} \cdot \frac{FN + TN}{TP + TN + FP + FN} = \frac{63}{125} \cdot \frac{61}{125} = 0.246$$

Therefore the Cohen's Kappa coefficient becomes this:

$$\kappa = 1 - \frac{1 - p_{observed}}{1 - (p_{positive} + p_{negative})} = 1 - \frac{1 - 0.968}{1 - (0.254 + 0.246)} = 0.936$$

Global classification report

scikit-learn provides a helpful built-in function to generate a global classification report based on the most common evaluation metrics. This is an example using binary logistic regression:

```
from sklearn.metrics import classification_report

print(classification_report(Y_test, lr.predict(X_test)))

             precision    recall  f1-score   support

          0       0.98      0.95      0.97        63
          1       0.95      0.98      0.97        62

avg / total       0.97      0.97      0.97       125
```

The first column represents the classes and, for each of them, `precision`, `recall`, `f1-score`, and `support` (number of assigned samples) are computed. The last row shows the average values corresponding to every column. I highly recommend this function instead of the single metrics because of its compactness and completeness.

Learning curve

In many tasks, it's helpful to check how the number of samples impacts training performance. This can be achieved by plotting a **learning curve**, which is normally based on both training and validation scores (preferably CV score). Let's consider the Wine dataset and a simple logistic regression. scikit-learn provides the built-in `learning_curve()` function, which can automatically compute the scores for a different number of training samples:

```
import numpy as np

from sklearn.datasets import load_wine
from sklearn.model_selection import learning_curve
from sklearn.linear_model import LogisticRegression
from sklearn.utils import shuffle

wine = load_wine()

X, Y = shuffle(wine['data'], wine['target'])

tsize, training_score, test_score = learning_curve(LogisticRegression(), X, Y, cv=20, random_state=1000)
```

```
avg_tr_scores = np.mean(training_score, axis=1)
avg_test_scores = np.mean(test_score, axis=1)
```

The function requires an estimator (a `LogisticRegression` instance in our case), the X, Y arrays and an optional number of folds for the cross-validation (I also suggest to specify a fixed `random_state` in order to guarantee reproducibility). It returns three arrays containing respectively the number of training samples for each evaluation, the training accuracy and the cross-validation. As we want to plot the average curves, we need to compute the means of both training and test scores.

The resulting plot is shown in the following graph:

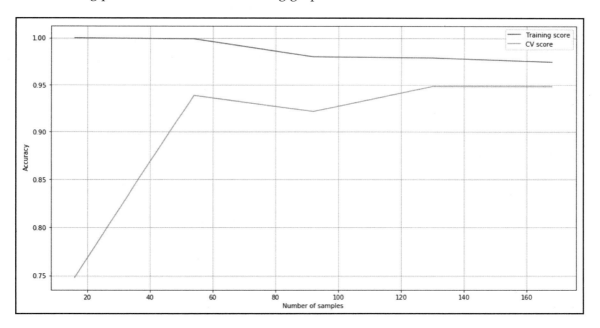

Training and CV learning curves

As you can see, the training accuracy is almost flat with a small number of samples, while it slightly decreases when other samples are added. This is mainly due to the increment in the complexity of the data structure; however, we can observe good model stability because the training accuracy saturates to an almost stable value of about 0.97. Conversely, the CV accuracy is normally subject to an increase because a larger number of training samples guarantees better generalization ability. However, the plot shows that the CV score remains constant starting from **140** samples. This means that further samples are redundant and their contribution cannot change the final configuration. In more complex scenarios where the datasets are very large, a learning curve is a valuable tool to check whether it's preferable to limit the number of samples or it's necessary to increase it to achieve the desired accuracy. The plot can be also extremely helpful to detect overfitting. In fact, when the training accuracy keeps increasing while the CV accuracy starts decreasing after a certain number of samples, it means that the variance of the model is becoming larger and larger, with a consequent loss of generalization ability (that is, the model is learning to perfectly predict only the training samples). In such cases, the solution can be to either increase the number of training samples or employ the standard regularization strategies previously discussed.

ROC curve

The **ROC curve** is a valuable tool to compare different classifiers that can assign a score to their predictions. In general, this score can be interpreted as a probability, so it's bounded between **0** and **1**. The plane is structured as shown in the following diagram:

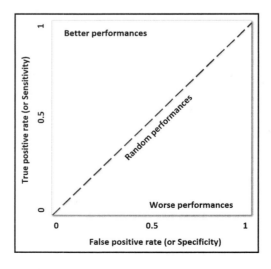

Standard structure of an ROC plane

The *x-axis* represents the increasing **false positive rate** (*1 - FPR*) also known as *1 - Specificity*, defined as follows:

$$FPR = \frac{FP}{N} = \frac{FP}{FP + TN}$$

The *y-axis* represents the **true positive rate** (**TPR**) also known as *Sensitivity*:

$$TPR = \frac{TP}{P} = \frac{TP}{TP + FN}$$

The dashed oblique line in the previous graph represents a perfectly random classifier (in a binary scenario, it's equivalent to tossing a fair coin to make every prediction), so all the curves below this threshold perform worse than a random choice, while the ones above it show better performance. Of course, the best classifier has an ROC curve split into the segments *[0, 0] - [0, 1]* and *[0, 1] - [1, 1]*. This is equivalent to a classifier with *FPR = 0* and *TPR = 1*. Therefore, our goal is to find an algorithm whose performance is as close as possible to this limit.

To show how to create an ROC curve with scikit-learn, we're going to train a model to determine the scores for the predictions (this can be achieved using the decision_function() or predict_proba() methods):

```
X_train, X_test, Y_train, Y_test = train_test_split(X, Y, test_size=0.25)

lr = LogisticRegression()
lr.fit(X_train, Y_train)

LogisticRegression(C=1.0, class_weight=None, dual=False,
fit_intercept=True,
          intercept_scaling=1, max_iter=100, multi_class='ovr', n_jobs=1,
          penalty='l2', random_state=None, solver='liblinear', tol=0.0001,
          verbose=0, warm_start=False)

Y_scores = lr.decision_function(X_test)
```

Linear Classification Algorithms

Now, we can compute the ROC curve:

```
from sklearn.metrics import roc_curve

fpr, tpr, thresholds = roc_curve(Y_test, Y_scores)
```

The output is made up of increasing true and false positive rates and decreasing thresholds (which aren't normally used for plotting the curve). Before proceeding, it's also useful to compute the **Area Under the Curve** (**AUC**), whose value is bounded between *0* (worst performance) and *1* (best performance), with a perfectly random value corresponding to *0.5* (which is the area of one of the two right triangles):

```
from sklearn.metrics import auc

print(auc(fpr, tpr))
0.97
```

We already know that our performance is rather good because the AUC is close to 1, so *FPR* << *TPR*. Now we can plot the ROC curve using matplotlib. As this book is not dedicated to this framework, I'm going to use a snippet that can be found in several examples:

```
import matplotlib.pyplot as plt

plt.figure(figsize=(8, 8))
plt.plot(fpr, tpr, color='red', label='Logistic regression (AUC: %.2f)' % auc(fpr, tpr))
plt.plot([0, 1], [0, 1], color='blue', linestyle='--')
plt.xlim([0.0, 1.0])
plt.ylim([0.0, 1.01])
plt.title('ROC Curve')
plt.xlabel('False Positive Rate')
plt.ylabel('True Positive Rate')
plt.legend(loc="lower right")

plt.show()
```

The resulting ROC curve is the following plot:

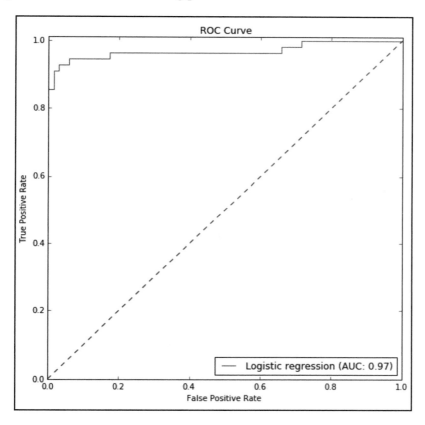

ROC curve of binary logistic regression

As confirmed by the AUC, our ROC curve shows almost optimal performance. In later chapters, we're going to use the ROC curve to visually compare different algorithms, which is probably the most helpful application of ROC curves. The last open question is this: how is an ROC curve generated? At the beginning of the section, we said that we need classifiers that are able to associate a probability to a prediction (for example, logistic regression). Normally, threshold λ to output class 0 or 1 is *0.5*, so if *p(x) < 0.5* → *y = 0* and viceversa. Let's suppose, for simplicity (even if it's a common criterion), that the probability density function for output *0* is *p(x)* and the one for output *1* is *1 - p(x)*. The curves are calculated as discrete integrals parameterized with threshold λ:

$$FPR(\lambda) = \int_{\lambda}^{1.0} p(x)dx \quad and \quad TPR(\lambda) = \int_{0.0}^{\lambda} (1 - p(x))dx$$

So, for each $\lambda \in \{0.0, 1.0\}$, we can compute two values (which are clearly bounded between *0.0* and *1.0* as they are both integrals of probability density functions) that determine a point on the curve. In practice, the integrals are approximated with sums and the set of λ values is finite. As an exercise, I invite the reader to try different parameters of the same model and plot all the ROC curves to immediately understand which global setting is preferable.

I suggest visiting `http://matplotlib.org` for further information and tutorials. One extraordinary tool is Jupyter (`http://jupyter.org`), which allows you to work with interactive notebooks where you can immediately try your code and visualize inline plots.

Summary

A linear model classifies samples using separating hyperplanes, so a problem is linearly separable if it's possible to find a linear model whose accuracy overcomes a predetermined threshold. Logistic regression is one of most famous linear classifiers, based on the principle of maximizing the probability of a sample belonging to the right class. SGD classifiers are a more generic family of algorithms, identified by the different loss functions that are adopted. SGD allows partial fitting, particularly when the amount of data is too large to be loaded in memory. A Perceptron is a particular instance of SGD, representing a linear neural network that cannot solve the `XOR` problem (for this reason, multi-layer perceptrons became the first choice for non-linear classification). However, in general, its performance is comparable to a logistic regression model.

The performances of all classifiers must be measured using different approaches, to be able to optimize their parameters or to change them when the results don't meet our requirements. We discussed different metrics and, in particular, the ROC curve, which graphically shows how the different classifiers are performing.

In the next chapter, `Chapter 6`, *Naive Bayes and Discriminant Analysis*, we're going to discuss **Naive Bayes classifiers**, which are another very famous and powerful family of algorithms. Thanks to this simple approach, it's possible to build spam filtering systems and solve apparently complex problems using only probabilities and the quality of results. Even after decades, it's still superior to, or atleast comparable to, much more complex solutions.

6
Naive Bayes and Discriminant Analysis

Naive Bayes algorithms are a family of powerful and easy-to-train classifiers that determine the probability of an outcome given a set of conditions using Bayes' theorem. The dynamic is based on the inversion of the conditional probabilities (that are associated with the causes) so that the query can be expressed as a function of measurable quantities. The approach is simple, and the adjective *naive* has been attributed not because these algorithms are limited or less efficient, but because of a fundamental assumption about the causal factors that we're going to discuss. Naive Bayes algorithms are multi-purpose classifiers, and it's easy to find their application in many different contexts. However, their performance is particularly good in all those situations, where the probability of a class is determined by the probabilities of some causal factors. Many ideal examples derive from **Natural Language Processing** (**NLP**), where a piece of text can be considered as a particular instance of a dictionary and the relative frequencies of all terms provide enough information to infer a belonging class.

In this chapter, we are going to discuss the following:

- Bayes' theorem and its applications
- Naive Bayes classifiers (Bernoulli, Multinomial, and Gaussian)
- Discriminant analysis (both linear and quadratic)

Bayes' theorem

Let's consider two probabilistic events, A and B. We can correlate the marginal probabilities $P(A)$ and $P(B)$ with the conditional probabilities $P(A|B)$ and $P(B|A)$, using the product rule:

$$\begin{cases} P(A \cap B) = P(A|B)P(B) \\ P(B \cap A) = P(B|A)P(A) \end{cases}$$

Considering that the intersection is commutative, the first members are equal, so we can derive Bayes' theorem:

$$P(A|B) = \frac{P(B|A)P(A)}{P(B)}$$

In the general discrete case, the formula can be re-expressed considering all possible outcomes for the random variable A:

$$P(A|B) = \frac{P(B|A)P(A)}{\sum_i P(B|A_i)P(A_i)}$$

As the denominator is a normalization factor, the formula is often expressed as a proportionality relationship:

$$P(A|B) \propto P(B|A)P(A)$$

This formula has very deep philosophical implications, and it's a fundamental element of statistical learning. First of all, let's consider the marginal probability, $P(A)$. This is normally a value that determines how probable a target event is, such as $P(Spam)$ or $P(Rain)$. As there are no other elements, this kind of probability is called **Apriori**, because it's often determined by mathematical or contextual considerations. For example, imagine we want to implement a very simple spam filter and we've collected 100 emails. We know that 30 are spam and that 70 are regular. So, we can say that $P(Spam) = 0.3$.

However, we'd like to evaluate using some criteria (for simplicity, let's consider a single one)—for example, email text is shorter than 50 characters. Therefore, our query becomes the following:

$$P(Spam|Text < 50chars) = \frac{P(Text < 50chars|Spam)P(Spam)}{p(Text < 50chars)}$$

The first term is similar to *P(Spam)* because it's the probability of spam given a certain condition. For this reason, it's called **A Posteriori** (in other words, it's a probability that we can estimate after knowing some additional elements). On the right-hand side, we need to calculate the missing values, but it's simple. Let's suppose that 35 emails have text shorter than 50 characters, so *P(Text < 50 chars) = 0.35*. Looking only into our spam folder, we discover that only 25 spam emails have short text, so that *P(Text < 50 chars | Spam) = 25/30 = 0.83*. The result is this:

$$P(Spam|Text < 50 chars) = \frac{0.83 \cdot 0.3}{0.35} = 0.71$$

So, after receiving a very short email, there is a 71% probability that it's spam. Now, we can understand the role of *P(Text < 50 chars | Spam)*; as we have actual data, we can measure how probable is our hypothesis given the query. In other words, we have defined a likelihood (compare this concept with the logistic regression), which is a weight between the Apriori probability and the A Posteriori one:

$$P_{APosteriori} \propto Likelihood \cdot P_{APriori}$$

The normalization factor is often represented using the letter α, so the original formula becomes this:

$$P(A|B) = \alpha P(B|A) P(A)$$

The last step is considering the case when there are more concurrent conditions (this is more realistic in real-life problems):

$$P(A|C_1 \cap C_2 \cap \ldots \cap C_n)$$

In general, this problem can become extremely complex when considering the joint prior probability. In fact, it's easy to consider the impact of single factors, but the problem can become intractable when it is expressed in the following way:

$$P(A|C_1 \cap C_2 \cap \ldots \cap C_n) = \alpha P(C_1 \cap C_2 \cap \ldots \cap C_n|A) P(A)$$

Hence, a common assumption is to consider the causes independent of one another when they are involved in causing the same effect. Formally, this assumption is called **conditional independence** and it allows employing a simplified expression:

$$P(A|C_1 \cap C_2 \cap \ldots \cap C_n) = \alpha P(C_1|A) P(C_2|A) \ldots P(C_n|A) P(A)$$

Obviously, it's easier to compute the singles $P(C_i|A)$ and multiply them instead of considering the joint probability. For example, considering the spam-detection problem, we could imagine a joint probability in this form:

$$P(Spam|Text < 50\ chars \cap "\ Shop\ " \in Text)$$

Even if the problem is very simple, we need to consider the intersection of all email messages whose text length is shorter than 50 characters and contains the word *shop*. In more complex scenarios, determining this probability is difficult because of the cross-influences among factors. The conditional independence is simply based on a null (or completely negligible) cross-influence and, as we are going to discuss, such a condition is very often met even when we expect an interaction among different causes.

Naive Bayes classifiers

A Naive Bayes classifier is so called because it's based on a naive condition, which implies the conditional independence of the causes. This can seem very difficult to accept in many contexts where the probability of a particular feature is strictly correlated to another one. For example, in spam filtering, a text shorter than 50 characters can increase the probability of the presence of an image, or if the domain has been already blacklisted for sending the same spam emails to million users, it's likely to find particular keywords. In other words, the presence of a cause isn't normally independent from the presence of other ones. However, in *The Optimality of Naive Bayes, AAAI 1, no. 2 (2004): 3, Zhang H.*, the author proved that, under particular conditions (not so rare to happen), different dependencies clear one another, and a Naive Bayes classifier succeeds in achieving very high performances even if the conditional independence is not assured.

Let's consider a dataset:

$$X = \{\bar{x}_1, \bar{x}_2, \ldots, \bar{x}_n\} \quad where \quad \bar{x}_i \in \mathbb{R}^m$$

Every feature vector, for simplicity, will be represented as follows:

$$\bar{x}_i = \left(x_i^{(1)}, x_i^{(2)}, \ldots, x_i^{(m)}\right)^T$$

We also need a target dataset (with P possible classes):

$$Y = \{y_1, y_2, \ldots, y_n\} \text{ where } y_i \in (0, 1, 2, \ldots, P-1)$$

Here, each *y* can belong to one of the *P-1* different classes. Considering Bayes' theorem under conditional independence, we can write the following:

$$P(y_i | x_i^{(1)}, x_i^{(2)}, \ldots, x_i^{(m)}) = \alpha P(y_i) \prod_j P(x_i^{(j)} | y_i)$$

The values of the marginal Apriori probability $P(y_i)$ and of the conditional probabilities $P(x_i^{(j)}|y_i)$ are generally obtained through a frequency count or a maximum likelihood estimation, therefore, given an input vector *x*, the predicted class is the one for which the A Posteriori probability is maximum. However, in many implementations (including scikit-learn), it's also possible to specify a prior for each class $P(y_i)$, so to focus the training on the optimization of the posterior distribution given the likelihood.

Naive Bayes in scikit-learn

scikit-learn implements three Naive Bayes variants based on the same number of different probabilistic distributions: Bernoulli, Multinomial, and Gaussian. The first one is a binary distribution, and is useful when a feature can be present or absent. The second one is a discrete distribution and is used whenever a feature must be represented by a whole number (for example, in NLP, it can be the frequency of a term), while the third is a continuous distribution characterized by its mean and variance.

Bernoulli Naive Bayes

If *X* is a Bernoulli-distributed random variable, it can have only two possible outcomes (for simplicity, let's call them *0* and *1*) and their probability is this:

$$p(X) = \begin{cases} p \text{ if } X = 1 \\ q \text{ if } X = 0 \end{cases} \text{ where } q = 1 - p \text{ and } 0 < p < 1$$

Naive Bayes and Discriminant Analysis

In general, the input vectors x_i are assumed to be multivariate Bernoulli distributed and each feature is binary and independent. The parameters of the model are learned according to a frequency count. Hence, if there are n samples with m features, the probability for the ith feature is this ($N_{x^{(i)}}$ counts the number of times the $i^{th} = 1$):

$$p_i = \frac{N_{\tilde{x}^{(i)}} = 1}{n}$$

To test this algorithm with scikit-learn, we're going to generate a dummy dataset. Bernoulli Naive Bayes expects binary feature vectors; however, the `BernoulliNB` class has a `binarize` parameter, which allows us to specify a threshold that will be used internally to transform the features:

```
from sklearn.datasets import make_classification

nb_samples = 300

X, Y = make_classification(n_samples=nb_samples, n_features=2, n_informative=2, n_redundant=0)
```

We have generated the bidimensional dataset shown in the following graph:

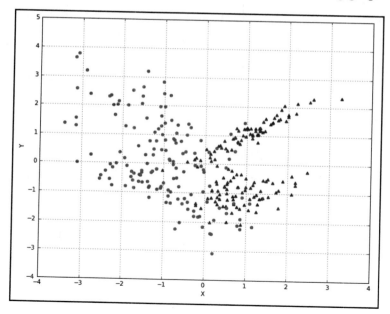

Dataset for Bernoulli Naive Bayes test

We have decided to use `0.0` as a binary threshold, so each point can be characterized by the quadrant where it's located. Of course, this is a rational choice for our dataset, but Bernoulli Naive Bayes is envisaged for binary feature vectors or continuous values, which can be precisely split with a predefined threshold:

```
from sklearn.naive_bayes import BernoulliNB
from sklearn.model_selection import train_test_split

X_train, X_test, Y_train, Y_test = train_test_split(X, Y, test_size=0.25)

bnb = BernoulliNB(binarize=0.0)
bnb.fit(X_train, Y_train)

print(bnb.score(X_test, Y_test))
0.85333333333333339
```

The score is rather good, but if we want to understand how the binary classifier worked, it's useful to see how the data has been internally binarized:

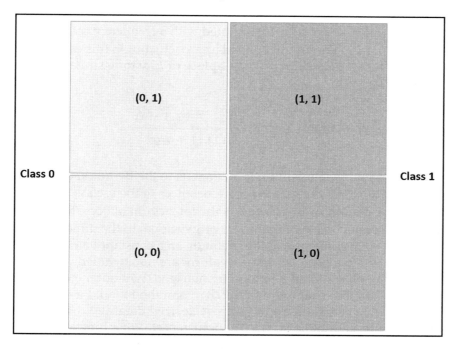

Structure of the binarized dataset

Now, checking the Naive Bayes predictions, we obtain the following:

```
data = np.array([[0, 0], [0, 1], [1, 0], [1, 1]])

print(bnb.predict(data))
[0, 0, 1, 1]
```

Multinomial Naive Bayes

A multinomial distribution is useful to model feature vectors where each value represents, for example, the number of occurrences of a term or its relative frequency. If the feature vectors have n elements and each of them can assume k different values with probability p_k, then:

$$p(X_1 = x_1 \cap X_2 = x_2 \cap \ldots \cap X_k = x_k) = \frac{n!}{\prod_i x_i!} \prod_j p_j^{x_j}$$

The conditional probabilities $P(x^{(i)}|y_j)$ are computed with a frequency count (which corresponds to applying a maximum likelihood approach), but in this case, it's important to consider also a correction parameter α (called **Laplace** or **Lidstone smoothing factor**), to avoid null probabilities:

$$p(\bar{x}^{(i)}|y_j) = \frac{(N_{\bar{x}^{(i)}} = y_j) + \alpha}{N(y_j) + m\alpha}$$

In the previous expression, the numerator is analogous to the Bernoulli case (with the addition of α), in fact, it counts the number of occurrences of the y_j class in the ith feature of the input samples. The denominator, instead, is the total count of occurrences of the y_j class (in all features) plus a correction factor which is proportional to the dimensionality of the inputs. In case of null counts, the probability defaults on a constant value. This correction is very helpful when evaluating samples whose features are not included in the training set. Even if this is a situation that should be properly managed, without the correction, their probability would be null. For example, in NLP, it's common to build a dictionary starting from a corpus of documents and then splitting a vectorized dataset (where each element contains n features representing the number of occurrences—or a function of it—of a specific word). If a word doesn't appear in any document belonging to the training set, a non-smoothed model wouldn't be able to properly manage the document in the majority of cases (that is, considering only the known elements to assign the most likely class).

The default value for α is *1.0* (in this case, it's called Laplace factor) and it prevents the model from setting null probabilities when the frequency is zero. It's possible to assign all non-negative values; however, larger values will assign higher probabilities to the missing features, and this choice could alter the stability of the model. When α < *1.0*, it's usually called the **Lidstone factor**. Clearly, if α → *0*, the effect becomes more and more negligible, returning to a scenario very similar to the Bernoulli Naive Bayes. In our example, we're going to consider the default value of *1.0*.

For our purposes, we're going to use `DictVectorizer`, already analyzed in Chapter 2, *Important Elements in Machine Learning*. There are automatic instruments to compute the frequencies of terms, but we're going to discuss them later. Let's consider only two records: the first one representing a city and the second one the countryside. Our dictionary contains hypothetical frequencies as if the terms were extracted from a text description:

```
from sklearn.feature_extraction import DictVectorizer

data = [
    {'house': 100, 'street': 50, 'shop': 25, 'car': 100, 'tree': 20},
    {'house': 5, 'street': 5, 'shop': 0, 'car': 10, 'tree': 500, 'river': 1}
]

dv = DictVectorizer(sparse=False)
X = dv.fit_transform(data)
Y = np.array([1, 0])

print(X)
[[ 100.,   100.,   0.,   25.,   50.,    20.],
 [  10.,     5.,   1.,    0.,    5.,   500.]]
```

Note that the term `'river'` is missing from the first set, so it's useful to keep α equal to `1.0` to give it a small probability. The output classes are 1 for city and 0 for the countryside. Now we can train a `MultinomialNB` instance:

```
from sklearn.naive_bayes import MultinomialNB

mnb = MultinomialNB()
mnb.fit(X, Y)

MultinomialNB(alpha=1.0, class_prior=None, fit_prior=True)
```

To test the model, we create a dummy city with a `river` and a dummy countryside place without any `river`:

```
test_data = data = [
    {'house': 80, 'street': 20, 'shop': 15, 'car': 70, 'tree': 10, 'river': 1},
    {'house': 10, 'street': 5, 'shop': 1, 'car': 8, 'tree': 300, 'river': 0}
]

print(mnb.predict(dv.fit_transform(test_data)))
[1, 0]
```

As expected, the prediction is correct. Later on, when discussing some elements of NLP, we're going to use a **Multinomial Naive Bayes** for text classification with larger corpora. Even if a multinomial distribution is based on the number of occurrences, it can be used successfully with frequencies or more complex functions.

An example of Multinomial Naive Bayes for text classification

In this example, we want to show how a Multinomial Naive Bayes model can be employed to efficiently classify text data. The dataset is called *20 Newsgroups* (http://qwone.com/~jason/20Newsgroups/) and it's already built-in in the scikit-learn (with all preprocessing steps already performed). There are 20,000 posts split into 20 categories, with a training set containing 11,314 posts and a test set with 7,532 posts. In `Chapter 13`, *Introduction to Natural Language Processing*, we are going to discuss the most common techniques that are necessary to transform a text into a numerical vector, however, for this example, it's enough to know that the dictionary contains 130,107 words and each document is a vector $x_i \in \Re^{130107}$, where every feature represents the frequency of a specific word. As a post is normally quite short (with respect to the dictionary), the vectors are extremely sparse, with the majority of values equal to *0.0* (meaning that a word is not present in the document). Our goal is to train a model that is able to assign an unknown post to the right category. In other words, we want to check whether a multinomial distribution can efficiently represent the dataset.

Let's start loading both train and test data blocks (already vectorized) using the `fetch_20newsgroups_vectorized()` function and setting the `subset` parameter equal to `'train'` and `'test'`:

```
from sklearn.datasets import fetch_20newsgroups_vectorized

train_data = fetch_20newsgroups_vectorized(subset='train')
test_data = fetch_20newsgroups_vectorized(subset='test')
```

At this point, we can instantiate and train the Multinomial Naive Bayes model (as the vectors are very sparse, we have preferred to set a Lidstone coefficient, `alpha=0.01`, because the absence of a word can yield wrong classification results. A higher value can smooth the vectors, but, in this case, many very unlikely words can negatively impact the accuracy):

```
from sklearn.naive_bayes import MultinomialNB

mnb = MultinomialNB(alpha=0.01)
mnb.fit(train_data['data'], train_data['target'])

print(mnb.score(test_data['data'], test_data['target']))
0.835103558152
```

Naive Bayes and Discriminant Analysis

The validation accuracy is rather high (83%), confirming that the documents can be modeled using a multinomial distribution, and, not surprisingly, the composition of specific words is enough to determine the right class in the majority of cases. To have a further confirmation, let's plot the confusion matrix:

Confusion matrix for the 20 Newsgroup dataset classified using a Multinomial Naive Bayes

A large number of categories are correctly detected with a minimum confusion. Moreover, we can also observe that a large percentage of errors are due to very similar categories (for example, **sci.electronics** and **comp.sys.ibm.pc.hardware** or **comp.sys.mac.hardware**). Even without analyzing the content of each post, we can assume that it's very likely to observe high similarities in all those cases and the classification probably depends only on a very small subset of discriminating words. I invite the reader to repeat the exercise using different Laplace/Lidstone coefficient values and a Bernoulli Naive Bayes. In this case, it's necessary to `binarize` the dataset using a threshold slightly higher than `0.0` (so as to remove all those elements with a very low probability), for example:

```
from sklearn.preprocessing import binarize

X_train_binary = binarize(train_data['data'], threshold=0.01)
X_test_binary = binarize(test_data['data'], threshold=0.01)
```

Gaussian Naive Bayes

Gaussian Naive Bayes is useful when working with continuous values whose probabilities can be modeled using Gaussian distributions whose means and variances are associated with each specific class (in this case, let's suppose j=1,2, ... P):

$$p(x_i|y_j) = \frac{1}{\sqrt{2\pi\sigma_j^2}} e^{-\frac{(x_i-\mu_j)^2}{2\sigma_j^2}}$$

Our goal is to estimate the mean and variance of each conditional distribution, using the maximum likelihood approach, which is quite easy, considering the mathematical nature of a Gaussian distribution. The likelihood for the whole dataset is as shown:

$$L(\mu, \sigma^2; X|Y) = \log \prod_i p(x_i|y_i) = \sum_i \log p(x_i|y_i)$$

Naive Bayes and Discriminant Analysis

Now, expanding the last term, we obtain the following expression (to avoid confusion, as the mean and the variance are associated with the y_i class, we are going to use the index j to indicate them, so they are excluded from the sum):

$$\sum_i \log p(x_i|y_i) = \sum_i \left(-\log \sqrt{2\pi} - \log \sigma_j - \frac{(x_i - \mu_j)^2}{2\sigma_j^2} \right)$$

To maximize the likelihood, we need to compute the partial derivatives with respect to μ_j and σ_j (the first term is constant and can be removed):

$$\frac{\partial L}{\partial \mu_j} = \frac{\partial}{\partial \mu_j} \sum_i \left(-\log \sigma_j - \frac{(x_i - \mu_j)^2}{2\sigma_j^2} \right) = -\sum_i \frac{(x_i - \mu_j)}{\sigma_j^2}$$

Setting the partial derivative equal to zero, we obtain a formula for the mean μ_j:

$$\mu_j = \frac{1}{n} \sum_i x_i$$

Analogously, we can compute the derivative with respect to σ_j:

$$\frac{\partial L}{\partial \sigma_j} = \frac{\partial}{\partial \sigma_j} \sum_i \left(-\log \sigma_j - \frac{(x_i - \mu_j)^2}{2\sigma_j^2} \right) = -\frac{n}{\sigma_j} + \sum_i \frac{(x_i - \mu_j)^2}{\sigma_j^2}$$

Hence, the expression for the variance σ_j^2 is this:

$$\sigma_j^2 = \frac{\sum_i (x_i - \mu_j)^2}{n}$$

It's important to remember that the index j has been introduced as an auxiliary term, but in the actual computation of the likelihood, it refers to the label assigned to the sample x_i.

As an example, we compare Gaussian Naive Bayes to a logistic regression using the ROC curves. The dataset has 300 samples with two features. Each sample belongs to a single class:

```
from sklearn.datasets import make_classification

nb_samples = 300
```

```
X, Y = make_classification(n_samples=nb_samples, n_features=2,
n_informative=2, n_redundant=0)
```

A plot of the dataset is shown in the following graph:

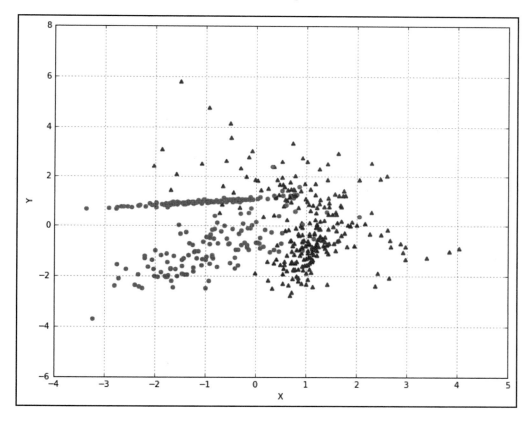

Dataset for Gaussian Naive Bayes test

Now we can train both models and generate the ROC curves (the Y scores for Naive Bayes are obtained through the `predict_proba` method):

```
from sklearn.naive_bayes import GaussianNB
from sklearn.linear_model import LogisticRegression
from sklearn.metrics import roc_curve, auc
from sklearn.model_selection import train_test_split

X_train, X_test, Y_train, Y_test = train_test_split(X, Y, test_size=0.25)

gnb = GaussianNB()
```

```
gnb.fit(X_train, Y_train)
Y_gnb_score = gnb.predict_proba(X_test)

lr = LogisticRegression()
lr.fit(X_train, Y_train)
Y_lr_score = lr.decision_function(X_test)

fpr_gnb, tpr_gnb, thresholds_gnb = roc_curve(Y_test, Y_gnb_score[:, 1])
fpr_lr, tpr_lr, thresholds_lr = roc_curve(Y_test, Y_lr_score)
```

The resulting ROC curves (generated in the same way shown in the previous chapter) are shown in the following graph:

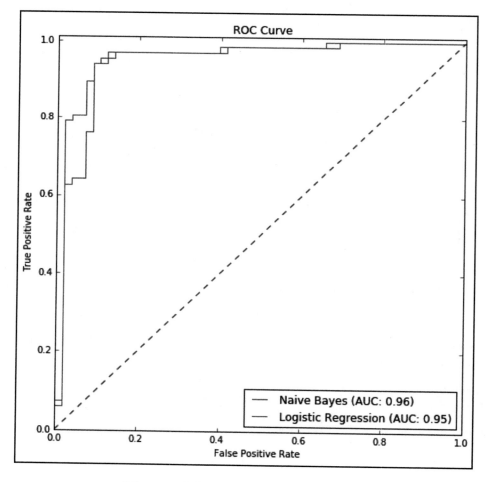

ROC curve comparing the scores of Naive Bayes versus a logistic regression

Naive Bayes' performance is slightly better than the logistic regression. However, the two classifiers have similar accuracy and **Area Under the Curve** (**AUC**). It's interesting to compare the performances of Gaussian and Multinomial Naive Bayes with the MNIST digit dataset. Each sample (belonging to 10 classes) is an 8 x 8 image encoded as an unsigned integer (0-255); therefore, even if each feature doesn't represent an actual count, it can be considered as a sort of magnitude or frequency:

```
from sklearn.datasets import load_digits
from sklearn.model_selection import cross_val_score

digits = load_digits()

gnb = GaussianNB()
mnb = MultinomialNB()

cross_val_score(gnb, digits.data, digits.target, scoring='accuracy',
cv=10).mean()
0.81035375835678214

cross_val_score(mnb, digits.data, digits.target, scoring='accuracy',
cv=10).mean()
0.88193962163008377
```

Multinomial Naive Bayes performs better than the Gaussian variant and the result is not really surprising. In fact, each sample can be thought of as a feature vector derived from a dictionary of 64 symbols and the effect of the Laplace coefficient can mitigate the deformations observed in a subset of the same digit class. The value of each feature (a pixel whose intensity is bounded between 0 and 16) is proportional to the count of each occurrence, so a multinomial distribution can better fit the data, while a Gaussian is slightly more limited by its mean and variance.

Discriminant analysis

Let's suppose we consider a multi-class classification problem where the conditional probability for a sample $x_i \in \mathcal{R}^m$ to belong to the y_j class can be modeled as a multivariate Gaussian distribution (X is assumed to be made up of **independent and identically distributed** (**i.i.d**) variables with extremely low collinearities):

$$p(\bar{x}_i | y_j) = \frac{1}{\sqrt{(2\pi)^m det(\Sigma_j)}} e^{-\frac{(\bar{x}_i - \bar{\mu}_j)^T \Sigma_j^{-1} (\bar{x}_i - \bar{\mu}_j)}{2}}$$

Naive Bayes and Discriminant Analysis

In this case, the class j is fully determined by the mean vector μ_j and the covariance matrix Σ_j. If we apply the Bayes' theorem, we can obtain the posterior probability $p(y_j|x_i)$:

$$p(y_j|\bar{x}_i) = \frac{p(\bar{x}_i|y_j)p(y_j)}{p(\bar{x}_i)}$$

Considering the discussion of Gaussian Naive Bayes, it's not difficult to understand how it's possible to estimate μ_j and Σ_j using the training set, in fact, they correspond to the sample mean and covariance and can be easily computed in closed form.

Now, for simplicity, let's consider a binary problem so that the parameters are (μ_0, Σ_0) and (μ_1, Σ_1). In the **Quadratic Discriminant Analysis (QDA)**, we predict the class a sample x_i by comparing the log-likelihoods and assigning them to the class 0 if the ratio is less than a pre-fixed threshold λ and to the class 1 if it's larger:

$$log\left(\frac{p(\bar{x}_i|y_1)}{p(\bar{x}_i|y_0)}\right) = log\left(\frac{\frac{1}{\sqrt{(2\pi)^m det(\Sigma_1)}}e^{-\frac{(\bar{x}_i-\bar{\mu}_1)^T\Sigma_1^{-1}(\bar{x}_i-\bar{\mu}_1)}{2}}}{\frac{1}{\sqrt{(2\pi)^m det(\Sigma_0)}}e^{-\frac{(\bar{x}_i-\bar{\mu}_0)^T\Sigma_0^{-1}(\bar{x}_i-\bar{\mu}_0)}{2}}}\right) =$$

$$= log\left(\sqrt{\frac{det(\Sigma_0)}{det(\Sigma_1)}}\right) + log\left(e^{-\frac{(\bar{x}_i-\bar{\mu}_1)^T\Sigma_1^{-1}(\bar{x}_i-\bar{\mu}_1)-(\bar{x}_i-\bar{\mu}_0)^T\Sigma_0^{-1}(\bar{x}_i-\bar{\mu}_0)}{2}}\right) > \lambda$$

In other words, once the parameters have been determined, a sample is assigned to the class corresponding the highest log-likelihood:

$$(\bar{x}_i - \bar{\mu}_0)^T \Sigma_0^{-1}(\bar{x}_i - \bar{\mu}_0) + log(det(\Sigma_0)) - (\bar{x}_i - \bar{\mu}_1)^T\Sigma_1^{-1}(\bar{x}_i - \bar{\mu}_1) - log(det(\Sigma_1)) > 2\lambda$$

The previous expression can be quickly computed and extended to a multi-class problem. The threshold can be also set equal to zero if we don't have any prior information that can increase the preference for a class. In case of equal covariance matrices $\Sigma_0 = \Sigma_1 = \Sigma$, the problem is called **Linear Discriminant Analysis (LDA)** and the computation can be further simplified. In fact, considering the previous expression, we can expand all the products:

$$\bar{x}_i^T\Sigma^{-1}\bar{x}_i - \bar{x}_i^T\Sigma^{-1}\bar{\mu}_0 - \bar{\mu}_0^T\Sigma^{-1}\bar{x}_i + \bar{\mu}_0^T\Sigma^{-1}\bar{\mu}_0 -$$
$$\bar{x}_i^T\Sigma^{-1}\bar{x}_i + \bar{x}_i^T\Sigma^{-1}\bar{\mu}_1 + \bar{\mu}_1^T\Sigma^{-1}\bar{x}_i - \bar{\mu}_1^T\Sigma^{-1}\bar{\mu}_1 > 2\lambda$$

As the covariance matrix is Hermitian (in the real case it simply means that $\Sigma = \Sigma^T$), we regroup the terms and cancel some elements out:

$$-2(\bar{x}_i^T \Sigma^{-1} \bar{\mu}_0) + 2(\bar{x}_i^T \Sigma^{-1} \bar{\mu}_1) + \bar{\mu}_0^T \Sigma^{-1} \bar{\mu}_0 - \bar{\mu}_1^T \Sigma^{-1} \bar{\mu}_1 > 2\lambda$$

If we keep on the left-hand side only the terms depending on x_i, we obtain a more practical expression:

$$\Sigma^{-1}(\bar{\mu}_1 - \bar{\mu}_0) \cdot \bar{x}_i > \lambda + \frac{1}{2}\left(\bar{\mu}_1^T \Sigma^{-1} \bar{\mu}_1 - \bar{\mu}_0^T \Sigma^{-1} \bar{\mu}_0\right)$$

Hence, the problem has become analogous to a perceptron (only formally); in fact, it can be expressed as the dot product $a \cdot x_i > b$, where the terms a and b are constant and need to be computed only once. Let's check now both algorithms with a bidimensional dataset where the variances of the classes are *1.0* and *100.0* respectively:

```
from sklearn.datasets import make_blobs

nb_samples = 1000

X, Y = make_blobs(n_samples=nb_samples, n_features=2, centers=2,
cluster_std=[1.0, 10.0], random_state=1000)
```

A plot of the dataset is shown in the following graph:

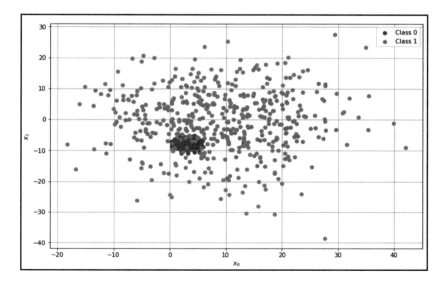

Dataset employed for discriminant analysis

The covariance matrices for the two classes is as follows:

```
import numpy as np

print(np.cov(X[Y==0].T))
[[ 1.00845543  0.0231022 ]
 [ 0.0231022   0.99716302]]

print(np.cov(X[Y==1].T))
[[ 109.48898805  -0.84146872]
 [  -0.84146872 104.89575589]]
```

As there's a large difference between the covariances (even if they are both decorrelated), we expect a higher accuracy using a QDA. Let's check both methods using the classes `LinearDiscriminantAnalysis` and `QuadraticDiscriminantAnalysis` and a 10-fold cross-validation:

```
from sklearn.discriminant_analysis import LinearDiscriminantAnalysis, QuadraticDiscriminantAnalysis

lda = LinearDiscriminantAnalysis()
print(cross_val_score(lda, X, Y, cv=10).mean())
0.831

qda = QuadraticDiscriminantAnalysis()
print(cross_val_score(qda, X, Y, cv=10).mean())
0.98
```

As expected, a QDA achieves 98% accuracy (while the linear variant is about 83%). I invite the reader to understand the dynamics of discriminant analysis with respect to other classifiers (for example, logistic regression). While in a standard linear model, we normally look for a separating hyperplane, in both LDA and QDA, we consider the likelihoods of belonging to each class. The concept is shown in the following graph (for a binary problem):

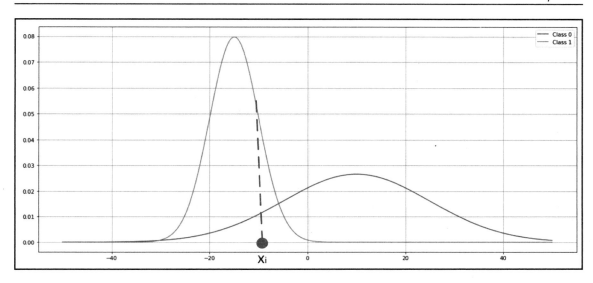

A binary classification problem where the dataset is modeled using multivariate Gaussian distributions

A point x_i has always a non-negative likelihood with respect to both Gaussians, but the likelihood for $y=1$ is quite higher than the one for $y=0$ (remember that the probabilities must be computed as definite integrals between $-\infty$ and x_i). In this way, the dataset can also be made up of partially (or even completely) overlapping subsets, and if the assumptions are met, the model is always able to distinguish with a high confidence. Clearly, there are a few drawbacks. First of all, the performances are extremely degraded by the presence of colinearities. Moreover, the Gaussianity of the dataset is often a requirement very hard to meet. Finally, even if we are in presence of a perfect dataset (in terms of assumptions), if the homogeneities of the two classes are different, the likelihood ratio can produce wrong estimations. If we consider the previous example and imagine a dense blob belonging to the class 1 and overlapping the class 0, all the points in this subset are classified as $y=0$, and the impact on the final accuracy can be dramatic. To better understand features and limitations, I invite the reader to repeat the example altering the dataset (for example, it's possible to create very dense overlapping regions or adding many samples in other areas so that the conditional probabilities are not Gaussian anymore).

Summary

In this chapter, we exposed the generic Naive Bayes approach, starting from the Bayes' theorem and its intrinsic philosophy. The naiveness of such algorithms is due to the choice to assume all the causes to be conditional independent. This means that each contribution is the same in every combination and the presence of a specific cause cannot alter the probability of the other ones. This is often unrealistic; however, under some assumptions, it's possible to show that internal dependencies clear one another so that the resulting probability appears unaffected by their relations.

scikit-learn provides three Naive Bayes implementations: Bernoulli, Multinomial, and Gaussian. The only difference between them is in the probability distribution adopted. The first one is a binary algorithm, which is particularly useful when a feature can be present or not. Multinomial assumes having feature vectors, where each element represents the number of times it appears (or, very often, its frequency). This technique is very efficient in NLP or whenever the samples are composed, starting from a common dictionary. Gaussian, instead, is based on a continuous distribution, and it's suitable for more generic classification tasks.

In the next chapter, `Chapter 7`, *Support Vector Machines* we're going to introduce a new classification technique called **Support Vector Machines**. These algorithms are very powerful for solving both linear and non-linear problems. They're often the first choice for more complex scenarios, because, despite their efficiency, the internal dynamics are very simple, and they can be trained in a very short period of time.

7
Support Vector Machines

In this chapter, we're going to introduce another approach to classification using a family of algorithms called **Support Vector Machines** (**SVMs**). They can work in both linear and non-linear scenarios, allowing high performance in many different contexts. Together with neural networks, SVMs probably represent the best choice for many tasks where it's not easy to find a good separating hyperplane. For example, for a long time, SVMs were the best choice for MNIST dataset classification, thanks to the fact that they can capture very high non-linear dynamics using a mathematical trick, without complex modifications to the algorithm. In the first part of this chapter, we're going to discuss the basics of linear SVM, which will then be used for their non-linear extensions. We'll also discuss some techniques to control the number of parameters and, in the end, the application of support vector algorithms to regression problems.

In particular, we are going to discuss the following:

- The maximum separation margin problem and linear SVM
- Kernel SVM
- Soft margins, C-SVM, and v-SVM
- **Support Vector Regression** (**SVR**)
- Introduction to **Semi-Supervised Support Vector Machines** (**S³VM**)

Linear SVM

Let's consider a dataset of feature vectors we want to classify:

$$X = \{\bar{x}_1, \bar{x}_2, \ldots, \bar{x}_n\} \ \ where \ \ \bar{x}_i \in \mathbb{R}^m$$

Support Vector Machines

For simplicity, we assume we are working with a bipolar classification (in all the other cases, it's possible to automatically use the one-versus-all strategy) and we set our class labels as -1 and 1:

$$Y = \{y_1, y_2, \ldots, y_n\} \text{ where } y_i \in \{-1, 1\}$$

Our goal is to find the best separating hyperplane, for which the equation is as follows:

$$\bar{\omega}^T \cdot \bar{x} + b = 0 \text{ where } \bar{\omega} = (\omega_1, \omega_2, \ldots, \omega_m)^T$$

In the following graph, there's a bidimensional representation of such a hyperplane:

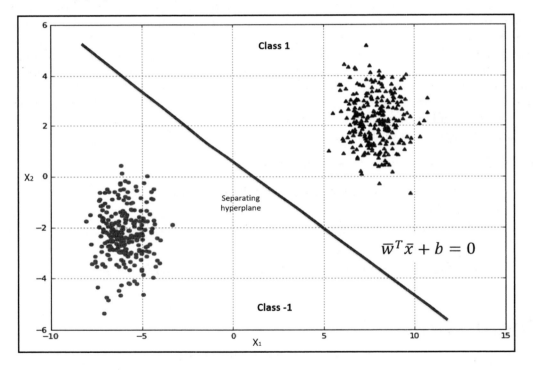

Structure of a linear SVM bipolar problem

In this way, our classifier can be written as follows:

$$\tilde{y} = f(\bar{x}) = sign\left(\bar{\omega}^T \cdot \bar{x} + b\right)$$

In a realistic scenario, the two classes are normally separated by a margin with two boundaries where a few elements lie. Those elements are called **support vectors** and the algorithm's name derives from their peculiar role. For a more generic mathematical expression, it's preferable to renormalize our dataset so that the support vectors will lie on two hyperplanes with the following equations:

$$\begin{cases} \bar{\omega}^T \cdot \bar{x} + b = -1 \\ \bar{\omega}^T \cdot \bar{x} + b = 1 \end{cases}$$

In the following graph, there's an example with two support vectors. The dashed line is the original separating hyperplane:

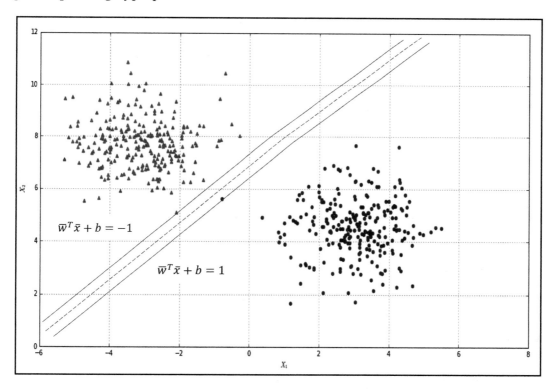

Separating hyperplane (dashed line) and support vector hyperplanes (solid lines)

Our goal is to maximize the distance between these two boundary hyperplanes to reduce the probability of misclassification (which is higher when the distance is short, and there aren't two well-defined blobs as in the previous graph).

Support Vector Machines

Considering that the boundaries are parallel, the distance between them is defined by the length of the segment perpendicular to both and connecting two points:

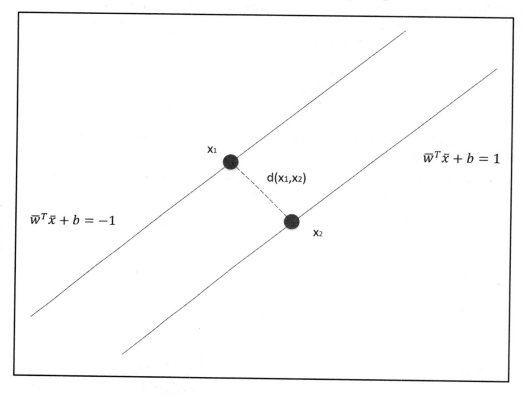

Distance between the support vector hyperplanes

Considering the points as vectors, therefore, we have the following:

$$\bar{x}_2 - \bar{x}_1 = t\bar{\omega}$$

Now, manipulating the boundary hyperplane equations, we get this:

$$\bar{\omega}^T \cdot \bar{x}_2 + b = \bar{\omega}^T \cdot (\bar{x}_1 + t\bar{\omega}) = 1 \Rightarrow (\bar{\omega}^T \cdot \bar{x}_1 + b) + t\|\bar{\omega}\|^2 = 1$$

The first term of the last part is equal to -1, so we solve for t, and we obtain this equation:

$$t = \frac{2}{\|\bar{\omega}\|^2}$$

The distance between x_1 and x_2 is the length 1 of segment t; substituting the previous expression, we can derive another equation:

$$d(\bar{x}_1, \bar{x}_2) = t\,\|\bar{\omega}\| = \frac{2}{\|\bar{\omega}\|}$$

Now, considering all the points of our dataset, we can impose the following constraint:

$$y_i(\bar{\omega}^T \cdot \bar{x}_i + b) \geqslant 1 \;\; \forall \; (\bar{x}_i, y_i)$$

This is guaranteed by using -1, 1 as class labels and boundary margins. The equality is true only for the support vectors, while for all the other points it will be greater than 1. It's important to consider that the model doesn't take into account vectors beyond this margin. In many cases, this can yield a very robust model, but in many datasets, this can also be a serious limitation. In the next paragraph, we're going to use a trick to avoid this rigidness while keeping the same optimization technique.

At this point, we can define the function to minimize (together with the constraints) to train an SVM (which is equivalent to maximizing the distance):

$$\begin{cases} min \frac{1}{2}\|\bar{\omega}\| \\ subject\ to\ y_i(\bar{\omega}^T \cdot \bar{x}_i + b) \geqslant 1 \;\; \forall \; (\bar{x}_i, y_i) \end{cases}$$

This can be further simplified (by removing the square root from the norm) in the following quadratic programming problem:

$$\begin{cases} min \frac{1}{2}\bar{\omega}^T\bar{\omega} \\ subject\ to\ y_i(\bar{\omega}^T \cdot \bar{x}_i + b) \geqslant 1 \;\; \forall \; (\bar{x}_i, y_i) \end{cases}$$

This problem is equivalent to the minimization of a hinge loss function (as already seen in passive-aggressive algorithms):

$$L = max\left(0, 1 - (\bar{\omega}^T \cdot \bar{x}_i + b)y_i)\right)$$

In fact, our goal is not only to find the optimal separating hyperplane, but also to maximize the distance between the support vectors (which are the *extreme delimiters*) when a sample x_i is correctly classified but its distance from the hyperplane is less than 1, $L > 0$ and the algorithm is forced to update the parameter vector ω, while it remains passive if $L = 0$ (a condition met by all those correctly classified samples whose distance from the separating hyperplane is larger than 1). To a certain extent, SVMs are very economic models, because they exploit the geometric properties of a dataset. As the support vectors are the closest different points (in terms of class), there's no need to care about all the other samples. When the best hyperplane has been found (only the support vectors contribute to its adjustment), $L = 0$ and no other corrections are needed.

SVMs with scikit-learn

To allow the model to have a more flexible separating hyperplane, all scikit-learn implementations are based on a simple variant that includes so-called **slack variables** ζ_i in the function to minimize (sometimes called **C-Support Vector Machines**):

$$min \frac{1}{2}\bar{\omega}^T \bar{\omega} + C \sum_i \zeta_i$$

In this case, the constraints become the following:

$$y_i(\bar{\omega}^T \cdot \bar{x}_i + b) \geqslant 1 - \zeta_i \;\; \forall \; (\bar{x}_i, y_i) \;\; and \;\; \zeta_i \geqslant 0$$

The introduction of slack variables allows us to create a flexible margin so that some vectors belonging to a class can also be found in the opposite part of the hyperspace and can be included in the model training. The strength of this flexibility can be set using the C parameter. Small values (close to zero) bring about very hard margins, while values greater than or equal to 1 allow more and more flexibility (while also increasing the misclassification rate). Equivalently, when $C \to 0$, the number of support vectors is minimized, while larger C values increase it.

The right choice of C is not immediate, but the best value can be found automatically by using a grid search, as seen in previous chapters. In our examples, we normally keep the default value of 1, even if in some cases we are going to reduce it, to increase the selectiveness.

Linear classification

Our first example is based on a linear SVM, as described in the previous section. We start by creating a dummy dataset with 500 vectors subdivided into two classes:

```
from sklearn.datasets import make_classification

nb_samples = 500

X, Y = make_classification(n_samples=nb_samples, n_features=2,
n_informative=2, n_redundant=0, n_clusters_per_class=1)
```

In the following graph, there's a plot of our dataset. Notice that some points overlap the two main blobs. For this reason, a positive C value is needed to allow the model to capture a more complex dynamic:

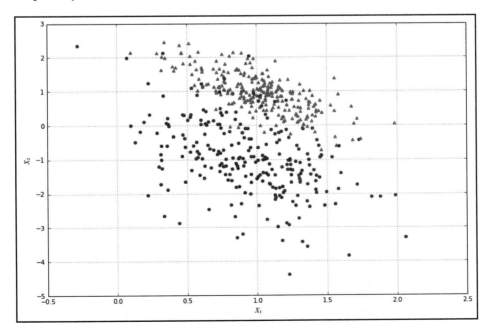

Test dataset for a linear SVM

Support Vector Machines

Scikit-learn provides the `SVC` class, which is a very efficient implementation that can be used in most cases. We're going to use it together with cross-validation to validate performance:

```
from sklearn.svm import SVC
from sklearn.model_selection import cross_val_score

svc = SVC(kernel='linear')
print(cross_val_score(svc, X, Y, scoring='accuracy', cv=10).mean())
0.93191356542617032
```

The `kernel` parameter must be set to `'linear'` in this example. In the next section, we're going to discuss how it works and how it can improve the SVM's performance dramatically in non-linear scenarios. As expected, the accuracy is comparable to a logistic regression, as this model tries to find an optimal linear separator. After training a model, it's possible to get an array of support vectors, through the `support_vectors_` instance variable.

A plot of them, for our example, is shown in the following graph:

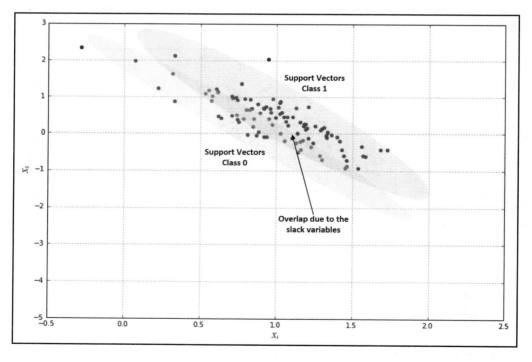

Support vectors of both classes

[216]

As it's possible to see, they are placed in a strip along the separating line. The effect of C and the slack variables determined a movable margin that partially captured the existing overlap. Of course, it's impossible to separate the sets in a perfect way with a linear classifier and most real-life problems are non-linear; for this reason, it's a necessary to employ a more complex strategy which is based on kernel functions.

Kernel-based classification

When working with non-linear problems, it's useful to transform the original vectors by projecting them into a (often higher-dimensional) space where they can be linearly separated. Let's suppose we consider a mapping function from the input sample space X to another one, V:

$$\phi(\bar{x}) : X \to V \quad \forall \, \bar{x} \in X$$

We saw a similar approach when we discussed polynomial regression. SVMs also adopt the same strategy, assuming that when the samples have been projected onto V they can be separated easily. However, now there's a complexity problem that we need to overcome. The mathematical formulation, in fact, becomes as follows:

$$\begin{cases} min \frac{1}{2}\bar{\omega}^T \bar{\omega} + C \sum_i \zeta_i \\ subject\ to\ y_i(\bar{\omega}^T \cdot \phi(\bar{x}_i) + b) \geqslant 1 - \zeta_i \ \forall \ (\bar{x}_i, y_i)\ and\ \zeta_i \geqslant 0 \end{cases}$$

Every feature vector is now filtered by a non-linear function that can completely reshape the scenario. However, the introduction of such a function generally increases the computational complexity in a way that may discourage you from using this approach. To understand what happens, it's necessary to express the quadratic problem using **Lagrange multipliers**. The entire procedure is beyond the scope of this book; however, the final formulation is as follows:

$$\begin{cases} max \left(\sum_i \alpha_i - \frac{1}{2} \sum_{i,j} \alpha_i \alpha_j y_i y_j \phi(\bar{x}_i)^T \cdot \phi(\bar{x}_j) \right) \\ subject\ to\ \sum_i \alpha_i y_i = 0 \end{cases}$$

Therefore, for every couple of vectors, it's necessary to compute the following dot product:

$$\phi(\bar{x}_i)^T \cdot \phi(\bar{x}_j)$$

Support Vector Machines

This procedure can be a bottleneck, which is unacceptable for large problems. However, it's now that the so-called **kernel trick** takes place. There are particular functions (called kernels) that have the following property:

$$K(\bar{x}_i, \bar{x}_j) = \phi(\bar{x}_i)^T \cdot \phi(\bar{x}_j)$$

In other words, the value of the kernel for two feature vectors is the product of the two projected vectors. The good news is that, according to **Mercer's theorem**, the $\Phi(x)$ function always exists whenever the kernel satisfies a particular condition (called **Mercer's condition**) on X. If we consider K as a matrix (which is a very common case), it's necessary that:

$$\sum_i \sum_j \bar{x}^{(i)} K_{ij} \bar{x}^{(j)} \geq 0 \quad \forall \bar{x} \in X$$

This simply means that K must be positive semi-definite. There's also an analogous condition for continuous functions, but in this context, it's useful to know that finding kernels it's not as complex as imagined; therefore, their application has become more and more widespread. Employing the kernel trick, the computational complexity remains almost the same, but it's possible to benefit from the power of non-linear projections, even in a very large number of dimensions.

Excluding the linear kernel, which is a simple product, scikit-learn supports three different kernels that can solve many real-life problems.

Radial Basis Function

The **Radial Basis Function** (RBF) kernel is the default value for SVC and is based on the following function:

$$K(\bar{x}_i, \bar{x}_j) = e^{-\gamma |\bar{x}_i - \bar{x}_j|^2}$$

Parameter γ determines the amplitude of the function, which is not influenced by the direction but only by the distance from the origin. This kernel is particularly helpful when the sets are concave and intersecting, for example, when a subset belonging to a class is surrounded by another one belonging to another class.

Polynomial kernel

The polynomial kernel is based on this function:

$$K(\bar{x}_i, \bar{x}_j) = (\gamma \bar{x}_i^T \cdot \bar{x}_j + r)^c$$

Exponent *c* is specified through the parameter degree, while the constant term *r* is called coef0. This function can easily expand the dimensionality with a large number of support variables and overcome very non-linear problems. The requirements in terms of resources are normally higher, but considering that a non-linear function can often be approximated quite well for a bounded area (by adopting polynomials), it's not surprising that many complex problems become easily solvable using this kernel.

Sigmoid kernel

The sigmoid kernel is based on this function:

$$K(\bar{x}_i, \bar{x}_j) = \frac{1 - e^{-2(\gamma \bar{x}_i^T \cdot \bar{x}_j + r)}}{1 + e^{-2(\gamma \bar{x}_i^T \cdot \bar{x}_j + r)}} = tanh(\gamma \bar{x}_i^T \cdot \bar{x}_j + r)$$

The constant term *r* is specified through the coef0 parameter. As pointed out in *A study on sigmoid kernels for SVM and the training of non-PSD kernels by SMO-type methods, Lin, Hsuan-Tien, and Chih-Jen Lin, Submitted to Neural Computation 3 (2003): 1-32*, if $\gamma \ll 1$ and $r < 0$, the sigmoid kernel behaves like an RBF one, however, in general, its performance is never dominant with respect to RBF or polynomial kernels. Hence, it's preferable to test the first two methods before trying this one.

Custom kernels

Normally, built-in kernels can efficiently solve most real-life problems; however, scikit-learn allows us to create custom kernels as normal Python functions:

```
import numpy as np

def custom_kernel(x1, x2):
    return np.square(np.dot(x1, x2) + 1)
```

The function can be passed to SVC through the `kernel` parameter, which can assume fixed string values (`'linear'`, `'rbf'`, `'poly'`, and `'sigmoid'`) or a callable (such as `kernel=custom_kernel`).

Non-linear examples

To show the power of kernel SVMs, we're going to solve two problems. The first one is simpler but purely non-linear and the dataset is generated through the `make_circles()` built-in function:

```
from sklearn.datasets import make_circles

nb_samples = 500
X, Y = make_circles(n_samples=nb_samples, noise=0.1)
```

A plot of this dataset is shown in the following graph:

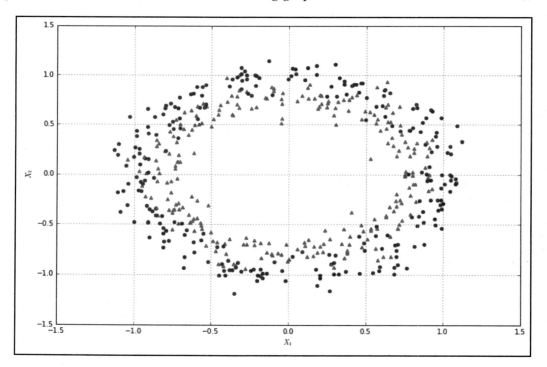

Test dataset for an RBF SVM

As it's possible to see, a linear classifier can never separate the two sets and every approximation will contain on average 50% misclassifications. A logistic regression example is shown here:

```
from sklearn.linear_model import LogisticRegression

lr = LogisticRegression()
print(cross_val_score(lr, X, Y, scoring='accuracy', cv=10).mean())
0.438
```

As expected, the accuracy is below 50% and no other optimizations can increase it dramatically. Let's consider, instead, a grid search with an SVM and different kernels (keeping the default values of each one):

```
import multiprocessing
from sklearn.model_selection import GridSearchCV

param_grid = [
    {
        'kernel': ['linear', 'rbf', 'poly', 'sigmoid'],
        'C': [ 0.1, 0.2, 0.4, 0.5, 1.0, 1.5, 1.8, 2.0, 2.5, 3.0 ]
    }
]

gs = GridSearchCV(estimator=SVC(), param_grid=param_grid,
            scoring='accuracy', cv=10,
n_jobs=multiprocessing.cpu_count())

gs.fit(X, Y)

GridSearchCV(cv=10, error_score='raise',
      estimator=SVC(C=1.0, cache_size=200, class_weight=None, coef0=0.0,
  decision_function_shape=None, degree=3, gamma='auto', kernel='rbf',
  max_iter=-1, probability=False, random_state=None, shrinking=True,
  tol=0.001, verbose=False),
      fit_params={}, iid=True, n_jobs=8,
      param_grid=[{'kernel': ['linear', 'rbf', 'poly', 'sigmoid'], 'C':
[0.1, 0.2, 0.4, 0.5, 1.0, 1.5, 1.8, 2.0, 2.5, 3.0]}],
      pre_dispatch='2*n_jobs', refit=True, return_train_score=True,
      scoring='accuracy', verbose=0)

print(gs.best_estimator_)
SVC(C=2.0, cache_size=200, class_weight=None, coef0=0.0,
  decision_function_shape=None, degree=3, gamma='auto', kernel='rbf',
  max_iter=-1, probability=False, random_state=None, shrinking=True,
  tol=0.001, verbose=False)
```

Support Vector Machines

```
print(gs.best_score_)
0.87
```

As expected from the geometry of our dataset, the best kernel is an RBF, which yields 87% accuracy. Further refinements on `gamma` could slightly increase this value, but as there is a partial overlap between the two subsets, it's very difficult to achieve an accuracy close to 100%. However, our goal is not to overfit our model; it is to guarantee an appropriate level of generalization. So, considering the shape, a limited number of misclassifications is acceptable, to ensure that the model captures sub-oscillations in the boundary surface.

Another interesting example is provided by the MNIST handwritten digit dataset. We have already seen it and classified it using linear models. Now, we can try to find the best kernel with an SVM:

```
from sklearn.datasets import load_digits

digits = load_digits()

param_grid = [
    {
        'kernel': ['linear', 'rbf', 'poly', 'sigmoid'],
        'C': [ 0.1, 0.2, 0.4, 0.5, 1.0, 1.5, 1.8, 2.0, 2.5, 3.0 ]
    }
]

gs = GridSearchCV(estimator=SVC(), param_grid=param_grid,
                  scoring='accuracy', cv=10,
n_jobs=multiprocessing.cpu_count())

gs.fit(digits.data, digits.target)

GridSearchCV(cv=10, error_score='raise',
       estimator=SVC(C=1.0, cache_size=200, class_weight=None, coef0=0.0,
   decision_function_shape=None, degree=3, gamma='auto', kernel='rbf',
   max_iter=-1, probability=False, random_state=None, shrinking=True,
   tol=0.001, verbose=False),
       fit_params={}, iid=True, n_jobs=8,
       param_grid=[{'kernel': ['linear', 'rbf', 'poly', 'sigmoid'], 'C':
[0.1, 0.2, 0.4, 0.5, 1.0, 1.5, 1.8, 2.0, 2.5, 3.0]}],
       pre_dispatch='2*n_jobs', refit=True, return_train_score=True,
       scoring='accuracy', verbose=0)

print(gs.best_estimator_)

SVC(C=0.1, cache_size=200, class_weight=None, coef0=0.0,
  decision_function_shape=None, degree=3, gamma='auto', kernel='poly',
  max_iter=-1, probability=False, random_state=None, shrinking=True,
```

```
    tol=0.001, verbose=False)

print(gs.best_score_)
0.97885364496382865
```

Hence, the best classifier (with almost 98% accuracy) is based on a polynomial kernel and a very low C value. This means that a non-linear transformation with very hard margins can easily capture the dynamics of all digits. Indeed, SVMs (with various internal alternatives) have always shown excellent performance with this dataset, and their usage can easily be extended to similar problems.

Another interesting example is based on the Olivetti face dataset, which is not part of scikit-learn but can be automatically downloaded and set up using the built-in fetch_olivetti_faces() function:

```
from sklearn.datasets import fetch_olivetti_faces

faces = fetch_olivetti_faces(data_home='/ML/faces/')
```

Through the optional data_home parameter, it is possible to specify in which local folder the dataset must be placed. A subset of samples is shown in the following screenshot:

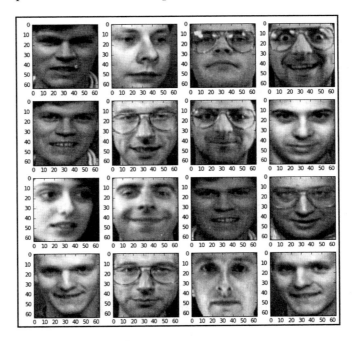

Subset of Olivetti faces dataset

Support Vector Machines

There are 40 different people, and each of them is represented by 10 grayscale pictures of 64 x 64 pixels. The number of classes (40) is not high, but considering the similarity of many photos, a good classifier should be able to capture some specific anatomical details. Performing a grid search with non-linear kernels, we get the following:

```
param_grid = [
    {
        'kernel': ['rbf', 'poly'],
        'C': [ 0.1, 0.5, 1.0, 1.5 ],
        'degree': [2, 3, 4, 5],
        'gamma': [0.001, 0.01, 0.1, 0.5]
    }
]

gs = GridSearchCV(estimator=SVC(), param_grid=param_grid,
scoring='accuracy', cv=8, n_jobs=multiprocessing.cpu_count())

gs.fit(faces.data, faces.target)

GridSearchCV(cv=8, error_score='raise',
       estimator=SVC(C=1.0, cache_size=200, class_weight=None, coef0=0.0,
    decision_function_shape=None, degree=3, gamma='auto', kernel='rbf',
    max_iter=-1, probability=False, random_state=None, shrinking=True,
    tol=0.001, verbose=False),
       fit_params={}, iid=True, n_jobs=8,
       param_grid=[{'kernel': ['rbf', 'poly'], 'C': [0.1, 0.5, 1.0, 1.5],
'gamma': [0.001, 0.01, 0.1, 0.5], 'degree': [2, 3, 4, 5]}],
       pre_dispatch='2*n_jobs', refit=True, return_train_score=True,
       scoring='accuracy', verbose=0)

print(gs.best_estimator_)
SVC(C=0.1, cache_size=200, class_weight=None, coef0=0.0,
    decision_function_shape=None, degree=2, gamma=0.1, kernel='poly',
    max_iter=-1, probability=False, random_state=None, shrinking=True,
    tol=0.001, verbose=False)
```

So, the best estimator is polynomial-based with `degree=2`, and the corresponding accuracy is the following:

```
print(gs.best_score_)
0.96999999999999997
```

This confirms the ability of SVMs to capture non-linear dynamics even with simple kernels that can be computed in a very limited amount of time. It would be interesting for the reader to try different parameter combinations or preprocess the data and apply **Principal Component Analysis (PCA)** to reduce its dimensionality.

ν-Support Vector Machines

With real datasets, SVMs can extract a very large number of support vectors to increase accuracy, and this strategy can slow down the whole process. To find a trade-off between precision and the number of support vectors, it's possible to employ a slightly different model called ν-SVM. The problem (with kernel support and n samples denoted by x_i) becomes the following:

$$\begin{cases} min \frac{1}{2}\bar{\omega}^T\bar{\omega} - \nu\tau + \frac{1}{n}\sum_i \zeta_i \\ subject\ to\ y_i(\bar{\omega}^T \cdot \phi(\bar{x}_i) + b) \geq \tau - \zeta_i\ \forall\ (\bar{x}_i, y_i)\ and\ \zeta_i \geq 0,\ \tau \geq 0 \end{cases}$$

Parameter ν is bounded between 0 (excluded) and 1, and can be used to control at the same time the number of support vectors (greater values will increase their number) and training error (lower values reduce the fraction of errors). The formal proof of these results requires us to express the problem using a Lagrangian; however, it's possible to understand the dynamics intuitively, considering the boundary cases. When $\nu \to 0$, the τ variable has no more effect on the target function. If $n > 1$, the slack variables are penalized like in C-SVM with $C < 1$. In this case, the number of support vectors becomes the minimum and, at the same time, the error increases. On the other side, when $\nu > 0$, the penalty on the target function becomes the following:

$$\alpha = \frac{1}{n}\sum_i \zeta_i - \nu\tau$$

As both ζ_i and τ are non-negative, if there's a final $\tau > 0$, it's possible to prove that the problem is equivalent to C-SVM with $C=1/n\tau$ (the formal proof is omitted); hence, when $\tau < n^{-1} \Rightarrow C > 1$ and the number of support vectors increases proportionally, one of the main advantages of ν-SVM is the possibility to tune up a single bounded parameter instead of an unbounded one. Therefore, a grid search can better explore the whole space, letting the data scientist focus only on the region where the trade-off between the number of support vectors and the margin error is optimal.

Scikit-learn provides an implementation called `NuSVC`, where parameter ν is called `nu`. To test this model, let's consider an example with a linear kernel and a simple dataset with a non-negligible overlap.

In the following graph, there's a scatter plot of our set:

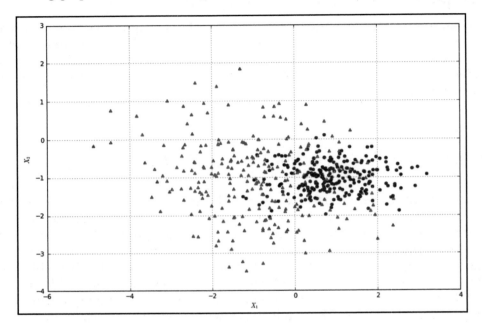

Test dataset for controlled SVM

Let's start checking the number of support vectors for a standard SVM:

```
svc = SVC(kernel='linear')
svc.fit(X, Y)

print(svc.support_vectors_.shape)
(242L, 2L)
```

So, the model has found 242 support vectors. Let's now try to optimize this number using cross-validation. The default value for v is 0.5, which is an acceptable trade-off:

```
from sklearn.svm import NuSVC

nusvc = NuSVC(kernel='linear', nu=0.5)
nusvc.fit(X, Y)

print(nusvc.support_vectors_.shape)
(251L, 2L)

print(cross_val_score(nusvc, X, Y, scoring='accuracy', cv=10).mean())
0.80633213285314143
```

As expected, the behavior is similar to a standard SVC. Let's now reduce the value of nu:

```
nusvc = NuSVC(kernel='linear', nu=0.15)
nusvc.fit(X, Y)

print(nusvc.support_vectors_.shape)
(78L, 2L)

print(cross_val_score(nusvc, X, Y, scoring='accuracy', cv=10).mean())
0.67584393757503003
```

In this case, the number of support vectors is less than before and the accuracy has also been affected by this choice. Instead of trying different values, we can look for the best choice with a grid search:

```
import numpy as np

param_grid = [
    {
        'nu': np.arange(0.05, 1.0, 0.05)
    }
]

gs = GridSearchCV(estimator=NuSVC(kernel='linear'), param_grid=param_grid,
            scoring='accuracy', cv=10,
n_jobs=multiprocessing.cpu_count())
gs.fit(X, Y)

GridSearchCV(cv=10, error_score='raise',
       estimator=NuSVC(cache_size=200, class_weight=None, coef0=0.0,
    decision_function_shape=None, degree=3, gamma='auto', kernel='linear',
    max_iter=-1, nu=0.5, probability=False, random_state=None,
    shrinking=True, tol=0.001, verbose=False),
       fit_params={}, iid=True, n_jobs=8,
       param_grid=[{'nu': array([ 0.05,   0.1 ,   0.15,  0.2 ,  0.25,  0.3 ,
 0.35,   0.4 ,   0.45,
         0.5 ,  0.55,  0.6 ,  0.65,  0.7 ,  0.75,  0.8 ,  0.85,  0.9 ,
 0.95])}],
       pre_dispatch='2*n_jobs', refit=True, return_train_score=True,
       scoring='accuracy', verbose=0)

print(gs.best_estimator_)
NuSVC(cache_size=200, class_weight=None, coef0=0.0,
    decision_function_shape=None, degree=3, gamma='auto', kernel='linear',
    max_iter=-1, nu=0.5, probability=False, random_state=None,
    shrinking=True, tol=0.001, verbose=False)
```

Support Vector Machines

```
print(gs.best_score_)
0.80600000000000005

print(gs.best_estimator_.support_vectors_.shape)
(251L, 2L)
```

Therefore, in this case the default value of 0.5 yielded the most accurate results. Normally, this approach works quite well, but when it's necessary to reduce the number of support vectors, it can be a good starting point for progressively reducing the value of nu until the result is acceptable.

Support Vector Regression

SVMs can also be efficiently employed for regression tasks. However, it's necessary to consider a slightly different loss function that can take into account the maximum discrepancy between prediction and the target value. The most common choice is the ε-insensitive loss (which we've already seen in passive-aggressive regression):

$$L = \begin{cases} 0 & if \ |y_i - \tilde{y}_i| - \epsilon \leqslant 0 \\ |y_i - \tilde{y}_i| - \epsilon & otherwise \end{cases}$$

In this case, we consider the problem as one of a standard SVM where the separating hyperplane and the (soft) margins are built sequentially to minimize the prediction error. In the following diagram, there's a schema representing this process:

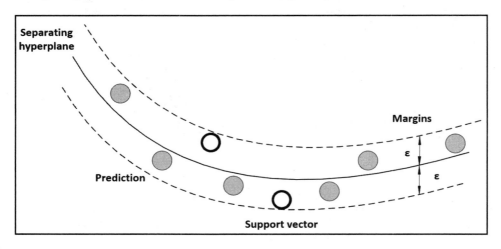

Example of Support Vector Regression; the empty circles represent two support vectors

The goal is to find the optimal parameters so that all predictions lie inside the margins (which are controlled by parameter ε). This condition minimized the ε-insensitive loss and guarantees a tolerable absolute error between target values and predictions. However, as already discussed for classification problems, the choice of hard margins can often be too restrictive. For this reason, it's preferable to introduce two sets of slack variables that also allow a prediction to be accepted as correct if the absolute error exceeds ε. Hence, the formulation of the problem becomes the following:

$$\begin{cases} min \frac{1}{2}\bar{\omega}^T\bar{\omega} + C\sum_i \left(\zeta_i + \zeta_i^*\right) \\ subject\ to \begin{cases} y_i - \bar{\omega}^T \cdot \phi(\bar{x}_i) + b \leq \epsilon - \zeta_i \\ -y_i + \bar{\omega}^T \cdot \phi(\bar{x}_i) - b \leq \epsilon - \zeta_i^* \end{cases} \forall\ (\bar{x}_i, y_i)\ and\ \zeta_i \geq 0,\ \zeta_i^* \geq 0 \end{cases}$$

Scikit-learn provides an SVR based on this algorithm. The real power of this approach (with respect to many other techniques) resides in the use of non-linear kernels (in particular, RBF and polynomials). However, the user is advised to evaluate the polynomial degree progressively because the complexity can grow rapidly, together with the training time.

For our first example, we are going to employ a dummy dataset based on a second-order noisy function:

```
import numpy as np

nb_samples = 50

X = np.arange(-nb_samples, nb_samples, 1)
Y = np.zeros(shape=(2 * nb_samples,))

for x in X:
    Y[int(x)+nb_samples] = np.power(x*6, 2.0) / 1e4 + np.random.uniform(-2, 2)
```

The dataset is plotted in the following graph:

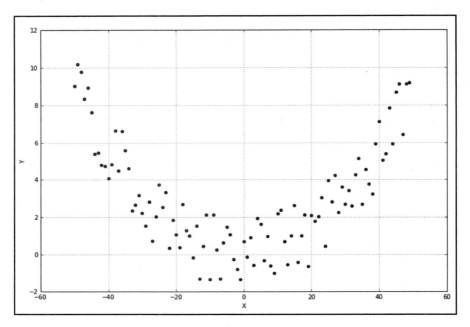

A non-linear noisy dataset for a polynomial SVR

To avoid a very long training process, the model is evaluated with `degree` set to 2. The `epsilon` parameter allows us to specify a soft margin for predictions; if a predicted value is contained in the ball centered on the target value and the radius is equal to `epsilon`, no penalty is applied to the function to be minimized. The default value is 0.1:

```
from sklearn.svm import SVR

svr = SVR(kernel='poly', degree=2, C=1.5, epsilon=0.5)

print(cross_val_score(svr, X.reshape((nb_samples*2, 1)), Y,
scoring='neg_mean_squared_error', cv=10).mean())
-1.4641683636397234
```

As it's possible to see, the CV negative mean squared error is lower than the standard deviation of the noise. To get confirmation, we can predict the regression value of all samples (the model must be fitted because the `cross_val_score` function creates internal instances):

```
svr.fit(X.reshape(-1, 1), Y.ravel())
Y_pred = svr.predict(X.reshape(-1, 1))
```

The result is shown in the following graph:

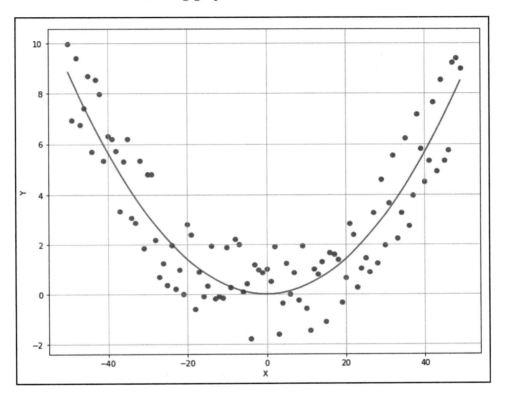

The dataset with the interpolating line

As expected, the interpolating line is a parabola that describes the high-level oscillation of the data. Employing higher degrees allows us to also capture the sub-oscillations, but this choice has two potential drawbacks: the computational cost increases and the model is more prone to overfitting the training set. The optimal choice of the hyperparameter, as usual, should be always made considering a validation set and a trade-off between training accuracy and generalization ability. In particular, when working with time series (as well as many non-stationary regressions), it's important to have an insight into the global trend, periodicities, and seasonalities. Training the model only on a small portion can hide important pieces of information, leading to an over-specialization that is not globally acceptable. The next example better exposes this problem.

An example of SVR with the Airfoil Self-Noise dataset

In this example, we are going to employ the Airfoil Self-Noise dataset freely provided by Brooks, Pope, Marcolini, and Lopez through the UCI website (for download information, please read the infobox at the end of the section). The dataset is made up of 1,503 samples (x_i), with five parameters describing the wind tunnel configuration and a dependent variable (y_i) that represents the scaled sound pressure (in dB). In this case, we want to train a regressor with 1,203 samples and test it using the remaining ones. The dataset is stored in a single TSV file; hence, we can easily load and inspect it using pandas:

```
import pandas as pd

file_path = '<DATA_PATH>\airfoil_self_noise.dat'

df = pd.read_csv(file_path, sep='\t', header=None)
```

The first operation checks the statistical properties:

```
print(df.describe())
```

The pretty output of the previous command is shown in the following table:

	0	1	2	3	4	5
count	1503.000000	1503.000000	1503.000000	1503.000000	1503.000000	1503.000000
mean	2886.380572	6.782302	0.136548	50.860745	0.011140	124.835943
std	3152.573137	5.918128	0.093541	15.572784	0.013150	6.898657
min	200.000000	0.000000	0.025400	31.700000	0.000401	103.380000
25%	800.000000	2.000000	0.050800	39.600000	0.002535	120.191000
50%	1600.000000	5.400000	0.101600	39.600000	0.004957	125.721000
75%	4000.000000	9.900000	0.228600	71.300000	0.015576	129.995500
max	20000.000000	22.200000	0.304800	71.300000	0.058411	140.987000

Statistical properties of the Airfoil Self-Noise dataset

The six attributes have very different means and standard deviations; therefore, it's preferable to scale (null mean and unitary standard deviation) them before training the model:

```
from sklearn.preprocessing import StandardScaler

X = df.iloc[:, 0:5].values
Y = df.iloc[:, 5].values

ssx, ssy = StandardScaler(), StandardScaler()

Xs = ssx.fit_transform(X)
Ys = ssy.fit_transform(Y.reshape(-1, 1))
```

At this point, we can split the dataset and train the SVM. In this example, we are going to employ an RBF SVR with `gamma=0.75`, `C=2.8`, `cache_size=500` (this value can also be kept to its default, but the performance will be slightly worse) and `epsilon=0.1`. The choice of the parameters has been made by evaluating different configurations, but I invite the reader to repeat the example using a grid search. Moreover, we have chosen to shuffle the samples using `train_test_split`, because this is not a time-series. The reader can try to use only the initial portion of the data and predict the remaining part after setting `shuffle=False` in the `split` function:

```
from sklearn.model_selection import train_test_split
from sklearn.svm import SVR

X_train, X_test, Y_train, Y_test = train_test_split(Xs, Ys.ravel(),
test_size=300, random_state=1000)

svr = SVR(kernel='rbf', gamma=0.75, C=2.8, cache_size=500, epsilon=0.1)

svr.fit(X_train, Y_train)
```

After the SVR has been trained, it's helpful to compute the R^2 scores for both training and test sets:

```
print(svr.score(X_train, Y_train))
0.896817214182

print(svr.score(X_test, Y_test))
0.871901045625
```

Support Vector Machines

 As we are not performing a CV, the R^2 scores can be influenced by the random seed. I suggest testing different values and employing the `cross_val_score()` function to get a deeper understanding of the impact of randomness on final performance.

The performance on the training set is quite good, and on the test set is only 2% worse. This confirms that the model has successfully learned the dynamics of the dataset. The plots of the original dataset and the predicted one are shown in the following graph (the values have been re-transformed into the original ones using the `StandardScaler` function `inverse_transform()`):

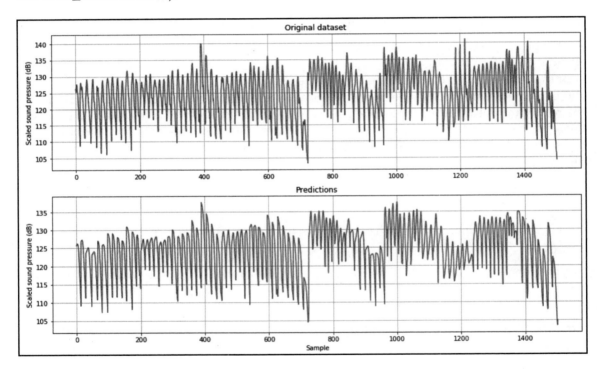

Original dataset (upper plot) and SVR predictions (lower plot)

As it's possible to see, the predictions are extremely reliable with a larger discrepancy only around sample **1200** (where the decreasing sub-trend observed is altered by three spikes). For further confirmation, it's possible to plot the absolute errors:

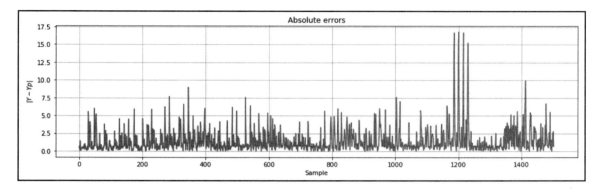

Prediction absolute errors

In the majority of cases, the error is lower than **5.0**, which is about three percent of the sound pressure (whose mean is about 125). The three spikes around sample **1200** yield an error of about **17.5**, which is *12 ÷ 14%* of the sound pressure. I invite the reader to test other hyperparameter configurations, reducing and increasing the test set size.

The Airfoil Self-Noise dataset can be downloaded from the UCI website (`https://archive.ics.uci.edu/ml/datasets/Airfoil+Self-Noise`). The `data` folder contains a single **Tab-Separated Value** (**TSV**) file with a `.dat` extension that can be easily parsed using pandas or any other standard method. The UCI website is possible thanks to the contribution of Dua, D. *and* Karra Taniskidou, E. (2017). UCI Machine-Learning Repository [`http://archive.ics.uci.edu/ml`]. Irvine, CA: University of California, School of Information and Computer Science.

Introducing semi-supervised Support Vector Machines (S³VM)

Let's suppose we have a dataset made up of N labeled points and M (normally $M \gg N$) unlabeled ones (a very common situation that arises when the labeling cost is very high). In a semi-supervised learning framework (for further details, please refer to *Mastering Machine Learning Algorithms, Bonaccorso G., Packt Publishing, 2018*), it's possible to assume that the information provided by the labeled samples is enough to understand the structure of the underlying data generating process. Clearly, this is not always true, in particular when the labeling has been done only on a portion of specific samples. However, in many cases, the assumption is realistic and, therefore, it's legitimate to ask whether it's possible to perform a full classification using only a poorly-labeled dataset. Of course, the reader must bear in mind that we don't want to train a model with only the labeled samples (this scenario defaults to a standard SVM), but we want to merge the *strong* contribution of labeled samples together with the structure of the unlabeled ones. The model we are going to present (without too many details as they are beyond the scope of this book) is called S³VM, and it has been proposed to address this kind of problem.

Without any loss of generality, let's consider a bipolar classification so that the labeled samples x_i ($i=1..N$) have a label $y_i \in \{-1, +1\}$, while the unlabeled samples x_i ($i=N+1...N+M$) are simply identified by $y_i = 0$. The structure of the target function is very similar to what we have already analyzed for a linear SVM:

$$min \frac{1}{2} \bar{\omega}^T \bar{\omega} + C \left(\sum_{i=1}^{N} \eta_i + \sum_{i=N+1}^{N+M} min(\xi_i, z_i) \right)$$

In this case, we have three sets of slack variables. The first one (η_i) defines the structure of the soft margins, considering only the labeled samples. The other two (ξ_j, z_j) impact only the unlabeled samples and are chosen using the min (•) function because our goal is to classify the sample in the half-plane when the error is the minimum. This strategy can be better understood by considering these constraints:

$$\begin{cases} y_i(\bar{\omega}^T \cdot \bar{x}_i + b) \geqslant 1 - \eta_i & \forall\ (\bar{x}_i, y_i)\ with\ i = 1..N\ and\ \eta_i \geqslant 0 \\ (\bar{\omega}^T \cdot \bar{x}_j + b) \geqslant 1 - \xi_j & \forall\ (\bar{x}_j, y_j)\ with\ j = N+1..N+M\ and\ \xi_j \geqslant 0 \\ -(\bar{\omega}^T \cdot \bar{x}_j + b) \geqslant 1 - z_j & \forall\ (\bar{x}_j, y_j)\ with\ j = N+1..N+M\ and\ z_j \geqslant 0 \end{cases}$$

The first constraint is the same as a standard SVM. The remaining ones are responsible for the choice of the optimal half-plane. In fact, if we consider a sample x_j whose optimal half-plane is identified by the label y_j = +1, the second constraint becomes $1 - \xi_j \leq K$ ($K > 0$), so $\xi_j \geq 1 - K$. The third one, instead, becomes $1 - z_j \leq -K$, which leads to $z_j \geq 1 + K$. Hence, as $z_j > \xi_j$, the slack variable to be minimized becomes $min(\xi_j, z_j) = \xi_j$, which corresponds to the +1 label.

Unfortunately, scikit-learn doesn't implement this algorithm, so we need to use SciPy to optimize the target function. As the problem is non-convex, the computational cost is high, and this can limit its usage to small datasets. However, using native libraries together with the computational power of modern personal computers allows us to solve medium to large problems in reasonable amounts of time (but always quite a bit longer than a standard SVM).

To test this algorithm, we are going to build a simple dataset (to limit the computational cost) with 50 labeled samples and 150 unlabeled ones:

```
from sklearn.datasets import make_classification

nb_samples = 200
nb_unlabeled = 150

X, Y = make_classification(n_samples=nb_samples, n_features=2,
n_redundant=0, random_state=1000)
Y[Y == 0] = -1
Y[nb_samples - nb_unlabeled:nb_samples] = 0
```

A plot of the dataset is shown in the following graph:

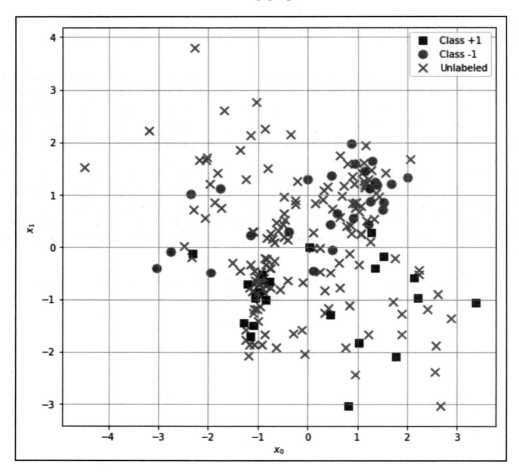

S³VM dataset; the crosses represent unlabeled samples

To optimize the target function, we are employing the SciPy `minimize()` function with the **Sequential Least SQuares Programming (SLSQP)** algorithm. Therefore, we need to initialize all the variables (we have decided to set C=0.5, but I invite the reader to test different configurations and compare the results), stack them into a single vector, and create a utility vectorized function to obtain the element-wise minimum between two arrays:

```
import numpy as np

w = np.random.uniform(-0.1, 0.1, size=X.shape[1])
```

```
eta = np.random.uniform(0.0, 0.1, size=nb_samples - nb_unlabeled)
xi = np.random.uniform(0.0, 0.1, size=nb_unlabeled)
zi = np.random.uniform(0.0, 0.1, size=nb_unlabeled)
b = np.random.uniform(-0.1, 0.1, size=1)
C = 0.5

theta0 = np.hstack((w, eta, xi, zi, b))

vmin = np.vectorize(lambda x1, x2: x1 if x1 <= x2 else x2)
```

 The `minimize()` function is very sensitive to the initial conditions, the underlying native libraries, the operating system, and the random seed. Therefore, I always suggest testing different values and selecting the optimal configuration for each specific environment.

At this point, we can create the `svm_target` function and all the constraints (three for the main inequalities and three for the non-negativity of the slack-variables):

```
import numpy as np

def svm_target(theta, Xd, Yd):
    wt = theta[0:2].reshape((Xd.shape[1], 1))
    
    s_eta = np.sum(theta[2:2 + nb_samples - nb_unlabeled])
    s_min_xi_zi = np.sum(vmin(theta[2 + nb_samples - nb_unlabeled:2 + nb_samples],
                              theta[2 + nb_samples:2 + nb_samples + nb_unlabeled]))
    
    return C * (s_eta + s_min_xi_zi) + 0.5 * np.dot(wt.T, wt)

def labeled_constraint(theta, Xd, Yd, idx):
    wt = theta[0:2].reshape((Xd.shape[1], 1))
    
    c = Yd[idx] * (np.dot(Xd[idx], wt) + theta[-1]) + \
        theta[2:2 + nb_samples - nb_unlabeled][idx] - 1.0
    
    return (c >= 0)[0]

def unlabeled_constraint_1(theta, Xd, idx):
    wt = theta[0:2].reshape((Xd.shape[1], 1))
    
    c = np.dot(Xd[idx], wt) - theta[-1] + \
        theta[2 + nb_samples - nb_unlabeled:2 + nb_samples][idx - nb_samples + nb_unlabeled] - 1.0
    
    return (c >= 0)[0]
```

```
def unlabeled_constraint_2(theta, Xd, idx):
    wt = theta[0:2].reshape((Xd.shape[1], 1))

    c = -(np.dot(Xd[idx], wt) - theta[-1]) + \
        theta[2 + nb_samples:2 + nb_samples + nb_unlabeled][idx -
nb_samples + nb_unlabeled] - 1.0

    return (c >= 0)[0]

def eta_constraint(theta, idx):
    return theta[2:2 + nb_samples - nb_unlabeled][idx] >= 0

def xi_constraint(theta, idx):
    return theta[2 + nb_samples - nb_unlabeled:2 + nb_samples][idx -
nb_samples + nb_unlabeled] >= 0

def zi_constraint(theta, idx):
    return theta[2 + nb_samples:2 + nb_samples+nb_unlabeled ][idx -
nb_samples + nb_unlabeled] >= 0
```

As we are working with a single vector containing all the variables, we always need to slice it to select only the appropriate values. The `minimize()` function allows us to add any number of constraints using a dictionary. In our case, they are all inequalities, so all of them have the `'type': 'ineq'` attribute, together with the function and the parameters to pass:

```
svm_constraints = []

for i in range(nb_samples - nb_unlabeled):
    svm_constraints.append({
        'type': 'ineq',
        'fun': labeled_constraint,
        'args': (X, Y, i)
    })
    svm_constraints.append({
        'type': 'ineq',
        'fun': eta_constraint,
        'args': (i,)
    })

for i in range(nb_samples - nb_unlabeled, nb_samples):
    svm_constraints.append({
        'type': 'ineq',
        'fun': unlabeled_constraint_1,
        'args': (X, i)
    })
    svm_constraints.append({
        'type': 'ineq',
        'fun': unlabeled_constraint_2,
```

```
            'args': (X, i)
    })
    svm_constraints.append({
            'type': 'ineq',
            'fun': xi_constraint,
            'args': (i,)
    })
    svm_constraints.append({
            'type': 'ineq',
            'fun': zi_constraint,
            'args': (i,)
    })
```

At this point, we run the optimization, setting a (`tol`) tolerance equal to `0.0001` and a maximum of `1000` iterations:

```
from scipy.optimize import minimize

result = minimize(fun=svm_target,
                  x0=theta0,
                  constraints=svm_constraints,
                  args=(X, Y),
                  method='SLSQP',
                  tol=0.0001,
                  options={'maxiter': 1000})
```

This operation (which is not very optimized) can be quite long when the number of samples is very high, so I suggest starting with smaller numbers and increasing them step by step. After the convergence, we need to extract all final parameters from the `theta` vector (which is returned as `result['x']`) and label all the unlabeled samples using the standard SVM approach, based on the sign of the dot product:

```
theta_end = result['x']
w = theta_end[0:2]
b = theta_end[-1]

Xu = X[nb_samples - nb_unlabeled:nb_samples]
yu = -np.sign(np.dot(Xu, w) + b)
```

The final result (compared with the original dataset) is shown in the following graph:

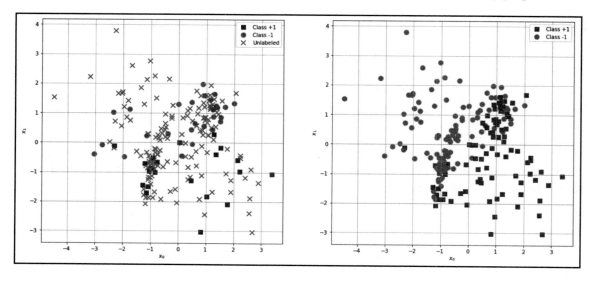

Original dataset (left); completely classified dataset (right)

As you can see, the algorithm has successfully labeled all the unlabeled samples in the most reasonable way, and the majority of potential misclassifications are due to the existing original overlap. This algorithm is not very efficient because of the high number of constraints (for n samples, there are $6n$ inequalities) and the limited level of optimization provided by SciPy. I invite the reader who is interested in this technique to check out the NLopt library (https://nlopt.readthedocs.io/en/latest/), which provides native optimization functions for the majority of operating systems. Starting from this example, it's not difficult to reproduce the structure of the problem using more performant frameworks.

Summary

In this chapter, we discussed how an SVM works in both linear and non-linear scenarios, starting with the basic mathematical formulation. The main concept is to find the hyperplane that maximizes the distance between the classes by using a limited number of samples (called support vectors) that are closest to the separation margin.

We saw how to transform a non-linear problem using kernel functions, which allows the remapping of the original space to another high-dimensional one where the problem becomes linearly separable. We also saw how to control the number of support vectors and how to use SVMs for regression problems.

In the next chapter, `Chapter 8`, *Decision Trees and Ensemble Learning*, we're going to introduce another classification method called **decision trees**, which is the last one explained in this book.

8
Decision Trees and Ensemble Learning

In this chapter, we're going to discuss Binary Decision Trees and ensemble methods. Even though they're probably not the most common methods for classification, they offer a good level of simplicity and can be adopted for many tasks that don't require a high level of complexity. They're also quite useful when it's necessary to show how a decision process works because they are based on a structure that can be easily shown in presentations and described step by step.

Ensemble methods are a powerful alternative to complex algorithms because they try to exploit the statistical concept of a majority vote. Many weak learners can be trained to capture different elements and make their own predictions, which are not globally optimal, but using a sufficient number of elements, it's statistically probable that a majority will evaluate correctly. In particular, we're going to discuss Random Forests of Decision Trees and some boosting methods that are for slightly different algorithms that can optimize the learning process by focusing on misclassified samples, or by continuously minimizing a target loss function.

In particular, we are going to discuss the following:

- The main structure of a Binary Decision Tree
- The most common impurity measures
- Decision Tree regression
- An introduction to Ensemble Learning (Random Forests, AdaBoost, Gradient Tree Boosting, and voting classifiers)

Binary Decision Trees

A Binary Decision Tree is a structure based on a sequential decision process. Starting from the root, a feature is evaluated and one of the two branches is selected. This procedure is repeated until a final leaf is reached, which normally represents the classification target we're looking for. One of the first formulations of Decision Trees is called **Iterative Dichotomizer 3 (ID3)**, and it required categorical features. This condition restricted its use and led to the development of **C4.5**, which could also manage continuous (but binned and discretized) values. Moreover, C4.5 was also known because of its ability to transform a tree into a sequence of conditional expressions (if <condition> then <...> else <...>). In this book, we are going to address the most recent development, which is called **Classification and Regression Trees (CART)**. These kinds of trees can manage both categorical and numerical features, can be employed either in classification or in regression tasks, and don't use any rule set as an internal representation.

Considering other algorithms, Decision Trees seem to be simpler in their dynamics. However, if the dataset is splittable while keeping an internal balance, the overall process is intuitive and rather fast in its predictions. Moreover, Decision Trees can work efficiently with un-normalized datasets because their internal structure is not influenced by the values assumed by each feature. In the following graph, there are plots from an unnormalized bidimensional dataset, as well as the cross-validation scores obtained using logistic regression and a Decision Tree:

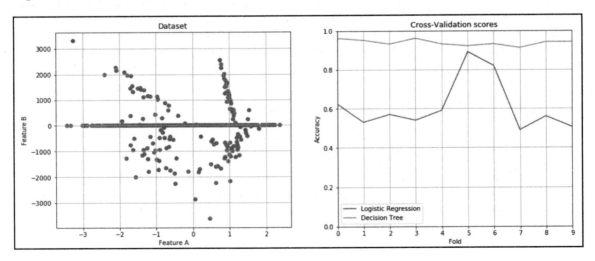

Example of a dataset with different variances (left), and cross-validation scores for a logistic regression and a Decision Tree (right)

The Decision Tree always achieves a score close to **1.0**, while the logistic regression has an average slightly greater than **0.6**. However, without proper limitations, a Decision Tree could potentially grow until a single sample (or a very low number) is present in every node. This situation drives the overfitting of the model, and the tree becomes unable to generalize correctly. Using a consistent test set or cross-validation, together with a maximum allowed depth, can help in avoiding this problem. In the section dedicated to *Decision Tree classification with scikit-learn* we're going to discuss how to limit the growth of the tree. Another important element to take into account is class balancing. Decision Trees are sensitive to unbalanced classes and can yield poor accuracy when a class is dominant. To mitigate this problem, it's possible to employ one of the resampling methods discussed in Chapter 2, *Important Elements in Machine Learning*, or use the class_weight parameter, which is provided by the scikit-learn implementations. In this way, a dominant class can be proportionally penalized, avoiding bias.

Binary decisions

Let's consider an input dataset, X:

$$X = \{\bar{x}_1, \bar{x}_2, \ldots, \bar{x}_n\} \quad where \quad \bar{x}_i \in \mathbb{R}^m$$

Every vector is made up of *m* features, so each of them is a candidate for the creation of a node based on a tuple (feature, threshold):

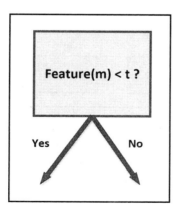

Single splitting node

According to the feature and the threshold, the structure of the tree will change. Intuitively, we should pick the feature that best separates our data. In other words, a perfectly separating feature will only be present in a node, and the two subsequent branches won't be based on it anymore. These conditions guarantee the convergence of the tree toward the final leaves, whose uncertainty is minimized. In real problems, however, this is often impossible, so it's necessary to find the feature that minimizes the number of decision steps that follow on from this.

For example, let's consider a class of students where all male students have dark hair and all females students have blonde hair, while both subsets have samples of different sizes. If our task is to determine the composition of the class, we can start with the following subdivision:

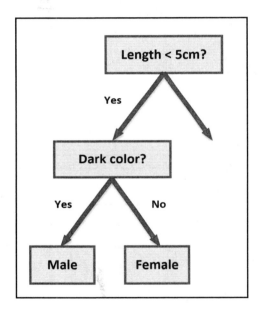

Splitting node (Length) with a branch containing another splitting node (Dark color)

However, the **Dark color?** block will contain both males and females (which are the targets we want to classify). This concept is expressed using the term *purity* (or, more often, its opposite concept, *impurity*). An ideal scenario is based on nodes where the impurity is null so that all subsequent decisions will be taken only on the remaining features. In our example, we can simply start from the color block:

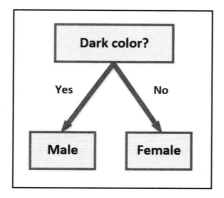

Example of an optimal splitting node – the impurity of the children is minimized

The two resulting sets are now pure according to the color feature, and this will be enough for our task. If we need further details, such as hair length, other nodes must be added; their impurity won't be null because we know that there are, for example, both **Male** and **Female** students with long hair.

More formally, suppose we define the selection tuple as follows:

$$\sigma = \langle i, t_k \rangle$$

Here, the first element is the index of the feature we want to use to split our dataset at a certain node (it will be the entire dataset only at the beginning; after each step, the number of samples decreases), while the second is the threshold that determines left and right branches. The choice of the best threshold is a fundamental element because it determines the structure of the tree and therefore its performance. The goal is to reduce the residual impurity in the least number of splits to have a very short decision path between the sample data and the classification result.

We can also define a total impurity measure by considering the two branches:

$$I(D, \sigma) = \frac{N_{left}}{N_D} I(D_{left}) + \frac{N_{right}}{N_D} I(D_{right})$$

Here, D is the whole dataset at the selected node, D_{left} and D_{right} are the resulting subsets (by applying the selection tuple), and I represents the impurity measures.

Impurity measures

To define the most frequently used impurity measures, we need to consider the total number of target classes:

$$Y = \{y_1, y_2, \ldots, y_n\} \text{ where } y_n \in \{0, 1, 2, \ldots, P-1\}$$

In a certain node, j, we can define the probability $p(y = i | Node = j)$ where i is an index [0, P-1] associated with each class. In other words, according to a frequentist approach, this value is the ratio between the number of samples belonging to class i and assigned to node j, and the total number of samples belonging to the selected node:

$$p(y = i | Node = j) = \frac{N_j^{y=i}}{N_j}$$

Gini impurity index

The Gini impurity index is defined as follows:

$$I_{Gini}(j) = \sum_i p(y = i | Node = j)(1 - p(y = i | Node = j))$$

Here, the sum is always extended to all classes. This is a very common measure and it's used as a default value by scikit-learn. Given a sample, the Gini impurity measures the probability of a misclassification if a label is randomly chosen using the probability distribution of the branch. The index reaches its minimum *(0.0)* when all the samples of a node are classified into a single category.

Cross-entropy impurity index

The cross-entropy measure is defined as follows:

$$I_{Cross-Entropy}(j) = -\sum_i p(y = i | Node = j) \log(p(y = i | Node = j))$$

This measure is based on information theory, and assumes null values when samples belonging to a single class are present in a split, while it is at its maximum when there's a uniform distribution among classes (which is one of the worst cases in Decision Trees because it means that there are still many decision steps until the final classification). This index is very similar to the Gini impurity, even though, more formally, cross-entropy allows you to select the split that minimizes uncertainty about the classification, while the Gini impurity minimizes the probability of misclassification.

In Chapter 2, *Important Elements in Machine Learning*, we defined the concept of mutual information, $I(X; Y) = H(X) - H(X|Y)$, as the amount of information shared by both variables, or the amount of information about X that we can obtain through knowledge of Y. It's also helpful to consider the data generating process, $p(x)$, and the conditional probability, $p(x|y)$. It's easy to prove that the mutual information is the expected value of the Kullback-Leibler divergence between them, with respect to the conditioned variable y:

$$I(X;Y) = E_{y \sim Y}[D_{KL}(p(x|y)||p(x))]$$

Therefore, this is when $I(X; Y) \to 0$, $p(x) \approx p(x|y)$ and knowledge of Y doesn't provide any useful information about X. Conversely, higher $I(X; Y)$ values imply an average strong divergence between the marginal distribution $p(x)$ and the conditional $p(x|y)$. The definition of conditional probability can provide better insight:

$$p(X|Y) = \frac{p(X \cap Y)}{p(Y)} \implies p(X \cap Y) \neq p(X)p(Y)$$

Here, the two variables are not statistically independent and knowledge of Y must provide proportional information about X. We can use this concept to define the **information gain** provided by a split (which is formally equivalent to the mutual information):

$$IG(\sigma) = H(Parent) - H(Parent|Children)$$

In this case, we are interested in the distributions of the parent node and the children. Therefore, when growing a tree, we start by selecting the split that provides the highest information gain. This guarantees to minimize the impurities of the subsequent nodes because, in general, the most relevant features are selected at the beginning and the next ones are reserved for fine-tuning. The growth of the tree proceeds until one of the following conditions is verified:

- All nodes are pure
- The information gain is null
- The maximum depth has been reached

Misclassification impurity index

The misclassification impurity index is the simplest index, and is defined as follows:

$$I_{Misclassification}(j) = 1 - max\ p(y = i|Node = j)$$

The interpretation is straightforward but, unfortunately, in terms of quality performance, this index is not the best choice because it's not particularly sensitive to different probability distributions. In all of these cases, Gini or cross-entropy indexes are the most natural choice.

Feature importance

When growing a Decision Tree with a multidimensional dataset, it can be useful to evaluate the importance of each feature in predicting the output values. In Chapter 3, *Feature Selection and Feature Engineering*, we discussed some methods to reduce the dimensionality of a dataset by selecting only the most significant features. Decision Trees offer a different approach based on the impurity reduction determined by every single feature. In particular, considering a feature, $x^{(i)}$, its importance can be determined as follows:

$$Importance\left(\bar{x}^{(i)}\right) = \sum_k \frac{N_k}{N} \Delta I_{\bar{x}^{(i)}}$$

The sum is extended to all nodes where $x^{(i)}$ is used, and N_k is the number of samples reaching the node, k. Therefore, the importance is a weighted sum of all impurity reductions computed, considering only the nodes where the feature is used to split them. If the Gini impurity index is adopted, this measure is also called **Gini importance**.

Decision Tree classification with scikit-learn

The scikit-learn library contains the `DecisionTreeClassifier` class, which can train a Binary Decision Tree with Gini and cross-entropy impurity measures. In our example, let's consider a dataset with 3 features and 3 classes:

```
from sklearn.datasets import make_classification

nb_samples = 500

X, Y = make_classification(n_samples=nb_samples, n_features=3,
n_informative=3, n_redundant=0, n_classes=3, n_clusters_per_class=1)
```

First, let's consider a classification with the default Gini impurity:

```
from sklearn.tree import DecisionTreeClassifier
from sklearn.model_selection import cross_val_score

dt = DecisionTreeClassifier()
print(cross_val_score(dt, X, Y, scoring='accuracy', cv=10).mean())
0.970
```

A very interesting feature is given by the ability to export the tree in graphviz format and convert it into a PDF.

Graphviz is a free tool that can be downloaded from http://www.graphviz.org. Alternatively, it's possible to use the free website http://www.webgraphviz.com, which allows you to plot graphs on the fly. In this case, however, it's necessary to paste the content of the exported file (which is plaintext) into the textbox.

To export a trained tree, it is necessary to use the built-in export_graphviz() function:

```
from sklearn.tree import export_graphviz

dt.fit(X, Y)
with open('dt.dot', 'w') as df:
    df = export_graphviz(dt, out_file=df,
                         feature_names=['A','B','C'],
                         class_names=['C1', 'C2', 'C3'])
```

In this case, we have used A, B, and C as feature names and C1, C2, and C3 as class names. Once the file has been created, it's possible to convert it into a PDF using the following command-line tool (we are assuming that the executable file has been installed in the <Graphviz Home> folder):

```
<Graphviz Home>/bindot -Tpdf dt.dot -o dt.pdf
```

The graph for our example is rather large, so in the following diagram, you can only see a part of a branch:

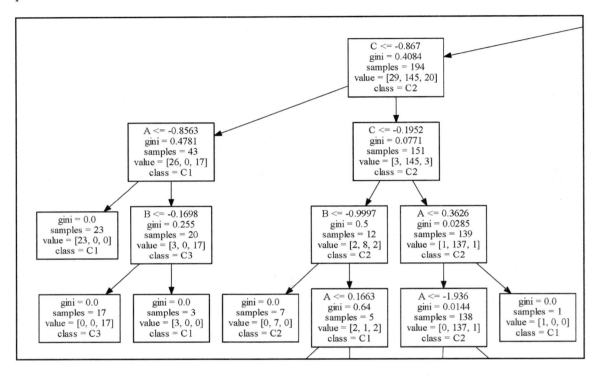

Example of a Decision Tree, visualized using Graphviz

As you can see, there are two kinds of nodes:

- Nonterminal, which contains the splitting tuple (as the <= *threshold* feature) and a positive impurity measure
- Terminal, where the impurity measure is null and a final target class is present

Chapter 8

In both cases, you can always check the number of samples. This kind of graph is very useful in understanding how many decision steps are needed. Unfortunately, even if the process is quite simple, the dataset structure can lead to very complex trees, while other methods can immediately find out the most appropriate class. Of course, not all features have the same importance. If we consider the root of the tree and the first nodes, we find features that separate a lot of samples; therefore, their importance must be higher than that of all terminal nodes, where the residual number of samples is at a minimum. In scikit-learn, it's possible to assess the Gini importance of each feature after training a model:

```
print(dt.feature_importances_)
[ 0.12066952,  0.12532507,  0.0577379 ,  0.14402762,  0.14382398,
  0.12418921,  0.14638565,  0.13784106]

print(np.argsort(dt.feature_importances_))
[2, 0, 5, 1, 7, 4, 3, 6]
```

The following graph shows a plot of the importance of every single feature:

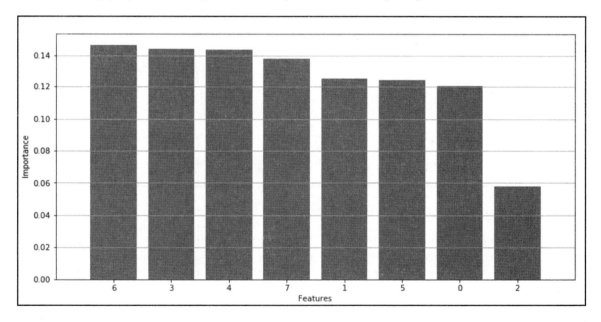

Plot of the importance of all features

The most important features are **6**, **3**, **4**, and **7**, while feature **2**, for example, separates a very small number of samples and can be considered non-informative for the classification task.

In terms of efficiency, a tree can also be pruned using the max_depth parameter. However, it's not always simple to understand which value is the best (grid searches and cross-validation can help in this task). Of course, it's extremely important to avoid over-specialization on the training set (which leads to overfitting the model), so it's often necessary to cut the tree at a certain level. In this way, it's easier to find a good trade-off between training and validation accuracy (if the model isn't too large, it's possible to test different values manually and pick the optimal one). The following graph shows common behavior of training and validation scores:

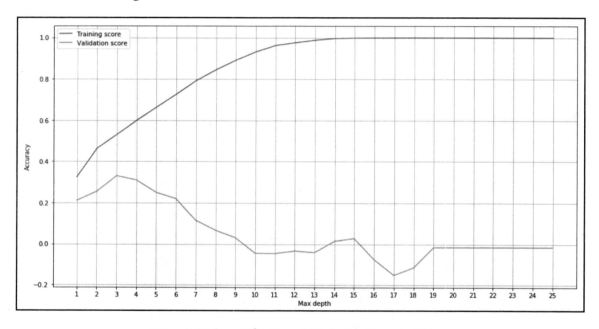

Training and validation scores (R^2 coefficients) plotted as functions of the max depth of the tree

It's easy to understand that the tree is prone to overfitting. In fact, when the **Max depth** reaches **14**, the training accuracy is **1.0**. Unfortunately, this condition corresponds to a proportional drop in the validation accuracy. The optimal value (given the overall existing configuration) is max_depth=3, which leads to the highest validation accuracy. Whenever performance is not acceptable, the data scientist should test different impurity measures, increase the training set size (also using data augmentation techniques), and tune up all other hyperparameters (possibly using a grid search). If none of these strategies yield the desired result, another model should be taken into account.

Sometimes, instead of working directly with the maximum depth, it's easier to decide what the maximum number of features to consider at each split should be. The max_features parameter can be used for this purpose:

- If it's a number, the value is directly taken into account at each split
- If it's 'auto' or 'sqrt', the square root of the number of features will be adopted
- If it's 'log2', the logarithm (base 2) will be used
- If it's 'None', all of the features will be used (this is the default value)

In general, when the number of total features is not too high, the default value is the best choice, although it's useful to introduce a small compression (via sqrt or log2) when too many features can interfere with each other, reducing the efficiency. Another parameter useful for controlling both performance and efficiency is min_samples_split, which specifies the minimum number of samples to consider for a split. Some examples are shown in the following snippet:

```
print(cross_val_score(DecisionTreeClassifier(), X, Y, scoring='accuracy', cv=10).mean())
0.77308070807080698

print(cross_val_score(DecisionTreeClassifier(max_features='auto'), X, Y, scoring='accuracy', cv=10).mean())
0.76410071007100711

print(cross_val_score(DecisionTreeClassifier(min_samples_split=100), X, Y, scoring='accuracy', cv=10).mean())
0.72999969996999692
```

As we already explained, finding the best parameters is generally a difficult task, and the best way to carry this out is to perform a grid search while including all of the values that could affect accuracy.

Using logistic regression on the previous set (only for comparison), we get the following:

```
from sklearn.linear_model import LogisticRegression

lr = LogisticRegression()
print(cross_val_score(lr, X, Y, scoring='accuracy', cv=10).mean())
0.9053368347338937
```

So the score is higher, as we expected. However, the original dataset was quite simple and based on the concept of having a single cluster per class. This allows a simpler and more precise linear separation. If we consider a slightly different scenario with more variables and a more complex structure (which is hard to capture by a linear classifier), we can compare a **Receiver Operating Characteristic** (**ROC**) curve for both linear regression and Decision Trees:

```
nb_samples = 1000

X, Y = make_classification(n_samples=nb_samples, n_features=8,
    n_informative=6, n_redundant=2, n_classes=2, n_clusters_per_class=4)
```

The resulting ROC curve is shown in the following screenshot:

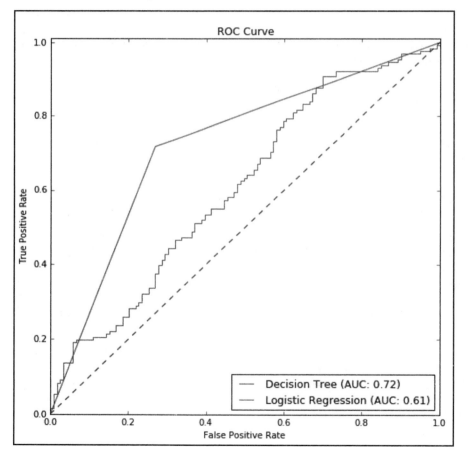

ROC curve comparing the performance of a Decision Tree and logistic regression

Using a grid search with the most common parameters on the MNIST digit dataset, we can get the following:

```
from sklearn.model_selection import GridSearchCV

param_grid = [
  {
    'criterion': ['gini', 'entropy'],
    'max_features': ['auto', 'log2', None],
    'min_samples_split': [ 2, 10, 25, 100, 200 ],
    'max_depth': [5, 10, 15, None]
  }
]

gs = GridSearchCV(estimator=DecisionTreeClassifier(),
param_grid=param_grid,
                  scoring='accuracy', cv=10,
n_jobs=multiprocessing.cpu_count())

gs.fit(digits.data, digits.target)
GridSearchCV(cv=10, error_score='raise',
       estimator=DecisionTreeClassifier(class_weight=None,
criterion='gini',        max_depth=None,
            max_features=None, max_leaf_nodes=None,
            min_impurity_split=1e-07, min_samples_leaf=1,
            min_samples_split=2, min_weight_fraction_leaf=0.0,
            presort=False, random_state=None, splitter='best'),
       fit_params={}, iid=True, n_jobs=8,
       param_grid=[{'max_features': ['auto', 'log2', None],
'min_samples_split': [2, 10, 25, 100, 200], 'criterion': ['gini',
'entropy'], 'max_depth': [5, 10, 15, None]}],
       pre_dispatch='2*n_jobs', refit=True, return_train_score=True,
       scoring='accuracy', verbose=0)

print(gs.best_estimator_)
DecisionTreeClassifier(class_weight=None, criterion='entropy',
max_depth=None,
            max_features=None, max_leaf_nodes=None,
            min_impurity_split=1e-07, min_samples_leaf=1,
            min_samples_split=2, min_weight_fraction_leaf=0.0,
            presort=False, random_state=None, splitter='best')

print(gs.best_score_)
0.8380634390651085
```

In this case, the element that impacted accuracy the most is the minimum number of samples to consider for a split. This is reasonable, considering the structure of this dataset and the need to have many branches to capture even small changes.

Decision Tree regression

Decision Trees can also be employed in order to solve regression problems. However, in this case, it's necessary to consider a slightly different way of splitting the nodes. Instead of considering an impurity measure, one of the most common choices is to pick the feature that minimizes the **mean squared error** (**MSE**), considering the average prediction of a node. Let's suppose that a node, *i*, contains *m* samples. The average prediction is as follows:

$$\bar{y}_i = \frac{1}{m} \sum_j y_j$$

At this point, the algorithm has to look for all of the binary splits in order to find the one that minimizes the target function:

$$MSE_i = \frac{1}{m} \sum_j (y_j - \bar{y}_i)^2$$

Analogous to classification trees, the procedure is repeated until the MSE is below a fixed threshold, λ. Even if it's not correct, we can think about an unacceptable impurity level when the prediction of a node has a low accuracy. In fact, in a classification tree, an impure node contains more than one class and the uncertainty could be too high to make a reasonable decision. In the same way, a node whose $MSE > \lambda$ is still too *uncertain* about the correct output means that further splits are required. An alternative approach is based on the **mean absolute error** (**MAE**), but in the majority of cases, the MSE is the optimal choice.

As usual, I always recommend employing both grid searches and cross-validation to find the optimal hyperparameters. In fact, when the number of features is large, it's sometimes difficult to immediately find the best configuration. Moreover, the structure of every dataset can lead to very different validation accuracies (as we are going to see), according to the selected training subset. In specific problems, it's helpful to understand whether the training set represents the actual data generation process (for example, with the help of a domain expert), to avoid exclusions that lead to different train and test distributions. Another common problem to address is overfitting. A very deep tree can perfectly map the training set while showing poor validation performance. Therefore, don't forget to include the `max_depth` parameter in the CV grid search and pick the solution that offers the optimal trade-off.

Example of Decision Tree regression with the Concrete Compressive Strength dataset

In this example, we are going to use the Concrete Compressive Strength dataset, which was freely provided by Yeh to the UCI repository (it was originally employed in the paper *Modeling of strength of high performance concrete using artificial neural networks, I-Cheng Yeh, Cement and Concrete Research, Vol. 28, No. 12, pp. 1797-1808 (1998)*). The dataset is made up of 1,030 samples with eight independent variables (Cement, Blast furnace slag, Fly ash, Water, Superplasticizer, Coarse aggregate, Fine aggregate, and Age) and a single dependent variable (Concrete Compressive Strength).

Once you have downloaded the Excel file (assuming that it has been stored in the <DATA_HOME> folder), it's possible to parse it using pandas in order to obtain the actual dataset (X, Y):

```
import pandas as pd

file_path = '<DATA_HOME>/Concrete_Data.xls'

df = pd.read_excel(file_path, header=0)

X = df.iloc[:, 0:8].values
Y = df.iloc[:, 8].values
```

The first five samples are shown in a pretty format in the following table:

	Cement (component 1)(kg in a m^3 mixture)	Blast Furnace Slag (component 2)(kg in a m^3 mixture)	Fly Ash (component 3)(kg in a m^3 mixture)	Water (component 4)(kg in a m^3 mixture)	Superplasticizer (component 5) (kg in a m^3 mixture)	Coarse Aggregate (component 6)(kg in a m^3 mixture)	Fine Aggregate (component 7) (kg in a m^3 mixture)	Age (day)	Concrete compressive strength(MPa, megapascals)
0	540.0	0.0	0.0	162.0	2.5	1040.0	676.0	28	79.986111
1	540.0	0.0	0.0	162.0	2.5	1055.0	676.0	28	61.887366
2	332.5	142.5	0.0	228.0	0.0	932.0	594.0	270	40.269535
3	332.5	142.5	0.0	228.0	0.0	932.0	594.0	365	41.052780
4	198.6	132.4	0.0	192.0	0.0	978.4	825.5	360	44.296075

The first five samples of the Concrete Compressive Strength dataset

Decision Trees and Ensemble Learning

The user is free to check the statistical properties, but we have already discussed the insensitivity of Decision Trees to different scales, and therefore we are not going to preprocess the data. This dataset contains very specific *regions* characterized by an uncommon behavior, and therefore the first step is to check the CV scores (R^2, as this is a regression) using an instance of `DecisionTreeRegressor` with `max_depth=11` (I invite the reader to test other values) and an MSE criterion (the default one):

```
from sklearn.tree import DecisionTreeRegressor
from sklearn.model_selection import cross_val_score

print(cross_val_score(DecisionTreeRegressor(criterion='mse', max_depth=11,
random_state=1000), X, Y, cv=20))

[ 0.20764059  0.55004291  0.13532372  0.94144166  0.53860366  0.67753093
  0.48176233  0.39555753  0.15537892  0.05613209  0.68861262  0.27333756
  0.05999872  0.74014659  0.78379972  0.8200266   0.88995849  0.9468295
  0.8325872   0.98061205]
```

As it's possible to see, 12 folds have $R^2 > 0.5$ (8 have $R^2 > 0.7$), while the remaining ones indicate more inaccurate predictions (the worst cases are the two folds with $R^2 \approx 0.05$). This means that not all random test subset selections have the same distribution as the training set. In these cases, it's important to gain a deeper understanding of the structure of the dataset. A statistical summary (obtained using the pandas `describe()` command invoked on the data frame) therefore becomes necessary:

	Cement (component 1)(kg in a m^3 mixture)	Blast Furnace Slag (component 2) (kg in a m^3 mixture)	Fly Ash (component 3)(kg in a m^3 mixture)	Water (component 4)(kg in a m^3 mixture)	Superplasticizer (component 5) (kg in a m^3 mixture)	Coarse Aggregate (component 6) (kg in a m^3 mixture)	Fine Aggregate (component 7) (kg in a m^3 mixture)	Age (day)	Concrete compressive strength(MPa, megapascals)
count	1030.000000	1030.000000	1030.000000	1030.000000	1030.000000	1030.000000	1030.000000	1030.000000	1030.000000
mean	281.165631	73.895485	54.187136	181.566359	6.203112	972.918592	773.578883	45.662136	35.817836
std	104.507142	86.279104	63.996469	21.355567	5.973492	77.753818	80.175427	63.169912	16.705679
min	102.000000	0.000000	0.000000	121.750000	0.000000	801.000000	594.000000	1.000000	2.331808
25%	192.375000	0.000000	0.000000	164.900000	0.000000	932.000000	730.950000	7.000000	23.707115
50%	272.900000	22.000000	0.000000	185.000000	6.350000	968.000000	779.510000	28.000000	34.442774
75%	350.000000	142.950000	118.270000	192.000000	10.160000	1029.400000	824.000000	56.000000	46.136287
max	540.000000	359.400000	200.100000	247.000000	32.200000	1145.000000	992.600000	365.000000	82.599225

Statistical summary of the Concrete Compressive Strength dataset

It's evident that a few components are null in a large portion of the samples. For example, **Fly Ash** has a 50th percentile equal to 0, while **Blast Furnace Slag** and **Superplasticizer** both have the 25th percentile equal to 0. A simple count showed that 566 samples have a null **Fly Ash** feature and 466 have a null **Blast Furnace Slag**. Hence, it's almost impossible to have uniform behavior for any random split, with some *extreme* scenarios where one or more features are constantly null and the predictions are consequently wrong. As we have already pointed out, in these cases it's necessary to obtain a test that represents the original data generation process. For our example, we are going to select 200 test samples using random_state=1000 (different values can yield other results), and we are going to be training a DecisionTreeRegressor instance with max_depth=11:

```
from sklearn.model_selection import train_test_split

X_train, X_test, Y_train, Y_test = train_test_split(X, Y, test_size=200,
random_state=1000)

dtr = DecisionTreeRegressor(criterion='mse', max_depth=11,
random_state=1000)
dtr.fit(X_train, Y_train)
```

Before evaluating the model, it's helpful to visualize a small portion of the tree using Graphviz (we are storing the output file in the <DATA_HOME> folder):

```
from sklearn.tree import export_graphviz

graphviz_path = '<DATA_HOME>/Concrete_Data.dot'

export_graphviz(dtr, out_file=graphviz_path,
                feature_names=['Cement','Blast furnace slag', 'Fly ash','Water',
                               'Superplasticizer','Coarse Aggregate','Fine Aggregate','Age'])
```

The tree is extremely wide, but a complete branch is shown in the following diagram:

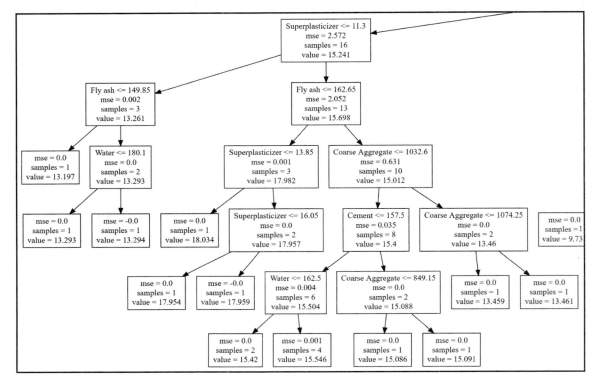

Branch of the final Decision Tree

As you can see, the top node (let's consider it as a root) has an **mse = 2.572**. All subsequent splits try to reduce the MSE or the number of samples belonging to the node. In an ideal scenario (without the risk of overfitting), each leaf should contain a single sample with an $MSE \approx 0$. In this case, there are many leaves that fulfill this condition, but there are also other ones whose MSE is null, but they contain **2** and **4** samples (in these cases, the value is averaged). This graph should also show the actual internal complexity of a Decision Tree (which is one of its characteristics). The root only has 16 samples, and five levels are needed to reach the leaves. That's why these kinds of models are very prone to overfitting: without an explicit control of their growth, they can become extremely deep, with an almost null generalization ability.

At this point, we can start evaluating the performance of the model:

```
print(dtr.score(X_train, Y_train))
0.988068139046

print(dtr.score(X_test, Y_test))
0.821126631614
```

The R^2 training score is extremely high (almost 99%), while the validation score is about 82% (which is quite a positive result). The **Original dataset** (Y) compared with the **Predictions** are shown in the following graph:

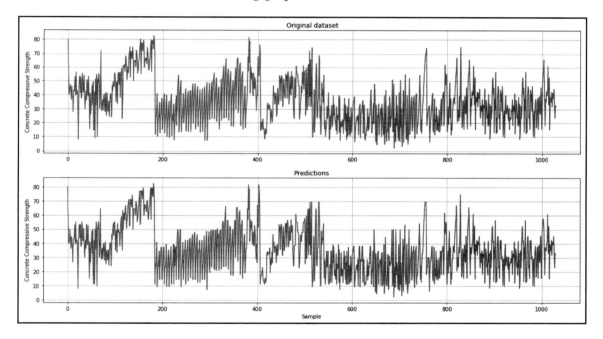

Original Compressive Strength (upper plot) and Predictions (lower plot)

It's clear that the discrepancies are minimal and cannot be caught through a visual inspection. The following graph shows the absolute error for each sample:

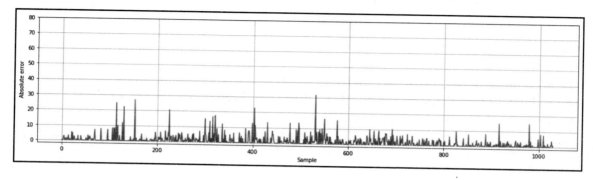

Absolute prediction errors

In the majority of cases, the errors are below **5**, with only a few high peaks whose corresponding prediction has an error of about 20-50%. To complete this example, let's plot the histogram (with *y* log-scale) of the absolute errors (which can provide us with the most valuable piece of information):

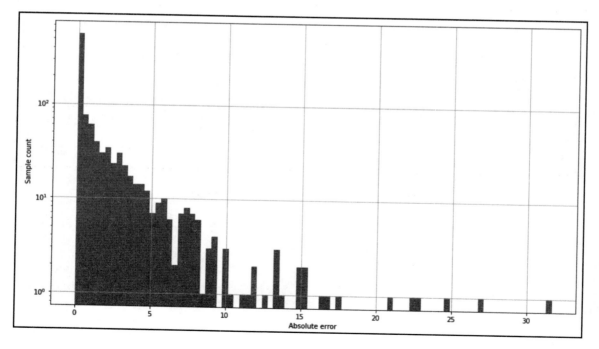

Histogram of the absolute errors

As expected, the distribution is an exponential (it can also be a Half-Gaussian with a small variance) with a very strong peak corresponding to *absolute error* ≈ *0* (in the plot, it seems shorter due to log-scale). The majority of errors are less than **5** and the tail contains only a few samples. This confirms that the regression (which was based on 200 test samples) successfully learned the structure of the dataset and only failed in a very limited number of cases. I invite the reader to repeat this example using a grid search and different test set dimensions.

The dataset can be downloaded from `https://archive.ics.uci.edu/ml/datasets/Concrete+Compressive+Strength`. It consists of a single Excel file that can be easily parsed using pandas.

Introduction to Ensemble Learning

Until now, we have trained models on single instances, iterating an algorithm in order to minimize a target loss function. This approach is based on so-called **strong learners**, or methods that are optimized to solve a specific problem by looking for the best possible solution (highest accuracy). Another approach is based on a set of **weak learners**, which, formally, are estimators that are able to achieve an accuracy slightly higher than 0.5. In the real world, the actual estimators used in Ensemble Learning are much more accurate than their theoretical counterparts, but generally they are able to specialize a single region of the sample space and show bad performance while considering the whole dataset. Moreover, they can be trained in parallel or sequentially (with slight modifications to the parameters) and used as an ensemble (group) based on a majority vote or the averaging of results. In this context, we are assuming that the weak learners are always Decision Trees, but this must not be considered as a limitation or a constraint (for further details about the algorithms and the theoretical part, please refer to *Mastering Machine Learning Algorithms, Bonaccorso G., Packt Publishing, 2018*).

These methods can be classified into three main categories:

- **Bagging (or bootstrapping)**: In this case, the ensemble is completely built. Each classifier is trained independently, considering a subset (X_i, Y_i) of the original dataset (X, Y). There are many possible implementation strategies; however, the main goal is to avoid two or more estimators specializing on the same subset. In general, according to the algorithm, a varying level of randomness is included in the process. This can lead to suboptimal partial solutions, but it allows the ensemble to efficiently explore the whole sample space. The most common example of bagging is the Random Forest algorithm.
- **Boosting**: In this case, the ensemble is built sequentially, focusing on the samples that have been previously misclassified. This process is normally achieved by reweighting the dataset. If, in the beginning, each sample has the same probability of being selected, after an iteration the distribution becomes more peaked to not increase the probability of sampling the samples that require more specialization. Examples of boosted trees are AdaBoost and Gradient Tree Boosting.
- **Stacking**: This method is based on a heterogeneous set of weak learners (for example, SVM, logistic regression, and Decision Trees). Every classifier is trained autonomously and the final choice is made by employing a majority vote, averaging the results, or using another auxiliary classifier that takes all of the intermediate predictions and outputs a final one. In the following sections, we are going to discuss the idea of a voting classifier.

Random Forests

A Random Forest is a bagging ensemble method based on a set of Decision Trees. If we have N_c classifiers, the original dataset is split into N_c subsets (with replacement) called bootstrap samples:

$$(X_i, Y_i) = \{(\bar{x}_j, y_j) \ for \ j = 1..K \ and \ (\bar{x}_j, y_j) \sim (X, Y)\}$$

Contrary to a single Decision Tree, in a Random Forest the splitting policy is based on a medium level of randomness. In fact, instead of looking for the best choice, a random subset of features (for each tree) is used (in general, the number of features is computed using *sqrt(•)* or *log(•)*), trying to find the threshold that best separates the data. As a result, there will be many trees that are trained in a weaker way, and each of them will produce a different prediction. At the same time, every tree will be more specialized on a portion of the sample space, while yielding inaccurate predictions in other regions.

There are two ways to interpret these results. The most common approach is based on a majority vote (the most voted-for class, obtained through a binned *argmax(•)*, will be considered correct):

$$\tilde{y}_i = argmax_j \left(c_j(\bar{x}_i)\right)$$

However, scikit-learn implements an algorithm based on averaging the results, which yields very accurate predictions:

$$\tilde{y}_i = \frac{1}{Nc} \sum_j c_j(\bar{x}_i)$$

Even if they are theoretically different, the probabilistic average of a trained Random Forest cannot be very different from the majority of predictions (otherwise, there should be different stable points), and therefore the two methods often lead to comparable results.

As an example, let's consider the MNIST dataset with Random Forests made of a different number of trees:

```
from sklearn.ensemble import RandomForestClassifier

nb_classifications = 100
accuracy = []

for i in range(1, nb_classifications):
    a = cross_val_score(RandomForestClassifier(n_estimators=i),
digits.data, digits.target,  scoring='accuracy', cv=10).mean()
    rf_accuracy.append(a)
```

The resulting plot is shown in the following graph:

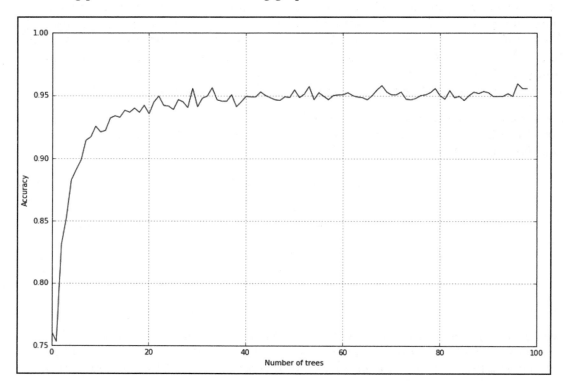

Accuracy of a Random Forest as a function of the number of Decision Trees

As we expected, the accuracy is low when the number of trees is under a minimum threshold. However, it starts increasing rapidly when this are fewer than 10 trees. A value between 20 and 30 trees yields the optimal result (95%), which is higher than for a single Decision Tree. When the number of trees is low, the variance of the model is very high and the averaging process produces many incorrect results; however, increasing the number of trees reduces the variance and allows the model to converge to a very stable solution.

scikit-learn also offers a variance that enhances the randomness in selecting the best threshold. Using the ExtraTreesClassifier class, it's possible to implement a model that randomly computes thresholds and picks the best one. This choice allows you to further reduce the variance and, very often, to achieve better final validation accuracy:

```
from sklearn.ensemble import ExtraTreesClassifier

nb_classifications = 100
```

```
for i in range(1, nb_classifications):
    a = cross_val_score(ExtraTreesClassifier(n_estimators=i), digits.data,
digits.target,   scoring='accuracy', cv=10).mean()
    et_accuracy.append(a)
```

The results (with the same number of trees) in terms of accuracy are slightly better, as shown in the following graph:

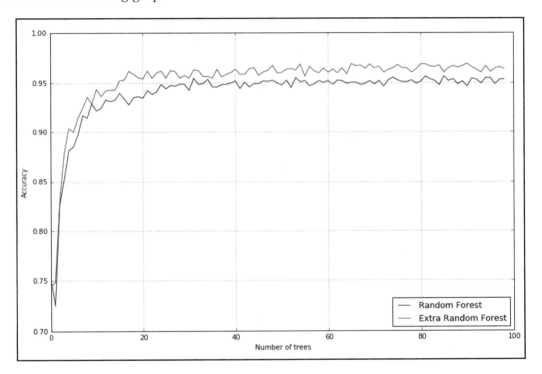

Comparison between the accuracy of Random Forest and Extra Random Forest

Feature importance in Random Forests

The concept of feature importance that we previously introduced can also be applied to Random Forests, computing the average over all of the trees in the forest:

$$Importance\left(\bar{x}^{(i)}\right) = \frac{1}{N_{trees}} \sum_i \sum_k \frac{N_k}{N} \Delta I_{\bar{x}^{(i)}}$$

We can easily test the evaluation of importance with a dummy dataset that contains 50 features with 20 noninformative elements:

```
nb_samples = 1000

X, Y = make_classification(n_samples=nb_samples, n_features=50,
       n_informative=30, n_redundant=20, n_classes=2, n_clusters_per_class=5)
```

The importance of the first 50 features according to a Random Forest with 20 trees is plotted in the following graph:

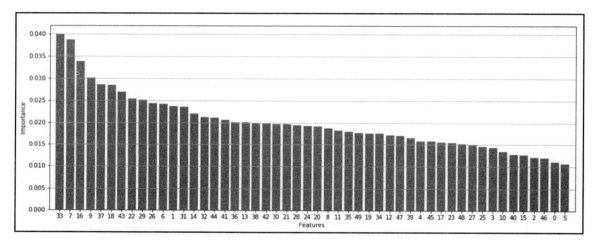

Feature importance in a Random Forest

As expected, there are a few *very* important features, a block of features with medium importance, and a tail containing features that have little influence on the predictions. This type of plot is also useful during the analysis stage to better understand how the decision process is structured. With multidimensional datasets, it's rather difficult to understand the influence of every factor, and sometimes many important business decisions are made without complete awareness of their potential impact. Using Decision Trees or Random Forests, it's possible to assess the real importance of all features and exclude all the elements under a fixed threshold. In this way, a complex decision process can be simplified and, at the same time, partially denoised.

AdaBoost

Another technique is called **AdaBoost** (short for **Adaptive Boosting**) and it works in a slightly different way to many other classifiers. The basic structure behind this can be a Decision Tree, but the dataset used for training is continuously adapted to force the model to focus on those samples that are misclassified. Moreover, the classifiers are added sequentially, so a new one boosts the previous one by improving performance in those areas where it was not as accurate as expected. At each iteration, a weight factor is applied to each sample to increase the importance of samples that are wrongly predicted and decrease the importance of others. In other words, the model is repeatedly boosted, starting as a very weak learner until the maximum n_estimators number is reached. The predictions, in this case, are always obtained by majority vote.

The original version of this algorithm is called **AdaBoost.M1**, and it's based on a dynamically updated weight set (considering n samples):

$$W^{(t)} = \left\{ w_1^{(t)}, w_2^{(t)}, \ldots, w_n^{(t)} \right\} \; where \; w_i^{(t)} \geq 0$$

The initial value, $W^{(0)}$, is set equal to $1/n$ for all weights so that no preference is expressed for any sample. Before starting the training process, every classifier sample is set, considering the existing weights. After each training step, an indicator function is computed:

$$\epsilon^{(t)} = \frac{\sum_{c_t(\bar{x}_i) \neq y_i} w_i}{\sum_i w_i}$$

If no misclassifications have occurred, $\varepsilon^{(t)} = 1$, while it's equal to 0 if all samples have been assigned to the wrong class. In order to reweight the dataset for the subsequent iteration, a special function is employed:

$$\alpha^{(t)} = log \left(\frac{1 - \epsilon^{(t)}}{\epsilon^{(t)} + \nu} \right) \; where \; \nu \ll \epsilon^{(t)}$$

Decision Trees and Ensemble Learning

The constant v is normally added in order to improve numerical stability (avoiding a division by zero); however, for our discussion, it can be considered null. When $\varepsilon^{(t)} \to 0$, $\alpha^{(t)} \to +\infty$, $\varepsilon^{(t)} \to 1$, $\alpha^{(t)} \to -\infty$. In the case of a binary random guess, $\varepsilon^{(t)} = 0.5$ and $\alpha^{(t)} = 0$. Even if it's not intuitive, our main goal is to exclude classifiers that are random oracles ($\varepsilon^{(t)} = 0.5$), which represent the worst situation, and their prediction is completely excluded. When $\varepsilon^{(t)} \to 1$, the classifier should receive a negative boost, but this would be in contrast to the initial purpose of the algorithm. Therefore, in this case, the output is inverted (also if it's a misclassification) and the boost becomes positive (for example, if $\varepsilon^{(t)} = 0.75$, it is transformed into $\varepsilon^{(t)} = 0.25$), which doesn't change the absolute value of $\alpha^{(t)}$.

This can be achieved by using a weighted global decision function:

$$d(\bar{x}_i) = sign\left(\sum_{j=1}^{N_c} \alpha^{(j)} c_j(\bar{x}_i)\right)$$

In this way, the classifiers that remain stuck at an accuracy close to 0.5 are automatically discarded, while all the others are boosted. The boosting procedure is very simple and it's based on the result of each classification. Let's consider sample x_i. We can introduce auxiliary variable o_i:

$$o_i = \begin{cases} 1 & if\ c^{(t)}(\bar{x}_i) \neq y_i \\ -1 & if\ c^{(t)}(\bar{x}_i) = y_i \end{cases}$$

The corresponding weight, w_i, is updated by considering the following rule:

$$w_i^{(t+1)} = w_i^{(t)} e^{\alpha_i^{(t)} o_i}$$

It's straightforward to understand that a weight is only increased when a misclassification has occurred, and it's decreased if the sample has been correctly classified. Moreover, parameter $\alpha^{(t)}$ takes into account the overall behavior of the estimator. If $\varepsilon^{(t)} \to 0.5$, the reweighting doesn't alter the existing configuration, while it becomes more aggressive when larger $\alpha^{(t)}$ values have been obtained. The result of this process is a new distribution, where the most problematic samples will become more and more likely to be sampled, while the simplest ones could be discarded after the first iterations.

Chapter 8

In the scikit-learn implementations (which are **AdaBoost.SAMME** for classification and **AdaBoost.SAMME.R** for regression, as both are designed to work in multiclass scenarios), there's also a parameter called `learning_rate` that weighs the effect of each classifier. The default value is 1.0, so all estimators are considered to have the same importance. However, as we can see with the MNIST dataset, it's useful to decrease this value so that each contribution is weakened:

```
from sklearn.ensemble import AdaBoostClassifier

accuracy = []

nb_classifications = 100

for i in range(1, nb_classifications):
    a = cross_val_score(AdaBoostClassifier(n_estimators=i,
learning_rate=0.1), digits.data, digits.target, scoring='accuracy',
cv=10).mean()
    ab_accuracy.append(a)
```

The result is shown in the following graph:

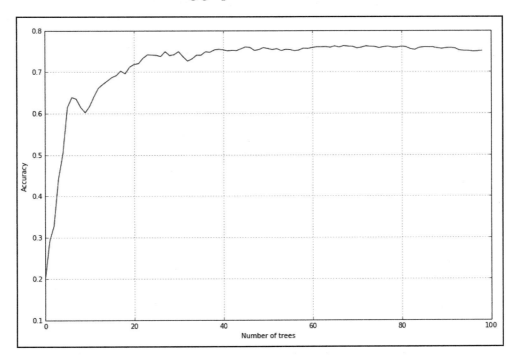

AdaBoost accuracy as a function of the number of Decision Trees

[275]

The accuracy is not as high as in the previous examples. However, it's possible to see that when the boosting adds about 20-30 trees, it reaches a stable value. A grid search on `learning_rate` could allow you to find the optimal value; however, the sequential approach, in this case, is not preferable. A classic Random Forest, which works with a fixed number of trees since the first iteration, performs better. This may well be due to the strategy adopted by AdaBoost. In this set, increasing the weight of the correctly classified samples and decreasing the strength of misclassifications can produce an oscillation in the loss function, with a final result that is not the optimal minimum point. Repeating the experiment with the Iris dataset (which is structurally much simpler) yields better results:

```
from sklearn.datasets import load_iris

iris = load_iris()

ada = AdaBoostClassifier(n_estimators=100, learning_rate=1.0)
print(cross_val_score(ada, iris.data, iris.target, scoring='accuracy',
cv=10).mean())
0.94666666666666666
```

In this case, a learning rate of 1.0 is the best choice, and it's easy to understand that the boosting process can be stopped after a few iterations. In the following graph, you can see a plot showing the accuracy of this dataset:

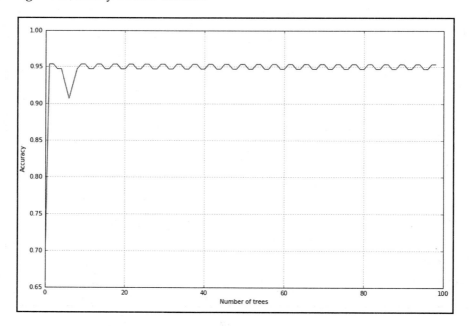

Accuracy of AdaBoost with the Iris dataset as a function of the number of Decision Trees

After about 10 iterations, the accuracy becomes stable (the residual oscillation can be discarded), reaching a value that is compatible with this dataset. The advantage of using AdaBoost can be appreciated in terms of resources; it doesn't work with a fully configured set of classifiers and the whole set of samples. Therefore, it can help save time when training on large datasets.

Gradient Tree Boosting

Gradient Tree Boosting is a technique that allows you to build a tree ensemble step by step (the method is also known as **forward stage-wise additive modeling**), with the goal of minimizing a target loss function. The generic output of the ensemble can be represented as follows:

$$y_E = \sum_j \alpha_j c_j(\bar{x}) = \sum_j \alpha_j f(\bar{x}; \bar{\theta}_i)$$

Here, $c_j(x)$ is a function representing a weak learner (in this particular case, it's always a Decision Tree that can be modeled as a single parametrized function, $f(\bullet)$, where the vector, θ_i, groups all the splitting tuples of the ith tree). The algorithm is based on the concept of adding a new Decision Tree at each step to minimize a global cost function (based on a predefined loss $L(\bullet)$) using the Steepest Gradient Descent method (for further information, see https://en.wikipedia.org/wiki/Method_of_steepest_descent). Considering that the classifiers are parametrized using a vector, θ, and all the weighting coefficients are grouped into a single array, the global cost function is defined as follows:

$$C(X, Y; \bar{\alpha}, \bar{\theta}) = \sum_i L\left(y_i, \alpha_i c(\bar{x}_i; \bar{\theta})\right)$$

Hence, the goal is to find the optimal tuple so that:

$$(\bar{\alpha}^*, \bar{\theta}^*) = argmin_{\bar{\alpha}, \bar{\theta}} C(X, Y; \bar{\alpha}, \bar{\theta})$$

Decision Trees and Ensemble Learning

Considering this goal, the incremental procedure can be rewritten as follows:

$$c_i(\bar{x}) = c_{i-1}(\bar{x}) + argmin_f \sum_j L\left(y_j, c_{i-1}(\bar{x}_j) + f(\bar{x}_j; \bar{\theta}_i)\right)$$

In the previous expression, the loss is computed by considering the previous contributions and optimized with respect to the new classifier. Unfortunately, although formally clear, this problem is extremely complex and requires an unacceptable computational cost. However, introducing the gradient, we can rewrite the previous expression to transform the additive model into a simpler optimization procedure:

$$c_i(\bar{x}) = c_{i-1}(\bar{x}) - \eta \alpha_i \sum_j \nabla_c L\left(y_j, c_{i-1}(\bar{x}_j)\right)$$

A reader who is familiar with **stochastic gradient descent (SGD)** algorithms can immediately understand that the goal of every step is to minimize the global cost function. In other words, by moving in the direction opposite to the gradient, the new classifier is built with the purpose of reducing the global cost function with respect to its predecessors. As usual in these kinds of tasks, parameter η is the learning rate, which must be chosen using a grid search in order to avoid either a very slow convergence or instability with consequent suboptimality. The weights, α_i, are instead computed using a line search algorithm (which is computationally affordable) after the computation of the gradient (considering α as an additional variable):

$$\alpha_i = argmin_\alpha \sum_j L\left(y_j, c_i(\bar{x}_j, \alpha)\right) = argmin_\alpha \sum_j L\left(y_j, c_{i-1}(\bar{x}) - \eta \alpha \sum_j \nabla_c L\left(y_j, c_{i-1}(\bar{x}_j)\right)\right)$$

The philosophy of this algorithm is not very different from AdaBoost; however, in this case we are focusing on a global goal without explicitly reweighting the dataset. A new estimator is chosen to improve the predecessor (in a perfect scenario, the cost function should always decrease until the minimum), but we don't know exactly which sample regions were more problematic.

scikit-learn implements the GradientBoostingClassifier class, which supports two classification loss functions:

- Binomial/multinomial negative log-likelihood (the default choice)
- Exponential (analogous to AdaBoost)

Let's evaluate the accuracy of this method by using a more complex dummy dataset made up of 500 samples with four features (three informative and one redundant) and three classes:

```
from sklearn.datasets import make_classification

nb_samples = 500

X, Y = make_classification(n_samples=nb_samples, n_features=4, n_informative=3, n_redundant=1, n_classes=3)
```

Now, we can collect the cross-validation average accuracy for a number of estimators in the range (1, 50). The loss function is the default one (multinomial negative log-likelihood):

```
from sklearn.ensemble import GradientBoostingClassifier
from sklearn.model_selection import cross_val_score

a = []

max_estimators = 50

for i in range(1, max_estimators):
    score = cross_val_score(GradientBoostingClassifier(n_estimators=i, learning_rate=10.0/float(i)), X, Y, cv=10, scoring='accuracy').mean()
    a.append(score)
```

While increasing the number of estimators (by using the `n_estimators` parameter), it's important to decrease the learning rate (by using the `learning_rate` parameter). The optimal value cannot be easily predicted; therefore, it's often useful to perform a grid search. In our example, I've set a very high learning rate at the beginning (5.0), which converges to 0.05 when the number of estimators is equal to 100. This is not a perfect choice (unacceptable in most real cases!), and it has only been made to show the differences in accuracy performance. The results are shown in the following graph:

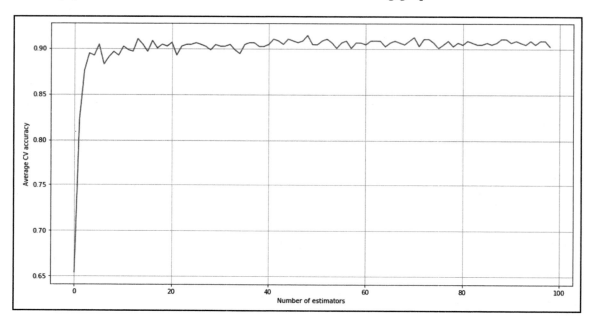

Average CV accuracy of Gradient Tree Boosting as a function of the number of estimators

As you can see, the optimal number of estimators is about 50, with a learning rate of 0.1. The reader can try different combinations and compare the performance of this algorithm with the other ensemble methods.

Voting classifier

A very interesting ensemble solution (which can be considered as part of the stacking subset) is offered by the `VotingClassifier` class, which isn't an actual classifier but a wrapper for a set of different ones that are trained and evaluated in parallel in order to exploit the different peculiarities of each algorithm.

The final decision on a prediction is taken by majority vote according to two different strategies:

- **Hard voting**: In this case, the class that received the highest number of votes, $N_c(y_t)$, will be chosen:

$$\tilde{y} = argmax\left(N_c(y_t^1), N_c(y_t^2), \ldots, N_c(y_t^n)\right)$$

- **Soft voting**: In this case, the probability vectors for each predicted class (for all classifiers) are summed up and averaged. The winning class is the one corresponding to the highest value:

$$\tilde{y} = argmax \frac{1}{N_{Classifiers}} \sum_{Classifier} (p_1, p_2, \ldots, p_n)$$

Let's consider a dummy dataset and compute the accuracy with a hard voting strategy:

```
from sklearn.datasets import make_classification

nb_samples = 500

X, Y = make_classification(n_samples=nb_samples, n_features=2,
n_redundant=0, n_classes=2)
```

For our examples, we are going to consider three classifiers: logistic regression, a Decision Tree (with default Gini impurity), and an SVM (with a polynomial kernel and `probability=True` in order to generate the probability vectors). This choice has only been made for didactic purposes and may not be the best one. When creating an ensemble, it's useful to consider the different features of each classifier involved and avoid "duplicate" algorithms (for example, a logistic regression and a linear SVM or a perceptron are likely to yield very similar performances). In many cases, it can be useful to mix non-linear classifiers with Random Forests or AdaBoost classifiers. The reader can repeat this experiment with other combinations, comparing the performance of every single estimator and the accuracy of the voting classifier:

```
from sklearn.linear_model import LogisticRegression
from sklearn.svm import SVC
from sklearn.tree import DecisionTreeClassifier
from sklearn.ensemble import VotingClassifier

lr = LogisticRegression()
svc = SVC(kernel='poly', probability=True)
dt = DecisionTreeClassifier()
```

Decision Trees and Ensemble Learning

```
classifiers = [('lr', lr),
               ('dt', dt),
               ('svc', svc)]

vc = VotingClassifier(estimators=classifiers, voting='hard')
```

Computing the cross-validation accuracies, we get the following:

```
from sklearn.model_selection import cross_val_score

a = []

a.append(cross_val_score(lr, X, Y, scoring='accuracy', cv=10).mean())
a.append(cross_val_score(dt, X, Y, scoring='accuracy', cv=10).mean())
a.append(cross_val_score(svc, X, Y, scoring='accuracy', cv=10).mean())
a.append(cross_val_score(vc, X, Y, scoring='accuracy', cv=10).mean())

print(np.array(a))
[ 0.90182873  0.84990876  0.87386955  0.89982873]
```

The average CV accuracies of every single classifier and of the ensemble are plotted in the following graph:

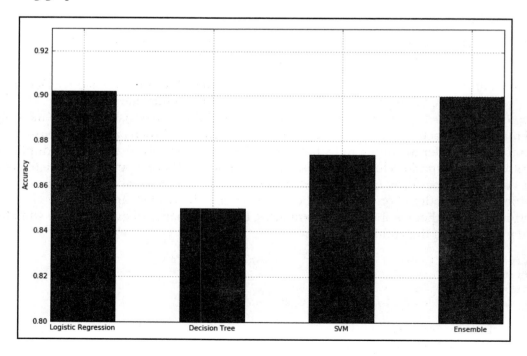

Average CV accuracy for each classifier belonging to the ensemble

As expected, the ensemble takes advantage of the different algorithms and yields better performance than any single one. We can now repeat the experiment with soft voting, considering that it's also possible to introduce a weight vector (through the `weights` parameter) to give more or less importance to each classifier:

$$\tilde{y} = argmax \frac{1}{N_{Classifiers}} \sum_{Classifier} w_c(p_1, p_2, \ldots, p_n)$$

For example, considering the previous graph, we can decide to give more importance to the logistic regression and less to the Decision Tree and SVM:

```
weights = [1.5, 0.5, 0.75]

vc = VotingClassifier(estimators=classifiers, weights=weights, voting='soft')
```

Repeating the same calculations for the cross-validation accuracy, we get the following:

```
print(np.array(a))
[ 0.90182873  0.85386795  0.87386955  0.89578952]
```

The resulting plot is shown in the following graph:

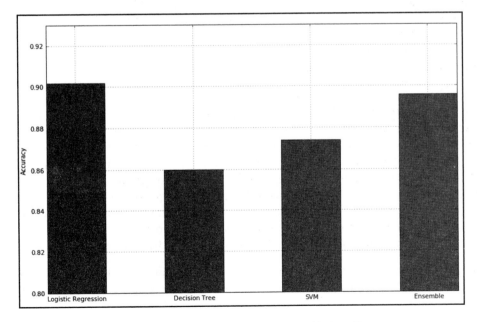

Average CV accuracy for each classifier belonging to the weighted ensemble

Weighting is not limited to the soft strategy. It can also be applied to hard voting, but in that case, it will be used to filter (reduce or increase) the number of actual occurrences:

$$\tilde{y} = argmax \left(N_c(y_t^1, \bar{w}), N_c(y_t^2, \bar{w}), \ldots, N_c(y_t^n, \bar{w}) \right)$$

Here, $N_c(y_t, w)$ is the number of votes for each target class, where each of them is multiplied by the corresponding classifier weighting factor.

A voting classifier can be a good choice whenever a single strategy is not able to reach the desired accuracy threshold. While exploiting the different approaches, it's possible to capture many microtrends using only a small set of strong (but sometimes limited) learners.

Summary

In this chapter, we introduced Decision Trees as a particular kind of classifier. The basic idea behind this concept is that a decision process can become sequential by using splitting nodes where, according to the sample we used, a branch is chosen until we reach a final leaf. In order to build such a tree, the concept of impurity was introduced; starting from a complete dataset, our goal was to find a split point that creates two distinct sets that should share the minimum number of features and, at the end of the process, should be associated with a single target class. The complexity of a tree depends on the intrinsic purity; in other words, when it's always easy to determine a feature that best separates a set, the depth will be reduced. However, in many cases, this is almost impossible, so the resulting tree needs many intermediate nodes to reduce the impurity until it reaches the final leaves.

We also discussed some Ensemble Learning approaches: Random Forests, AdaBoost, Gradient Tree Boosting, and voting classifiers. They are all based on the idea of training several weak learners and evaluating their predictions using a majority vote or an average. However, while a Random Forest creates a set of Decision Trees that are partially randomly trained, AdaBoost and Gradient Boost Trees adopt the technique of boosting a model by adding a new one, step after step, and focusing only on those samples that have been previously misclassified or on the minimization of a specific loss function. A voting classifier instead allows the mixing of different classifiers, adopting a majority vote to decide which class must be considered as the winning one during a prediction.

In the next chapter, `Chapter 9`, *Clustering Fundamentals* we're going to introduce the first unsupervised learning approach, k-means, which is one of the most diffuse clustering algorithms. We will concentrate on its strengths and weaknesses, and explore some alternatives that are offered by scikit-learn.

9
Clustering Fundamentals

In this chapter, we're going to introduce the basic concepts of clustering and the structure of some quite common algorithms that can solve many problems efficiently. However, their assumptions are sometimes too restrictive; in particular, those concerning the convexity of the clusters can lead to some limitations in their adoption. After reading this chapter, the reader should be aware of the contexts where each strategy can yield accurate results and how to measure the performances and make the right choice regarding the number of clusters.

In particular, we are going to discuss the following:

- The general concept of clustering
- The **k-Nearest Neighbors (k-NN)** algorithm
- Gaussian mixture
- The K-means algorithm
- Common methods for selecting the optimal number of clusters (inertia, silhouette plots, Calinski-Harabasz index, and cluster instability)
- Evaluation methods based on the ground truth (homogeneity, completeness, and Adjusted Rand Index)

Clustering basics

Let's consider a dataset of m-dimensional samples:

$$X = \{\bar{x}_1, \bar{x}_2, \ldots, \bar{x}_n\} \quad where \quad \bar{x}_i \in \mathbb{R}^m$$

Clustering Fundamentals

Let's assume that it's possible to find a criterion (not a unique) so that each sample can be associated with a specific group according to its peculiar features and the overall structure of the dataset:

$$g_k = G(\bar{x}_i) \ where \ k = \{0, 1, 2, \dots, t\}$$

Conventionally, each group is called a **cluster**, and the process of finding the function, G, is called **clustering**. Right now, we are not imposing any restriction on the clusters; however, as our approach is unsupervised, there should be a similarity criterion to join some elements and separate other ones. Different clustering algorithms are based on alternative strategies to solve this problem, and can yield very different results.

In the following graph, there's an example of clustering based on four sets of bidimensional samples; the decision to assign a point to a cluster depends only on its features and sometimes on the position of a set of other points (neighborhood):

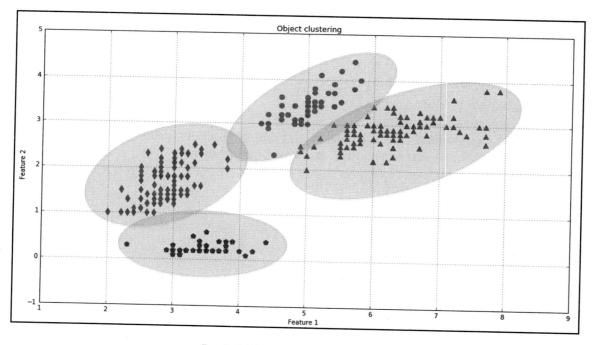

Example of a bidimensional dataset split into four clusters

[286]

In this book, we're going to discuss a large number of **hard-clustering** techniques, where each element must belong to a single cluster. The alternative approach, called **soft clustering** (sometimes called **fuzzy clustering**), is based on a membership score that defines how much the elements are *compatible* with each cluster. The generic clustering function becomes as follows:

$$\bar{m}_i = F(\bar{x}_i) \ where \ \bar{m}_i = \left(m_i^0, m_i^1, \ldots, m_i^t\right)^T \ and \ m_i^k \in [0, 1]$$

A vector, m_i, represents the relative membership of x_i, and it's often normalized as a probability distribution (that is, the sum is always forced to be equal to 1). In other scenarios, the single degrees are kept bounded between 0 and 1 and, hence, are considered as different probabilities. This is often a consequence of the underlying algorithm. As we are going to see in the *Gaussian mixture* section, a sample implicitly belongs to all distributions, so for each of them, we obtain a probability that is equivalent to a membership degree. In the majority of cases, hard clustering is the most appropriate choice, especially if the assigned cluster is a piece of information that's immediately employed in other tasks. In all of these cases, a soft approach whose output is a vector must be transformed into a single prediction (normally using the operator *argmax(•)*). However, there are particular applications where the vectorial output of a soft clustering algorithm is fed into another model (for example, together with other features) in order to produce a different final output (for example, a set of suggested items). In these scenarios, having several membership degrees (also normalized) can improve performance thanks to the potential relationships between secondary choices (that is, not the classes with the highest degrees/probabilities) and other parameters. An example of such kinds of systems is represented by the recommender engines. In these models, a soft clustering describing a user is often associated with other time-variant pieces of information, and the output is influenced by the dominant cluster (that can represent a user-segment) but also the secondary ones, which can become more and more important according to, for example, the navigation history of the user.

Clustering Fundamentals

k-NN

This method is intrinsically one of the simplest algorithms, belonging to the family of **instance-based learning** methods. Such a general approach is not based on a parameterized model that must be fit, for example, in order to maximize the likelihood. Conversely, instance-based algorithms rely completely on the data and their underlying structure. In particular, k-NN is a technique that can be employed for different purposes (even if we are going to consider it as a clustering algorithm), and it's based on the idea that samples that are close with respect to a predefined distance metric are also similar, so they can share their peculiar features. More formally, let's consider a dataset:

$$X = \{\bar{x}_1, \bar{x}_2, \ldots, \bar{x}_n\} \text{ where } \bar{x}_i \in \mathbb{R}^m$$

In order to measure the similarity, we need to introduce a distance function. The most common choice is the Minkowski metric, which is defined as follows:

$$d_p(\bar{x}_1, \bar{x}_2) = \left(\sum_j \left| \bar{x}_1^{(j)} - \bar{x}_2^{(j)} \right|^p \right)^{\frac{1}{p}}$$

$p = 1$, $d_1(\bullet)$ becomes the Manhattan (or city block) distance, while $p = 2$, $d_{2(\bullet)}$ is the classical Euclidean distance. Larger p values lead to shorter measures and, for $p \to \infty$, $d_p(\bullet)$, converges to the largest component absolute difference, $|x_1^{(k)} - x_2^{(k)}|$ (assuming that k is the index corresponding to the largest difference). In many applications, the Euclidean distance is the optimal choice; however, the value assigned to p can affect the *semantics* of the metric itself. In fact, when $p = 1$, all components are taken into account in the same way. This measure is called Manhattan because, given two points, the distance is equivalent to the path of a car moving along a piece-wise line (just like a taxi in New York). On the other side, increasing p will proportionally reduce the impact of all small component differences, forcing the measure to represent only the most relevant one. Finding the most appropriate value for p requires a pre-analysis of the dataset and full domain knowledge (for example, in a particular domain, two samples that have a large difference in a single component must be considered dissimilar, while in another one, the Euclidean distance is the most accurate way to measure their similarity). I always suggest testing different values, comparing the results, and picking the one which best represents the structure of the underlying problem.

Once a distance function has been chosen, it's easy to define the neighbors. In general, two approaches are employed. The first one is based on the number of nearest neighbors (which justifies the name of the algorithm), hence, given a sample (x_i), the neighborhood is defined as follows:

$$N_k(\bar{x}_i) = argmin_j^k \, d_p(\bar{x}_i, \bar{x}_j)$$

In the previous formula, the $argmin_j^k(\bullet)$ function selects the k last indexes j corresponding to the smallest distances from the center, x_i. In some cases, however, it's helpful to obtain the set of all neighbors whose distance is less than a prefixed radius, R:

$$N_R(\bar{x}_i) = \{\bar{x}_j : d_p(\bar{x}_i, \bar{x}_j) \leq R\}$$

Both approaches are perfectly compatible; however, the second one can only be employed when the data scientist has a complete awareness of the distances. In fact, for high-dimensional datasets, it's often simpler to set the maximum number of neighbors than finding the right radius. However, this is a context-sensitive problem and the right choice cannot be easily generalized.

Considering the previous formulas, the sharp-eyed reader should have noticed that they are extremely inefficient. In fact, if the dataset contains n samples $x_i \in \mathfrak{R}^m$, the computational complexity is $O(mn^2)$ because it's necessary to compute all pairwise distances. For this reason, the *vanilla* k-NNs are an affordable choice, but only for small low-dimensional sets, while, in all of the other cases, more efficient solutions have been designed.

The first approach is based on a **KD tree**, which is the extension of a binary tree to m-dimensional vectors. At each level, a feature is selected and a split is done. In the following diagram, there's an example with three-dimensional samples:

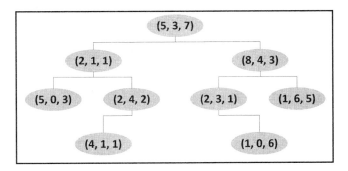

Example of a KD tree based on three-dimensional vectors

As it's possible to see, the first level is based on the first feature, and so on, until the leaves. If the tree is perfectly balanced, the computational complexity is analogous to a binary version and becomes $O(m \log n)$. Even if KD trees offer a valid solution in the majority of cases, they are prone to being uneasily unbalanced and, in particular, when $m \gg 1$, the complexity tends to be $O(mn)$ due to the *curse of dimensionality*. The first problem can be mitigated by always choosing the feature corresponding to the median of the corresponding subset, but the second problem is much more severe and, in many cases, there's no practical solution.

Another approach is strictly connected to the radius-based neighborhood, and this is based on a data structure called **ball tree**. A ball centered on a sample, x_i, is formally equivalent to $N_R(x_i)$, where R defines the radius. Therefore, the tree is built by nesting smaller balls (in terms of radius) into larger ones, as shown in the following diagram:

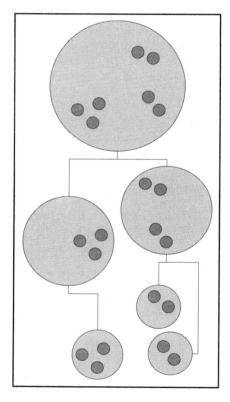

Example of a ball tree

An extremely important condition in a sample can belong to a single ball. This trick keeps the computational complexity equal to $O(m \log n)$, even when $m \gg 1$. Thanks to the property of the balls, it's enough to measure the distance between the point and the center to know whether it belongs to the ball. Hence, starting from the leaves (which are the smallest balls), it's possible to repeat this operation until the right ball is found. It's straightforward that such a data structure is a perfect choice when the dataset is partitioned into neighbors that can be found inside the same ball. This means that the number of samples contained in the leaves must not be too small, because this increases the depth of the tree and can lead to searches extended to more nodes (in particular, when the number of neighbors is fixed). Clearly, the opposite condition (nodes with too many samples) is very similar to the vanilla algorithm, and most of the advantages of this approach are lost. The choice of the leaf size is very problem dependent. Let's suppose that an application is based on cycles made up of *Nk* queries, and each cycle requires *1 k = 10* query and *20 k = 30* queries. In this case, the computational cost of the second block of queries can be dramatically reduced if the leaf size is close to 30 (slightly smaller or larger). Even if the first query is penalized, the more complex part is simplified. The reader should always consider the actual application and make the choice of whose negative impact is minimized. In any case, a leaf size that is optimal in every situation is extremely hard to find and, in the majority of real-life scenarios, requires a trade-off.

In order to test this algorithm, we are going to employ the MNIST handwritten digits dataset, which is available in the `scikit-learn` library. Contrary to the original version, in this case, there are 1,797 grayscale 8 × 8 images representing the digits from 0 to 9. Our goal is to cluster them using k-NNs and then find the neighbors of a noisy sample. The first step consists of, as usual, loading and scaling the dataset:

```
from sklearn.datasets import load_digits
from sklearn.preprocessing import StandardScaler

digits = load_digits()

ss = StandardScaler(with_std=False)
X = ss.fit_transform(digits['data'])
```

We have chosen to set the `with_std=False` attribute in the `StandardScaler` instance, so as to limit the process to the mean. After scaling the samples, the values are bound between -1.0 and 1.0.

Clustering Fundamentals

At this point, we can instantiate the `NearestNeighbors` class with `algorithm='ball_tree'`, `n_neighbors=25`, and `leaf_size=30`. As an exercise, I invite the reader to modify these parameters and to benchmark the performances with different queries (for example, as the leaf size is 30, it's possible to compare the computational time of 1,000 k = 10 queries and 1,000 k = 100 queries):

```
from sklearn.neighbors import NearestNeighbors

knn = NearestNeighbors(algorithm='ball_tree', n_neighbors=25, leaf_size=30)
knn.fit(X)
```

Once the model is fit, we expect that digits belonging to the same class will have very short distances and, in particular, as the digits are handwritten, the deformations spread uniformly around a mean value (which should be a perfect representation of each specific digit). Therefore, we want to query the model with a new sample, which is obtained by applying Gaussian noise to an existing one:

```
import numpy as np

X_noise = X[50] + np.random.normal(0.0, 1.5, size=(64, ))
```

The original version (representing a 2) and the noisy one are shown in the following screenshot:

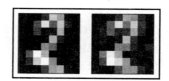

Original digit (left) and the noisy version (right)

We can now use the `kneighbors` function to retrieve the 25 nearest neighbors of the sample (the function returns the indexes, not the samples themselves). If the `n_neighbors` attribute is missing, the default value (declared in the constructor) will be employed. Moreover, setting the `return_distance=True` attribute allows us to also obtain the distances in ascending order:

```
distances, neighbors = knn.kneighbors(X_noise.reshape(1, -1),
return_distance=True)

print(distances[0])

[ 11.12060333  20.28609914  23.02542331  27.44598861  27.59396014
  29.31642387  30.05927362  31.04779967  31.67898264  32.00076453
```

```
   32.9237453    33.12438956   33.77814375   34.09565415   34.16874265
   34.36520324   34.59085311   35.12394042   35.37141415   35.42838394
   35.59343185   35.72420622   35.79115518   35.86460867   36.07930963]
```

The smallest distance is not very small because of the noise, but it's helpful to plot the samples so that we have a confirmation of the result:

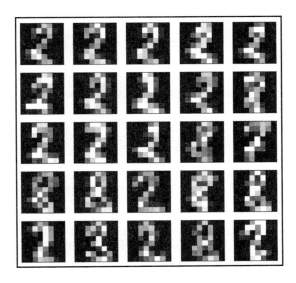

Nearest neighbors of the noisy sample

The set also contains some spurious digits (in particular, belonging to the classes 8 and 3), but the algorithm successfully found enough similar samples. This example showed the simplicity of the process and highlighted the importance of a distance metric function, which is the only real discriminating element. In this particular case, the Euclidean distance worked quite effectively, but in some cases, the results can be worse. In particular, this could be when the complexity of the samples can lead to similar distances, even when in the presence of strong differences. I invite the reader to repeat this example using the Olivetti Faces dataset (available in the scikit-learn library), adding some noise or excluding a test face from the training process. Understanding when k-NN is reliable is one of the most important tasks in many machine learning fields (for example, recommendation engines), and so the user must be aware of all the situations where the similarity between two feature vectors can be obtained by using a distance function and when, instead, more complex steps are necessary (for example, in natural language processing, two words cannot be compared, considering that their ASCII or UTF representations and algorithms such as Word2Vec have been designed to solve this problem efficiently).

Gaussian mixture

Let's suppose that we have a dataset made up of n m-dimensional points drawn from a data generating process, p_{data}:

$$X = \{\bar{x}_1, \bar{x}_2, \ldots, \bar{x}_n\} \quad \text{where} \quad \bar{x}_i \in \mathbb{R}^m$$

In many cases, it's possible to assume that the blobs (that is, the densest and most separated regions) are symmetric around a mean (in general, the symmetry is different for each axis), so that they can be represented as multivariate Gaussian distributions. Under this assumption, we can imagine that the probability of each sample is obtained as a weighted sum of k (the number of clusters) multivariate Gaussians parametrized by the mean vector, μ_j and the covariance matrix, Σ_j:

$$p(\bar{x}_i) = \sum_{j=1}^{k} p(N=j) N(\bar{x}_i; \bar{\mu}_j, \Sigma_j) = \sum_{j=1}^{k} w_j N(\bar{x}_i; \bar{\mu}_j, \Sigma_j)$$

This model is called **Gaussian mixture** and can be employed either as a soft- or a hard-clustering algorithm. The former option is clearly the *native* way because each point is associated with a probability vector representing the membership degree with respect to each cluster (that is, each Gaussian distribution). However, in many cases, it's preferable to apply an *argmax(•)* function to obtain the most likely cluster. As we are going to discuss in the next section, this choice leads to the behavior of another very famous algorithm, K-means (it's not strange at all that Gaussian mixture is also known as **soft K-means**).

In the previous formula, the term $p(N=j) = w_j$ is the relative weight of each Gaussian (their sum must be equal to 1 because we want to represent a probability distribution), and hence the parameters that must be learned are k weights, k mean vectors, and k covariance matrices. Considering the introduction to statistical learning discussed in Chapter 2, *Important Elements in Machine Learning*, the reader should have already understood the best way to find the parameters is the **maximum likelihood estimation** (MLE). In fact, starting from random values, we want to adjust the weights, means, and covariances to enhance the probability that each sample is generated by the model. Therefore, the standard way to solve this problem is based on the EM algorithm and requires many calculations (the whole derivation can be found in *Mastering Machine Learning Algorithms, Bonaccorso G, Packt Publishing, 2018*). As the complexity is non-trivial, we omit the proof and directly show the final results (which are extremely easy to compute).

For simplicity, let's group all of the parameters into a vector $\theta=(w_j, \mu_j, \Sigma_j)$, and let's also suppose a sequence, $\theta_0 \to \theta_1 \to ... \to \theta_t \to \theta_\infty$, that starts from a random initial guess, θ_0 and converges after a sufficiently large number of steps to the optimal value, θ_∞, which maximizes the likelihood. This is the iterative approach followed by the EM algorithm, and therefore we are going to indicate the generic parameter set at time t with θ_t. Moreover, let's define the probability of a Gaussian, j, given a sample, x_i, and the parameter set, θ_t, as $p(j|x_i; \theta_t)$. The generic value for each parameter after the iteration, t, is as follows:

$$\begin{cases} \bar{\mu}_j = \dfrac{\sum_i p(j|\bar{x}_i; \bar{\theta}_t)\bar{x}_i}{\sum_i p(j|\bar{x}_i; \bar{\theta}_t)} \\ \Sigma_j = \dfrac{\sum_i p(j|\bar{x}_i; \bar{\theta}_t)\left[(\bar{x}_i - \bar{\mu}_j)(\bar{x}_i - \bar{\mu}_j)^T\right]}{\sum_i p(j|\bar{x}_i; \bar{\theta}_t)} \\ w_j = \dfrac{\sum_i p(j|\bar{x}_i; \bar{\theta}_t)}{n} \end{cases}$$

Each iteration is based on the computation of $p(j|x_i; \theta_t)$, which can be easily obtained using the Bayes' theorem:

$$p(j|\bar{x}_i; \bar{\theta}_t) = \alpha p(\bar{x}_i; j, \bar{\theta}_t) p(N=j, \bar{\theta}_t)$$

The first term on the right-hand side is the probability of the sample under the Gaussian, j, while the second is the relative weight of the Gaussian, j, assuming the parameter is set to θ_t. Once this probability has been computed, it's possible to obtain the new values for the parameters (the covariance must be computed after the mean as it requires it). The process is iterated until convergence or after setting a maximum number of iterations. In the general case, which is also the default one, the covariance matrices are independent, and no diagonal constraints are imposed. However, scikit-learn allows setting the `covariance_type` attribute (the default value is `full`), but the user can also choose `tied`, which forces all of the Gaussians to share the same covariance matrix, `diag`, which enforces the diagonality condition (uncorrelated covariances) and `spherical`, where the covariance matrix degenerates into a single variance.

In this example, we are going to create a synthetic bidimensional dataset with three partially overlapped blobs:

```
from sklearn.datasets import make_blobs

nb_samples = 800
```

```
X, Y = make_blobs(n_samples=nb_samples, n_features=2, centers=3,
cluster_std=2.2, random_state=1000)
```

A plot of the dataset is shown in the following graph:

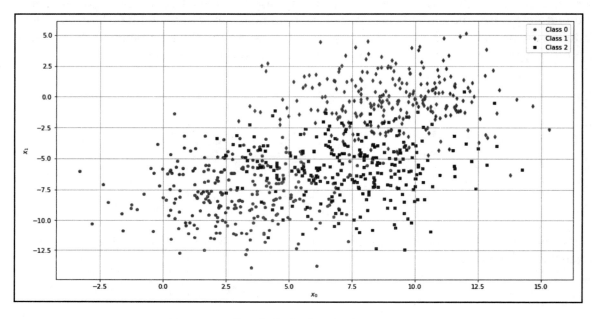

Dataset for Gaussian mixture clustering

I invite the reader to implement the algorithm directly using the previous formulas; however, in this case, we are going to employ the scikit-learn implementation `GaussianMixture` with `n_components=3` (we expect three generating Gaussian distributions) and `max_iter=1000`. We are keeping the default full covariance because we don't want to impose any restriction on the final matrices:

```
from sklearn.mixture import GaussianMixture

gm = GaussianMixture(n_components=3, max_iter=1000, random_state=1000)
gm.fit(X)
```

Once the model has been fitted, we can check means, covariances, and weights by using the instance variables `means_`, `covariances_`, and `weights_`:

```
print(gm.means_)

[[ 2.98469906 -7.43734851]
 [ 9.07336044 -0.42240226]
```

```
    [ 7.73445792 -6.05424097]]

print(gm.covariances_)

[[[ 4.11923384  0.46244723]
  [ 0.46244723  5.99491645]]

 [[ 4.7835118   0.07778618]
  [ 0.07778618  4.70847923]]

 [[ 4.16916545  0.41469122]
  [ 0.41469122  5.52496132]]]

print(gm.weights_)

[ 0.34896872  0.34073081  0.31030047]
```

As it's possible to see, the three final Gaussians are almost decorrelated and we can plot them, assuming that the diagonal elements are zero. The result is shown in the following graph:

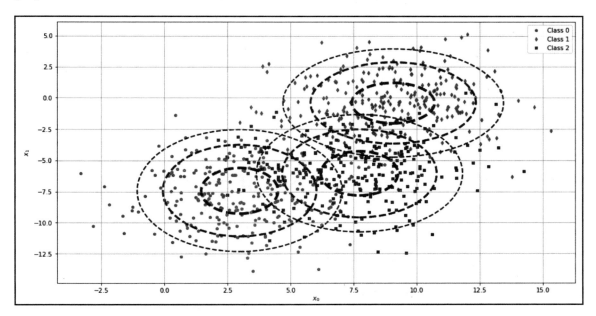

Dataset with the final three Gaussians distributions

The result confirms the hypothesis that the data-generating process can be approximated with a Gaussian mixture (at the end of this chapter, the reader will also learn how to measure the quality of the clustering when the ground truth is known). Now, let's check the probabilities for two test points:

```
import numpy as np

X_test_1 = np.array([10.0, -2.5])
print(gm.predict_proba(X_test_1.reshape(1, -1)))

[[ 0.00078008  0.76294815  0.23627177]]

X_test_2 = np.array([5.0, -6.0])
print(gm.predict_proba(X_test_2.reshape(1, -1)))

[[ 0.5876569   0.00776548  0.40457762]]
```

In the first case, the sample has a 76% membership degree with respect to the top cluster and 23% with respect to the central one (check the means for a confirmation). This *asymmetry* is due to the higher variances of the top Gaussian, which forces the distribution to capture more potential outliers. In the second case, instead, the point is on the boundary between the lower and central Gaussians, with a higher likelihood for the former. In both cases, it's possible to see how a soft-clustering approach can be used to manage situations where there's an intrinsic uncertainty. In some specific applications, the two test points can represent boundary samples that can inherit the common features of the two assigned clusters proportionally to their degree. For example, in a recommender system, a user vector belonging to more than one cluster can receive suggestions from each of them with a probability that's proportional to its membership degree.

Finding the optimal number of components

When the desired number of components is not known or there are no specific constraints, it's necessary to evaluate different models in order to decide which configuration is the best. As Gaussian mixture is a model based on the maximization of likelihood, a helpful method is provided by the **Akaike Information Criterion** (AIC). Let's suppose that the total number of parameters is n_p (of course, we are not considering the hyperparameters, but only the means, covariances, and weights) and that the maximum log-likelihood achieved after fitting the model is L_{opt}. If this is the case, then the AIC is defined as follows:

$$AIC(n_p; X) = 2n_p - 2L_{opt}$$

As it considers the negative log-likelihood, the smaller the *AIC* is, the higher the score. However, contrary to a method based only on the likelihood, *AIC* is based on the Occam's razor principle and penalizes the model with a very large number of parameters. Hence, the optimal value is always a trade-off between the MLE and the complexity. Another similar method is the **Bayesian Information Criterion** (**BIC**), which also considers the number of training samples, n:

$$BIC(n, n_p; X) = log(n)n_p - 2L_{opt}$$

The main difference between the two methods is the penalty, which is generally higher for the *BIC*, forcing the choice of even simpler models. In many cases, the two indexes show similar properties; however, it has been proven that the *AIC* is generally more reliable, even if it normally requires a large number of samples in order to avoid an overfitting. On the other side, the minimum value of the BIC when the number of training samples is sufficiently large (in theory, when $n \to \infty$) corresponds to a model whose probability distribution, p_m, satisfies $D_{KL}(p_m || p_{data}) \to 0$. In other words, the *BIC* guarantees to find a model that can perfectly reproduce the data-generating process.

To have a better understanding of this, let's compute both values considering the previously defined dataset and a set of a different number of components:

```
nb_components = [2, 3, 4, 5, 6, 7, 8]

aics = []
bics = []

for n in nb_components:
    gm = GaussianMixture(n_components=n, max_iter=1000, random_state=1000)
    gm.fit(X)
    aics.append(gm.aic(X))
    bics.append(gm.bic(X))
```

The plots for both indexes are shown in the following graph:

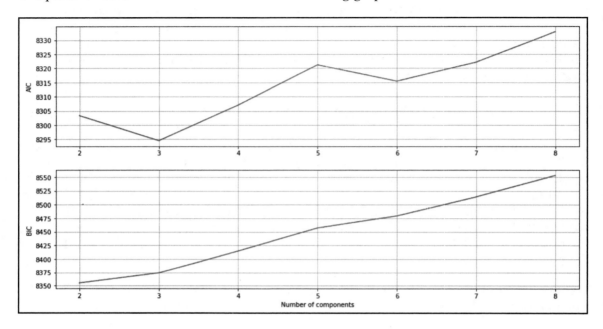

AIC (top) and BIC (bottom) indexes as functions of the number of components

The **AIC** has a minimum corresponding to **3** components (which is the ground truth). On the other side, the penalty term of the **BIC** (which is proportional to $log(800) \approx 6.7$ instead of **2**) doesn't allow you to reach the same minimum. However, it's possible to see a slight change in the slope, indicating that when the number of components becomes larger than **3**, the resulting **BIC** increases more quickly. I suggest analyzing the AIC before the BIC and, if there are small discrepancies, selecting the minimum provided by the AIC unless the number of training samples is extremely small. If the data scientist has at least a partial knowledge of the ground truth, the research can be immediately directed in the neighborhood of the expected number of components. Otherwise, a simpler model is generally preferable than a complex one (in terms of the number of parameters) in order to avoid a loss of inter-cluster separation (for example, in our example, a mixture of 6 Gaussians necessarily needs to split the blobs into two partially overlapping parts).

K-means

In the previous section, we discussed an algorithm based on the assumption that the data-generating process can be represented as a weighted sum of multivariate Gaussian distributions. What happens when the covariance matrices are shrunk towards zero? As it's easy to imagine, when $\Sigma_i \to 0$, the corresponding distribution degenerates to a Dirac's Delta centered on the mean. In other words, the probability will become almost *1* if the sample is extremely close to the mean, and *0* otherwise. In this case, the membership to a cluster becomes binary and it's determined only by the distance between the sample and the mean (the shortest distance will determine the winning cluster).

The K-means algorithm is the natural hard extension of Gaussian mixture and it's characterized by *k* (pre-determined) **centroids** or **means** (which justifies the name):

$$K = \{\bar{\mu}_1, \bar{\mu}_2, \ldots, \bar{\mu}_k\}$$

The goal of the algorithm, similarly to Gaussian mixture, is to find the optimal centroids in order to maximize the following:

- The intra-cluster cohesion
- The inter-cluster separation

In other words, we want to find the means so that all the points belonging to the same cluster are closer to each other much more than to any other point and, at the same time, the shortest distance between the closest points belonging to different clusters is always larger than the maximum intra-cluster distance.

The process is iterative and starts with a random guess of the centroids. At each step, the distance between every sample and every centroid is computed and the sample is assigned to the cluster where the distance is the minimum. This approach is often called **minimizing the inertia** of the clusters, which is defined as follows:

$$S = \sum_{j=1}^{k} \sum_{\bar{x}_i \in C_j} \|\bar{x}_i - \bar{\mu}_j\|^2$$

Considering the previous formula, it's possible to understand that when S is large, the internal cohesion is low because, probably, too many points have been assigned to clusters whose centroids are very far. To a certain extent, the previous one can be considered as a cost function because its minimization corresponds to the solution of the original problem. However, as we are going to discuss in the next section, the inertia is strictly related to the number of clusters and, therefore, it can be employed to pick the optimal value for *k* given a certain problem structure (for example, if $k = n$, each point represents a cluster and $S = 0$, but clearly this is an undesired situation).

Once all of the samples have been processed, a new set of centroids, $K^{(t)}$, is computed (now considering the actual elements belonging to the cluster), and all the distances are recomputed. The algorithm (formally known as **Lloyd's algorithm**) stops when the desired tolerance is reached, or, in other words, when the centroids become stable and, therefore, the inertia is minimized. Of course, this approach is quite sensitive to the initial conditions, and some methods have been studied to improve convergence speed.

One of them is called **K-means++** (*Robust Seed Selection Algorithm for K-Means Type Algorithms, Karteeka Pavan K., Allam Appa Rao, Dattatreya Rao A. V., and Sridhar G.R., International Journal of Computer Science and Information Technology 3, no. 5, October 30, 2011*), which selects the initial centroids so that they are statistically close to the final ones. This mathematical explanation is discussed in *Mastering Machine Learning Algorithms, Bonaccorso G., Packt Publishing, 2018*; however, this method is the default choice for scikit-learn, and it's normally the best choice for any clustering problem that's solvable with this algorithm. In general, considering the randomness, the process is repeated a fixed number of times (in scikit-learn, this value can be set considering the parameter `n_init`, whose default value is 10), and the configuration with the lowest inertia is chosen as the initial guess.

Now, let's consider an example with a dataset similar to one that's employed for Gaussian mixture (the example is very easy in order to allow an immediate visual confirmation of the results):

```
from sklearn.datasets import make_blobs

nb_samples = 1000
X, _ = make_blobs(n_samples=nb_samples, n_features=2, centers=3,
cluster_std=1.5, random_state=1000)
```

We expect to have three clusters with bidimensional features and a partial overlap due to the standard deviation of each blob. In our example, we won't use the *Y* variable (which contains the expected cluster) because we only want to generate a set of locally coherent points to try our algorithms.

The resultant plot is shown in the following graph:

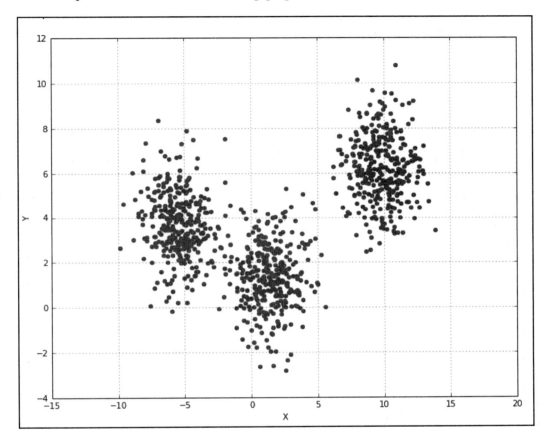

Dataset containing three blobs

 Slightly different datasets can be obtained using different random seeds. However, the final result of the make_blobs() function is only partially affected and the number of clusters should remain equal to three.

In this case, the problem is quite simple to solve, so we expect K-means to separate the three groups with a minimum error in the region of X bounded between [-5, 0]. Keeping the default values, we get the following:

```
from sklearn.cluster import KMeans

km = KMeans(n_clusters=3)
```

Clustering Fundamentals

```
km.fit(X)

KMeans(algorithm='auto', copy_x=True, init='k-means++', max_iter=300,
    n_clusters=3, n_init=10, n_jobs=1, precompute_distances='auto',
    random_state=None, tol=0.0001, verbose=0)

print(km.cluster_centers_)

[[ 1.39014517,  1.38533993]
 [ 9.78473454,  6.1946332 ]
 [-5.47807472,  3.73913652]]
```

Replotting the data using three different markers, it's possible to verify how K-means successfully separated the data:

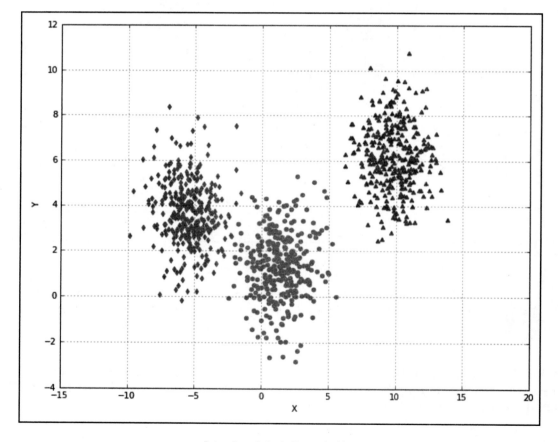

Dataset clustered using the K-means algorithm

In this case, the separation was very easy because K-means is based on Euclidean distance, which is radial, and therefore the clusters are expected to be convex. When this doesn't happen, the problem cannot be solved using this algorithm. Most of the time, even if the convexity is not fully guaranteed, K-means can produce good results, but there are several situations when the expected clustering is impossible, and letting K-means find out the centroid can lead to completely wrong solutions.

Let's consider the case of concentric circles. scikit-learn provides a built-in function to generate such datasets:

```
from sklearn.datasets import make_circles

nb_samples = 1000
X, Y = make_circles(n_samples=nb_samples, noise=0.05)
```

The plot of this dataset is shown in the following graph:

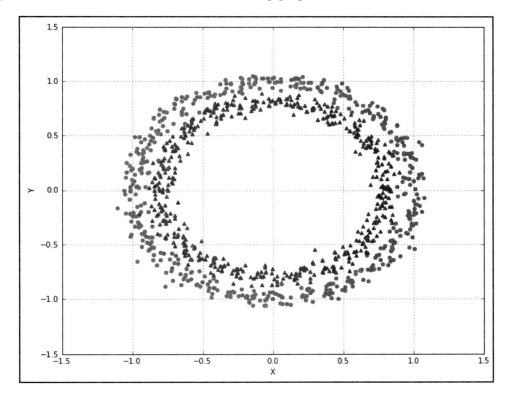

Example of a non-convex dataset

Clustering Fundamentals

We would like to have an internal cluster (corresponding to the samples depicted with triangular markers) and an external one (depicted by dots). However, such sets are not convex, and it's impossible for K-means to separate them correctly (the means should be the same!). In fact, suppose we try to apply two clusters to the algorithm:

```
km = KMeans(n_clusters=2)
km.fit(X)

KMeans(algorithm='auto', copy_x=True, init='k-means++', max_iter=300,
    n_clusters=2, n_init=10, n_jobs=1, precompute_distances='auto',
    random_state=None, tol=0.0001, verbose=0)
```

We get the separation that's shown in the following graph:

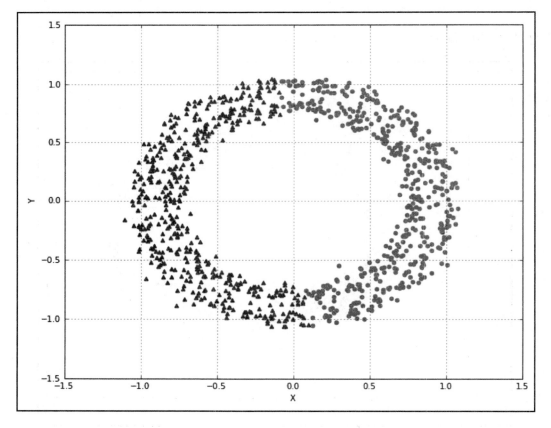

Non-convex dataset clustered using K-means

As expected, K-means converged on the two centroids in the middle of the two half-circles, and the resultant clustering is quite different from what we expected. Moreover, if the samples must be considered different according to the distance from the common center, this result will lead to completely wrong predictions. It's obvious that the convexity of the dataset is a fundamental condition and, whenever it's not met, K-means (as well as Gaussian mixture) will fail to cluster correctly. In all of these cases, another solution must be chosen. In the next chapter, Chapter 10, *Advanced Clustering*, we are going to discuss a few algorithms that are insensitive to this problem and can cluster more complex datasets in a more precise way. On the other hand, assessing the convexity is not a trivial task when the dimensionality is high, and so it's necessary to evaluate the clustering performances using one of the methods described at the end of this chapter. Unfortunately, they can produce valuable results, but only when the ground truth is accessible (that is, the labels of the training samples must be known). In all other cases, it's necessary to perform more complex evaluations, considering the similarity of randomly selected samples and their relative assignments. In the case of the nested circles, it's easy to verify that, when picking N random couples of samples from each cluster, the average distance between them is much lower than the average true intra-cluster distance. For example, take a look at the following:

```
import numpy as np

from scipy.spatial.distance import pdist

true_distances = pdist(X[Y == 0], metric='euclidean')
print(np.mean(true_distances))

1.281

Y_pred = km.predict(X)
sampled_X = np.random.choice(X[Y_pred == 0, 0], replace=False,
size=300).astype(np.int32)

distances = pdist(X[sampled_X], metric='euclidean')

print(np.mean(distances))

0.175
```

In this particular case, an average intra-cluster distance equal to 0.175 (with respect to the ground truth of 1.281) means that the non-convexity has not been managed correctly. In fact, on average, two points belonging to the same diameter should have a distance larger than 0.175 (see the preceding plot for a confirmation). Clearly, this approach can only be employed when the data scientist has a clear understanding of the dataset structure, but it's a good exercise in order to understand the limitations of K-means and the advantages of other algorithms.

Finding the optimal number of clusters

One of the most common disadvantages of K-means is related to the choice of the optimal number of clusters. An excessively small value will determine large groupings that contain heterogeneous elements, while a large number leads to a scenario where it can be difficult to identify the differences among clusters. Therefore, we're going to discuss some methods that can be employed to determine the appropriate number of splits and to evaluate the corresponding performance.

Optimizing the inertia

The first method we are going to consider is based on the assumption that an appropriate number of clusters must produce a small inertia. However, this value reaches its minimum (0.0) when the number of clusters is equal to the number of samples. Therefore, we can't look for the minimum, but for a value that is a trade-off between the inertia and a reasonable number of clusters.

Let's suppose we have a dataset of 1,000 elements (such as the one employed in the previous example):

```
nb_samples = 1000
X, _ = make_blobs(n_samples=nb_samples, n_features=2, centers=3,
cluster_std=1.5, random_state=1000)
```

We can compute and collect the inertias (scikit-learn stores these values in the `inertia_` instance variable) for a different number of clusters:

```
from sklearn.cluster import KMeans

nb_clusters = [2, 3, 5, 6, 7, 8, 9, 10]

inertias = []

for n in nb_clusters:
```

```
km = KMeans(n_clusters=n)
km.fit(X)
inertias.append(km.inertia_)
```

Plotting the values, we get the result that's shown in the following graph:

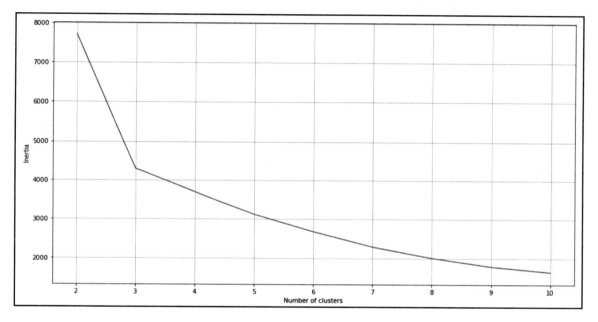

Inertias as a function of the number of clusters

As you can see, there's a dramatic reduction between **2** and **3** (the initial inertia is about **8,000**, and adding a single cluster allows us to reduce it to about **4,000**) and then the curve starts reducing the slope. We want to find a value that, if reduced, leads to a great inertial increase and, if increased, produces a very small inertial reduction. Therefore, a good choice could be **4** or **5**, while greater values are likely to produce unwanted intra-cluster splits (until the extreme situation where each point becomes a single cluster). On the other side, for $k = 4$, the inertia is about 3,800, and so there's no reason to reject the previous knowledge of the ground truth of the decision (corresponding to $k = 3$). However, let's suppose that we don't have any prior information and let's keep the potential values 3, 4, and 5 (less likely).

Clustering Fundamentals

This method is very simple and can be employed as the first approach to determine a potential range. In general, the optimal value is always a trade-off because the inertia decreases monotonically, and therefore the data scientist who decides to use this method always has to start with a minimum knowledge about the potential number of clusters and selects a value that is close to it. The following strategies are more complex and can be used to find the final number of clusters.

Silhouette score

The silhouette score is based on the principle of *maximum internal cohesion and maximum cluster separation*. In other words, we would like to find the number of clusters that produce a subdivision of the dataset into dense blocks that are well-separated from each other. In this way, every cluster will contain very similar elements and, selecting two elements belonging to different clusters, their distance should be greater than the maximum intra-cluster one.

After defining a distance metric (Euclidean is normally a good choice), we can compute the average intra-cluster distance for each element:

$$a(\bar{x}_i) = E_{\bar{x}_j \in C}[d(\bar{x}_i, \bar{x}_j)] \ \forall \ \bar{x}_i \in C$$

We can also define the average nearest-cluster distance (which corresponds to the lowest intercluster distance):

$$b(\bar{x}_i) = E_{\bar{x}_j \in D}[d(\bar{x}_i, \bar{x}_j)] \ \forall \ \bar{x}_i \in C \ where \ D = argmin\{d(C, D)\}$$

The silhouette score for an element, x_i, is defined as follows:

$$s(\bar{x}_i) = \frac{b(\bar{x}_i) - a(\bar{x}_i)}{max\left[a(\bar{x}_i), b(\bar{x}_i)\right]}$$

This value is bounded between *-1* and *1*, with the following interpretation:

- A value close to *1* is good (*1* is the best condition) because it means that $a(x_i) \ll b(x_i)$
- A value close to *0* means that the difference between the intra- and inter-cluster measures is almost null and therefore there's a cluster overlap
- A value close to *-1* means that the sample has been assigned to a wrong cluster because $a(x_i) \gg b(x_i)$

scikit-learn allows you to compute the average silhouette score so as to have an immediate overview for different numbers of clusters:

```
from sklearn.metrics import silhouette_score

nb_clusters = [2, 3, 5, 6, 7, 8, 9, 10]

avg_silhouettes = []

for n in nb_clusters:
    km = KMeans(n_clusters=n)
    Y = km.fit_predict(X)
    avg_silhouettes.append(silhouette_score(X, Y))
```

The corresponding plot is shown in the following graph:

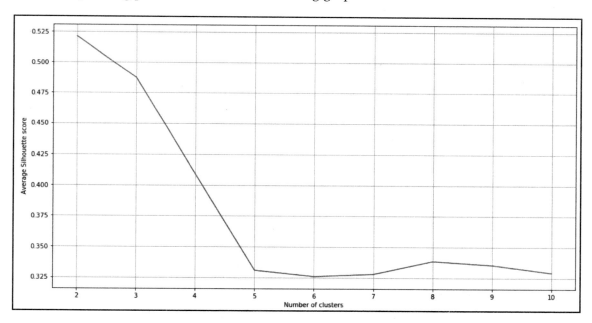

Average silhouette scores as a function of the number of clusters

Excluding the option *k* = 2, the best value is **3** (which corresponds to the ground truth, even if there are a few overlaps and wrong assignments). Between **2** and **5**, the score drops to its minimum, and so it does make sense to consider **5** anymore. However, the final decision between **3** and **4** is not immediate and should be evaluated by also considering the nature of the dataset. The silhouette score, together with the ground truth, indicates that there are three dense agglomerates (without the ground truth, a single cluster could be considered as the optimal choice), but the inertia diagram suggests that one of them (at least) can probably be split into two clusters. To have a better understanding of how the clustering is working, it's also possible to graph the silhouette plots, showing the sorted score for each sample in all clusters. In the following snippet, we create the plots for a number of clusters equal to **2, 3, 4,** and **8**:

```
import matplotlib.pyplot as plt
import matplotlib.cm as cm

from sklearn.metrics import silhouette_samples

fig, ax = plt.subplots(2, 2, figsize=(15, 10))

nb_clusters = [2, 3, 4, 8]
mapping = [(0, 0), (0, 1), (1, 0), (1, 1)]

for i, n in enumerate(nb_clusters):
    km = KMeans(n_clusters=n)
    Y = km.fit_predict(X)

    silhouette_values = silhouette_samples(X, Y)
    ax[mapping[i]].set_xticks([-0.15, 0.0, 0.25, 0.5, 0.75, 1.0])
    ax[mapping[i]].set_yticks([])
    ax[mapping[i]].set_title('%d clusters' % n)
    ax[mapping[i]].set_xlim([-0.15, 1])
    ax[mapping[i]].grid()
    y_lower = 20

    for t in range(n):
        ct_values = silhouette_values[Y == t]
        ct_values.sort()
        y_upper = y_lower + ct_values.shape[0]

        color = cm.Accent(float(t) / n)
        ax[mapping[i]].fill_betweenx(np.arange(y_lower, y_upper), 0,
                                     ct_values, facecolor=color,
                                     edgecolor=color)

        y_lower = y_upper + 20
```

The silhouette coefficients for each sample are computed using the `silhouette_values` function (which is always bounded between *-1* and *1*). In this case, we are limiting the graph between -0.15 and 1 because there are no smaller values. However, it's important to check the whole range before restricting it.

The resulting output is shown in the following graph:

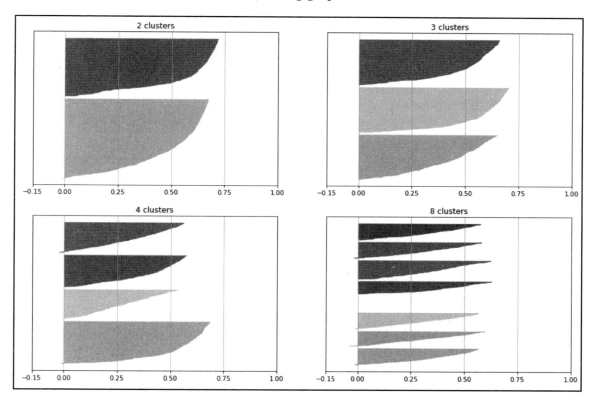

Silhouette plots for clusters 2, 3, 4, and 8

The width of each silhouette is proportional to the number of samples belonging to a specific cluster, and its shape is determined by the scores of each sample. An ideal plot should contain homogeneous blocks, without asymmetries. The silhouettes should be long and without sharp peaks (they must be similar to trapezoids rather than triangles) because we expect to have a very low score variance among samples in the same cluster. An ideal shape is more similar to a cigar, while a suboptimal one looks like a blade.

With **2** clusters, the shapes are acceptable, but one cluster is larger than the other one, and we know that this is an unlikely situation. A completely different situation is shown in the plot corresponding to 8 clusters. All the silhouettes are triangular and their maximum score is slightly greater than 0.6. This means that all the clusters are internally coherent, but the separation is unacceptable. With three clusters, the plot is almost perfect, except for the width of the third silhouette (a little bit too sharp, but better than any other configuration).

Without further metrics, we could consider this number as the best choice (which is also confirmed by the average score), but the inertia is lower for a higher number of clusters. With 4 clusters, the plot is slightly worse, with two silhouettes having a maximum score of about 0.5. This means that two clusters are perfectly coherent and separated, while the remaining two are rather coherent, but they probably aren't well separated. Right now, our choice should be made between 3 (which is currently the best choice) and 4. The following methods will help us in banishing all doubts.

Calinski-Harabasz index

Another method that is based on the concept of dense and well separated clusters is the Calinski-Harabasz index. To build it, we first need to define the inter-cluster dispersion. If we have *k* clusters with their relative centroids and the global centroid, the inter-cluster dispersion or **Between Cluster Dispersion (BCD)** is defined as follows:

$$BCD(k) = Tr(B_k) \quad \text{where} \quad B_k = \sum_i n_i (\bar{\mu}_i - \bar{\mu})(\bar{\mu}_i - \bar{\mu})^T$$

In the preceding expression, n_i is the number of elements belonging to the cluster, i, μ is the global centroid, μ_i is the centroid of cluster i, and $Tr(\bullet)$ is the trace of a square matrix. The intra-cluster dispersion or **Within Cluster Dispersion (WCD)** is defined as follows:

$$WCD(k) = Tr(X_k) \quad \text{where} \quad X_k = \sum_i \sum_{\bar{x} \in C_k} (\bar{x} - \bar{\mu}_i)(\bar{x} - \bar{\mu}_i)^T$$

The Calinski-Harabasz index is defined as the ratio between *BCD(k)* and *WCD(k)*:

$$CH(k) = \frac{n-k}{k-1} \cdot \frac{BCD(k)}{WCD(k)}$$

Chapter 9

As we look for a low intra-cluster dispersion (dense agglomerates) and a high intercluster dispersion (well separated agglomerates), we need to find the number of clusters that maximizes this index. We can obtain a graph in a way that's similar to what we have already done for the silhouette score:

```
from sklearn.metrics import calinski_harabaz_score

nb_clusters = [2, 3, 5, 6, 7, 8, 9, 10]

ch_scores = []

for n in nb_clusters:
    km = KMeans(n_clusters=n)
    Y = km.fit_predict(X)
    ch_scores.append(calinski_harabaz_score(X, Y))
```

The resulting plot is shown in the following graph:

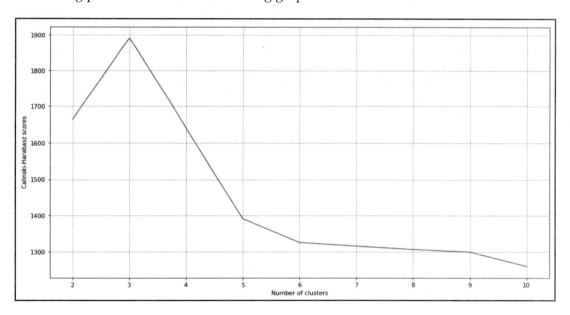

Calinski-Harabasz index has a function of the number of clusters

As expected, the highest value (about 1,900) is obtained with **3** clusters, while **4** clusters yield a value of about 1,650. Considering only this method, there's no doubt that the best choice is **3**, even if **4** is still a reasonable value. Let's consider the last strategy, which evaluates the overall stability.

Cluster instability

Another approach is based on the concept of cluster instability which is defined in *Cluster stability: an overview, Von Luxburg U., arXiv 1007:1075v1, 7 July 2010*. Intuitively, we can say that a clustering approach is stable if perturbed versions of the same dataset produce very similar results. More formally, if we have a dataset, X, we can define a set X_n of m perturbed (down-sampled or noisy) versions:

$$X_n = \{X_n^0, X_n^1, \ldots, X_n^m\}$$

Considering a distance metric, $d(C(X_1), C(X_2))$, between two clusterings with the same number (k) of clusters, the instability is defined as the average distance between couples of clusterings of noisy versions:

$$I(C) = E_{x_n^{i,j} \in X_n}\left[d\left(C(X_n^i), C(X_n^j)\right)\right]$$

For our purposes, we need to find the value of k that minimizes $I(C)$ (and, therefore, maximizes the stability). First of all, we need to produce some noisy versions of the dataset. Let's suppose that X contains 1,000 bidimensional samples with a standard deviation of 10.0. We can perturb X by adding a uniform random value (in the range [-2.0, 2.0]) with a probability of 0.25:

```
import numpy as np

nb_noisy_datasets = 10

X_noise = []

for _ in range(nb_noisy_datasets):
    Xn = np.ndarray(shape=(1000, 2))
    for i, x in enumerate(X):
        if np.random.uniform(0, 1) < 0.25:
            Xn[i] = X[i] + np.random.uniform(-2.0, 2.0)
        else:
            Xn[i] = X[i]

    X_noise.append(Xn)
```

Here, we are assuming that we have four perturbed versions. As a metric, we adopt the Hamming distance, which is proportional (if normalized) to the number of output elements that disagree.

 As the noisy datasets are based on random perturbations, the instability can be slightly different when changing the random seed. However, the overall trend should not be affected by the different random sequences when the noise has the same features (mean and standard deviation).

At this point, we can compute the instabilities for various numbers of clusters:

```
from sklearn.metrics.pairwise import pairwise_distances

instabilities = []

for n in nb_clusters:
    Yn = []
    for Xn in X_noise:
        km = KMeans(n_clusters=n)
        Yn.append(km.fit_predict(Xn))

    distances = []

    for i in range(len(Yn)-1):
        for j in range(i, len(Yn)):
            d = pairwise_distances(Yn[i].reshape(-1, 1), Yn[j].reshape(-1, 1), 'hamming')
            distances.append(d[0, 0])
    instability = (2.0 * np.sum(distances)) / float(nb_noisy_datasets ** 2)
    instabilities.append(instability)
```

As the distances are symmetrical, we can only compute them for the upper triangular part of the matrix. The result is shown in the following graph:

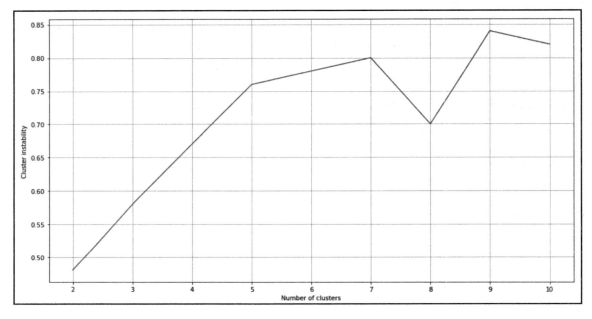

Cluster instability as a function of the number of clusters

Excluding the configuration with **2** clusters, where the inertia is very high, we have a minimum for **3** clusters, a value that has already been confirmed by the three previous methods. Therefore, we can finally decide to set `n_clusters=3`, excluding the options of 4 or more clusters. This method is very powerful, but it's important to evaluate the stability with a reasonable number of noisy datasets, taking care not to excessively alter the original geometry. A good choice is to use Gaussian noise with a variance set to a fraction (for example, 1/10) of the dataset variance. Alternative approaches are presented in *Cluster stability: an overview, Von Luxburg U., arXiv 1007:1075v1, 7 July 2010*.

 Even if we have presented these methods with K-means, they can be applied to any clustering algorithm to evaluate the performances and compare them.

Evaluation methods based on the ground truth

In this section, we will present some evaluation methods that require knowledge of the ground truth. This condition is not always easy to obtain because clustering is normally applied as an unsupervised method; however, in some cases, the training set has been manually (or automatically) labeled, and it's useful to evaluate a model before predicting the clusters of new samples.

Homogeneity

An important requirement for a clustering algorithm (given the ground truth) is that each cluster should only contain samples belonging to a single class. In Chapter 2, *Important Elements in Machine Learning*, we have defined the concepts of entropy *H(X)* and conditional entropy *H(X|Y)*, which measures the uncertainty of *X* given the knowledge of *Y*. Therefore, if the class set is denoted as *C* and the cluster set as *K*, *H(C|K)* is a measure of the uncertainty in determining the right class after having clustered the dataset. To have a homogeneity score, it's necessary to normalize this value considering the initial entropy of the class set, *H(C)*:

$$h = 1 - \frac{H(C|K)}{H(C)}$$

If we define the function $n(i_{true}, j_{pred})$, which corresponds to the number of samples with the true label, *i*, assigned to cluster, *j*, and the function $n_{pred}(j_{pred})$ that counts the number of samples assigned to the cluster, *j*, we can compute the conditional entropy as follows (*n* is the total number of samples):

$$H(C|K) = -\sum_{i_{true}} \sum_{j_{pred}} \frac{n(i_{true}, j_{pred})}{n} \log \frac{n(i_{true}, j_{pred})}{n_{pred}(j_{pred})}$$

In the same way, the entropy, H(C), is obtained as follows (the function, $n_{true}(i_{true})$, counts the number of samples belonging to the class, i):

$$H(C) = -\sum_{i_{true}} \frac{n_{true}(i_{true})}{n} \log \frac{n_{true}(i_{true})}{n}$$

In scikit-learn, there's the built-in `homogeneity_score()` function which can be used to compute this value. For this and the following few examples, we labeled dataset X (with true label Y) which we created in the previous section:

```
from sklearn.metrics import homogeneity_score

km = KMeans(n_clusters=3)
Yp = km.fit_predict(X)

print(homogeneity_score(Y, Yp))
0.430000122123
```

A value of 0.43 means that there's a residual uncertainty of about 57% because one or more clusters contains some points belonging to another class (this can be easily verified considering the high standard deviation of the blobs). As with the other methods shown in the previous section, it's possible to use the homogeneity score to determine the optimal number of clusters.

Completeness

A complementary requirement is that each sample belonging to a class is assigned to the same cluster. This measure can be determined by using the conditional entropy H(K|C), which is the uncertainty in determining the right cluster given the knowledge of the class. Like for the homogeneity score, we need to normalize this by using the entropy H(K):

$$c = 1 - \frac{H(K|C)}{H(K)}$$

The conditional entropy, H(K|C), and the entropy, H(K), can be computed using the frequency counts in the same way that was shown in the previous section.

We can compute this score (on the same dataset) by using
the `completeness_score()` function:

```
from sklearn.metrics import completeness_score

km = KMeans(n_clusters=3)
Yp = km.fit_predict(X)

print(completeness_score(Y, Yp))
0.897314145389
```

In this case, the value is rather high, meaning that the majority of samples belonging to a class have been assigned to the same cluster. This value can be further improved using a different number of clusters or by selecting another algorithm.

Adjusted Rand Index

The Adjusted Rand Index measures the similarity between the original class partitioning (Y) and the clustering. Bearing in mind the same notation adopted in the previous scores, we can define the following:

- **a**: The number of pairs of elements belonging to the same partition in the class set, C, and to the same partition in the clustering set, K
- **b**: The number of pairs of elements belonging to different partitions in the class set, C, and to different partitions in the clustering set, K

If the total number of samples in the dataset is n, the Rand Index is defined as follows:

$$R = \frac{a+b}{\binom{n}{2}}$$

The *corrected for chance* version is the Adjusted Rand Index, which is defined as follows:

$$AR = \frac{R - E[R]}{max(R) - E[R]}$$

We can compute the Adjusted Rand Index (or score) by using the `adjusted_rand_score()` function:

```
from sklearn.metrics import adjusted_rand_score

km = KMeans(n_clusters=3)
Yp = km.fit_predict(X)

print(adjusted_rand_score(Y, Yp))
0.374445476535
```

As the Adjusted Rand Index is bounded between *-1.0* and *1.0*, with negative values representing a bad situation (the assignments are strongly uncorrelated), a score of 0.37 means that the clustering is similar to the ground truth, even with some discrepancies (which is also confirmed by the homogeneity score and the high standard deviations of the blobs). Also, in this case, it's possible to optimize this value by trying different numbers of clusters or clustering strategies.

Summary

In this chapter, we introduced the fundamental clustering algorithms, starting with k-NN, which is an instance-based method that can be employed whenever it's helpful to retrieve the most similar samples given a query point. Then, we discussed the Gaussian mixture approach, focusing on its peculiarities and requirements, discussing how it's possible to use it whenever a soft-clustering is preferable than a hard method.

The natural evolution of Gaussian mixture with null covariances leads to the K-means algorithm, which is based on the idea of defining (randomly, or according to some criteria) *k* centroids that represent the clusters and optimize their position so that the sum of squared distances for every point in each cluster and the centroid is minimal. We have discussed different methods to find out the optimal number of clusters and, consequently, to evaluate the performance of an algorithm with and without the knowledge of the ground truth.

As the distance is a radial function, K-means (as well as Gaussian mixture) assumes that the clusters are convex and, hence, the algorithm cannot solve problems where the shapes have deep concavities (such as the half-moon problem). For this reason, in the next chapter, `Chapter 10`, *Advanced Clustering* we are going to introduce some algorithms that don't have such a limitation and can also be efficiently employed in non-convex scenarios.

10
Advanced Clustering

In this chapter, we're going to discuss some advanced clustering algorithms that can be employed when K-means (as well as other similar methods) fails to cluster a dataset. In Chapter 9, *Clustering Fundamentals*, we have seen that such models are based on the assumption of convex clusters that can be surrounded by a hyperspherical boundary. In this way, simple distance metrics can be employed to determine the correct labeling. Unfortunately, many real-life problems are based on concave and irregular structures that are wrongly split by K-means or a Gaussian mixture.

We will also explain two famous online algorithms that can be chosen whenever the dataset is too large to fit into the memory or when the data is streamed in a real-time flow. Surprisingly, even if these models work with a limited number of samples, their performance is only slightly worse than their standard counterparts trained with the whole dataset.

In particular, we are going to discuss the following:

- **Density-Based Spatial Clustering of Applications with Noise (DBSCAN)** clustering
- Spectral Clustering
- Online Clustering (mini-batch, K-means, and BIRCH)
- Biclustering (spectral biclustering algorithm)

DBSCAN

DBSCAN is a powerful algorithm that can easily solve non-convex problems where K-means fails. The main idea is quite simple: a cluster is a high-density area (there are no restrictions on its shape) surrounded by a low-density one. This statement is generally true and doesn't need an initial declaration about the number of expected clusters. The procedure is mainly based on a metric function (normally the Euclidean distance) and a radius, ε. Given a sample x_i, its boundary is checked for other samples. If it is surrounded by at least n_{min} points, it becomes a core point:

$$N(d(\bar{x}_i, \bar{x}_j) \leqslant \epsilon) \geqslant n_{min}$$

A sample x_j is defined as *directly reachable* from a core point x_i if:

$$d(\bar{x}_i, \bar{x}_j) \leqslant \epsilon$$

An analogous concept holds for sequences of directly reachable points. Hence, if there's a sequence $x_i \to x_{i+1} \to \ldots \to x_j$, then x_i and x_j are said to be *reachable*. Moreover, given a sample x_k, if x_i and x_j are reachable from x_k, they are said to be *density connected*. All the samples that don't meet these requirements are considered noisy.

All density-connected samples belong to a cluster whose geometrical structure has no limitations. After the first step, the neighbors are taken into account. If they also have a high density, they are merged with the first area; otherwise, they determine a topological separation. When all the areas have been scanned, the clusters have also been determined, and they appear like islands surrounded by empty space with a few noisy samples.

Scikit-learn allows us to control this procedure with two parameters:

- `eps`: This is responsible for defining the maximum distance between two neighbors. Higher values will aggregate more points, while smaller ones will create more clusters.
- `min_samples`: This determines how many surrounding points are necessary to define an area (also known as the core point).

Let's try DBSCAN with a very difficult clustering problem, called **half-moons**. The dataset can be created using a built-in function:

```
from sklearn.datasets import make_moons

nb_samples = 1000
X, Y = make_moons(n_samples=nb_samples, noise=0.05)
```

A plot of the dataset is shown in the following graph:

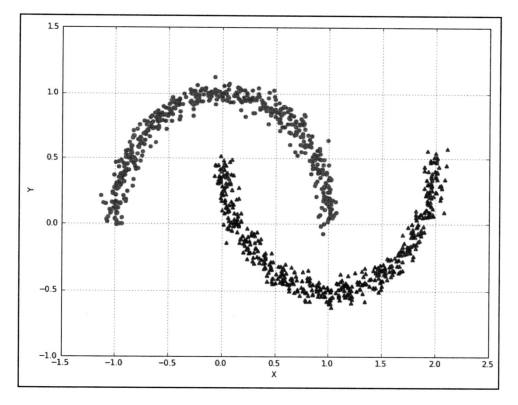

Half-moon dataset for DBSCAN clustering test

Advanced Clustering

To better understand the differences, we can check how K-means clusters this set. As explained in Chapter 9, *Clustering Fundamentals*, it works by finding the optimal convexity, and the result is shown in the following graph:

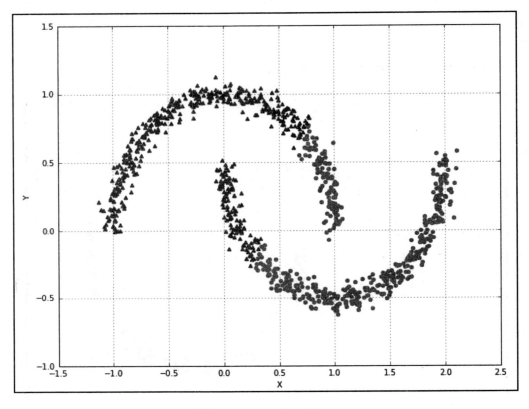

Half-moon dataset clustered using K-means

Of course, this separation is unacceptable and there's no way to improve the accuracy (it's impossible for K-means, as well as any other similar algorithm, to obtain a better result with concave clusters).

Let's try to cluster with DBSCAN (with eps set to 0.1 and the default value of 5 for min_samples):

```
from sklearn.cluster import DBSCAN

dbs = DBSCAN(eps=0.1)
Y = dbs.fit_predict(X)
```

In a different manner than other implementations, DBSCAN predicts the label during the training process, so we already have an array, Y, containing the cluster assigned to each sample. In the following graph, there's a representation with two different markers:

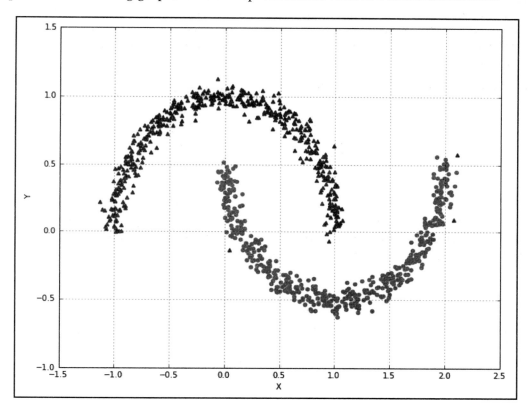

Half-moon dataset clustered using DBSCAN

As you can see, the accuracy is very high, and only three isolated points have been assigned to the wrong cluster because they are extremely noisy (that is, they are not density connected to any other sample). Of course, by performing a grid search, it's easy to find the best values that optimize the clustering process in every specific scenario. In general, it's important to tune up the parameters in order to avoid two common problems: a few big clusters and many small ones. This problem can be easily avoided using the **Spectral Clustering algorithm**.

Advanced Clustering

Spectral Clustering

Spectral Clustering is a more sophisticated approach based on the G={V, E} graph of the dataset. The set of vertices, V, is made up of the samples, while the edges, E, connecting two different samples are weighted according to an affinity measure, whose value is proportional to the distance of two samples in the original space or in a more suitable one (in a way analogous to Kernel SVMs).

If there are n samples, it's helpful to introduce a **symmetric affinity matrix**:

$$W = \begin{pmatrix} w_{11} & \cdots & w_{n1} \\ \vdots & \ddots & \vdots \\ w_{n1} & \cdots & w_{nn} \end{pmatrix}$$

Each element w_{ij} represents a measure of affinity between two samples. The most diffuse measures (also supported by scikit-learn) are the **Radial Basis Function (RBF)** and **k-Nearest Neighbors (k-NN)**. The former is defined as follows:

$$w_{ij} = e^{-\gamma \|\bar{x}_i - \bar{x}_j\|^2}$$

The latter is based on a parameter, k, defining the number of neighbors:

$$w_{ij} = \begin{cases} 1 & if\ \bar{x}_j \in kNN(\bar{x}_i) \\ 0 & otherwise \end{cases}$$

RBF is always non-null, while k-NN can yield singular affinity matrices if the graph is partially connected (and the sample x_i is not included in the neighborhood). This is in general not a severe issue because it's possible to correct the weight matrix to make it always invertible, but I invite the reader to test both methods, to check which is the most accurate. Whenever these approaches are not suitable, it's possible to employ any custom kernel that produces measures that have the same features of a distance (non-negative, symmetric, and increasing).

The algorithm is based on the normalized Laplacian graph:

$$L_n = I - D^{-1}W$$

Matrix *D* is called a *degree matrix*, and it's defined as this:

$$D = \left(\sum_j w_{ij} \; \forall \; i \in (1, n) \right)$$

Scikit-learn implements the Shi-Malik algorithm (*Normalized Cuts and Image Segmentation, Shi J, Malik J, IEEE Transactions on Pattern Analysis and Machine Intelligence, Vol 22, 08/2000*), also known as **normalized-cuts**, which partitions the samples into two sets (G_1 and G_2, which are formally graphs where each point is a vertex and the edges are derived from the normalized Laplacian matrix L_n) so that the weights corresponding to the points inside a cluster are higher than the one belonging to the cut. A complete mathematical explanation is beyond the scope of this book and can be found in *Mastering Machine Learning Algorithms, Bonaccorso G, Packt Publishing, 2018*. However, it's helpful to remark that the main role of L_n (thanks to its eigendecomposition) is to define a new subspace where the components are convex and can be easily separated using a standard algorithm such as K-means. Without any mathematical complications, the reader can imagine the process of a Kernel PCA, which, in this case, is a preparatory step. In fact, Spectral Clustering is a method that needs a concrete algorithm to assign the labels (for example, K-means).

Let's consider the previous half-moon example. In this case, the affinity (just like for DBSCAN) should be based on the nearest neighbors function; however, it's useful to compare different kernels. In the first experiment, we use an RBF kernel with different values for the gamma parameter:

```
import numpy as np

from sklearn.cluster import SpectralClustering

Yss = []
gammas = np.linspace(0, 12, 4)

for gamma in gammas:
    sc = SpectralClustering(n_clusters=2, affinity='rbf', gamma=gamma)
    Yss.append(sc.fit_predict(X))
```

In this algorithm, we need to specify how many clusters we want, so we set the value to 2. The resulting plots are shown in the following graph:

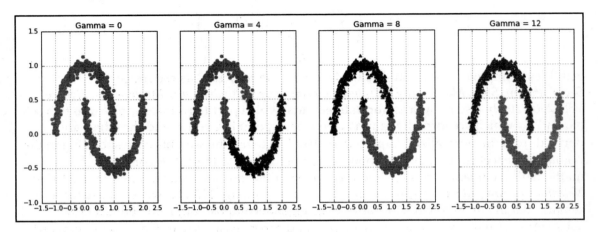

Half-moon dataset clustered using RBF Spectral Clustering with different γ values

As you can see, when the scaling factor **Gamma** is increased, the separation becomes more accurate. However, considering the dataset, the nearest neighbors kernel (with the default value n_neighbors=10) can also be a valid choice:

```
sc = SpectralClustering(n_clusters=2, affinity='nearest_neighbors')
Ys = sc.fit_predict(X)
```

The resulting plot is shown in the following graph:

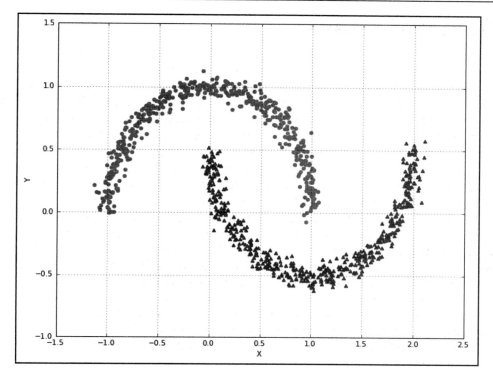

Half-moon dataset clustered using 10-NN Spectral Clustering

As with many other kernel-based methods, Spectral Clustering needs a previous analysis to detect which kernel can provide the best values for the affinity matrix. scikit-learn also allows us to define custom kernels for those tasks that cannot easily be solved using the standard ones. However, contrary to DBSCAN, in this case it's easy to avoid the wrong assignment of noisy samples.

Online Clustering

Sometimes, dataset X is too large, and the algorithms can become extremely slow, with a proportional need for memory. In these cases, it's preferable to employ a batch strategy that can learn while the data is streamed. As the number of parameters is generally very small, Online Clustering is quite fast and only a little bit less accurate than standard algorithms working with the whole dataset.

Mini-batch K-means

The first approach we are going to consider is a mini-batch version of the standard K-means algorithm. In this case, we cannot compute the centroids for all samples, and so the main problem is to define a criterion to reassign the centroids after a partial fit. The standard process is based on a streaming average, and therefore there will be centroids with a higher sample count and others with lower values. In scikit-learn, the fine-tuning of this process is achieved using the `reassigment_ratio` parameter (whose default value is `0.01`). Small values (for example, 0.0001) will let the centroids whose clusters have a fewer number of samples to the same value for a longer time, yielding a sub-optimal but faster solution. On the other hand, if the structure of the dataset is more complex and the batch size is very small, it could be preferable to increase this value, to force the process to reassign the *secondary* centroids more often. To better understand this process, let's consider a scenario with two large horizontal blobs. If the first batches contain 95% of samples belonging to a small area and the remaining ones are far away from this block, the process can create a spurious centroid to map the isolated samples. While the streaming process continues, such samples get closer and closer to the other ones; hence, with a low reassignment ratio the centroids are kept constant until a larger number of samples has been collected, while with a higher one the probability to update them increases, and even a single new batch can force a reassignment. It's straightforward that the latter approach will yield more accurate results, because the number of updates is higher and the process will need more iterations to converge.

Let's analyze this algorithm with a dataset containing `2000` samples that are streamed using `80` sample batches:

```
from sklearn.datasets import make_blobs

nb_samples = 2000
batch_size = 80

X, Y = make_blobs(n_samples=nb_samples, n_features=2, centers=5,
cluster_std=1.5, random_state=1000)
```

The `make_blobs` function shuffles the points automatically. In all the other cases, this is an extremely important step to guarantee that each batch represents the original data generation process in the most accurate way (even if it's always a partial representation). At this point, it's possible to instantiate the `MiniBatchKMeans` class with the default reassignment ratio and start the batch training process using the `partial_fit()` method. As we want to evaluate the overall performance, we are going to collect the predictions for the whole samples arrived till a certain moment:

```
from sklearn.cluster import MiniBatchKMeans

mbkm = MiniBatchKMeans(n_clusters=5, max_iter=1000, batch_size=batch_size,
random_state=1000)

X_batch = []
Y_preds = []

for i in range(0, nb_samples, batch_size):
    mbkm.partial_fit(X[i:i + batch_size])
    X_batch.append(X[:i + batch_size])
    Y_preds.append(mbkm.predict(X[:i + batch_size]))
```

The final result is shown in the following screenshot:

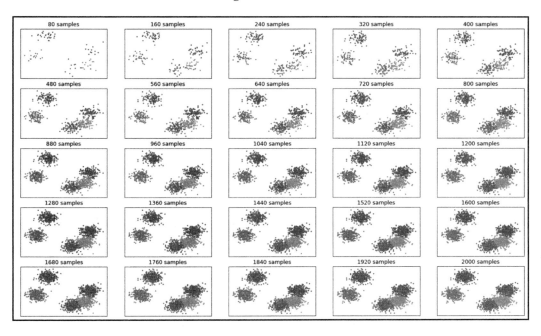

Partial clustering obtained using mini-batch K-means

Advanced Clustering

As you can see, the centroids show negligible modifications, and the most significant changes (which, however, are extremely small) happened only in the lower block, where the separation is almost null. Let's now compare the performance on the whole dataset to a standard K-means with the same parameters:

```
from sklearn.cluster import KMeans
from sklearn.metrics import adjusted_rand_score

print(adjusted_rand_score(mbkm.predict(X), Y))
0.916704982063      #Output
km = KMeans(n_clusters=5, max_iter=1000, random_state=1000)
km.fit(X)

print(adjusted_rand_score(km.predict(X), Y))
0.920328850257      #Output
```

As expected, the performance using the mini-batches is only slightly worse than a standard K-means, but the advantage in terms of computational and memory complexity is higher. I invite the reader to repeat the example with a larger and more complex dataset, trying to find out the optimal parameters (in particular, `reassignment_ratio` and `batch_size`) that yield an Adjusted Rand Index almost equal to the standard K-means one.

BIRCH

Another online algorithm is **Balanced Iterative Reducing and Clustering using Hierarchies** (**BIRCH**). The structure is a little more complex, but it's fundamentally based on a tree structure called the **Clustering-Feature Tree** (**CF-Tree**), sometimes also called the **Characteristic-Feature Tree**. In particular, let's suppose we have a dataset, X:

$$X = \{\bar{x}_1, \bar{x}_2, \ldots, \bar{x}_n\} \quad \text{where} \quad \bar{x}_i \in \mathbb{R}^m$$

A CF-Tree at level k is a tuple containing the following:

$$CF_k = \left(N_k, \sum_j \bar{x}_j, \sum_j \|\bar{x}_j\|^2\right)$$

The first element is the number of samples belonging to the sub-cluster, while the remaining two are the sum of the points and the square sum of the points. Without an excessively long mathematical explanation, we can describe the initial behavior of BIRCH as the creation of the CF-Tree, assuming a parameter B `branching_factor` (in scikit-learn, the default value is 50 samples). A non-terminal node must contain at most B values, which are stored together with a reference to their direct child. The leaves are instead stored directly (in the real implementations, references to adjacent leaves are inserted too, to speed up the computation). Each terminal node represents a sub-cluster where the distinctive features are the sum and the square sum of the points. Similar to Spectral Clustering, BIRCH doesn't perform the clustering autonomously, but after a compacting step (where the leaves are joined to avoid excessive fragmentation and increase internal cohesion), it employs a standard **Hierarchical Clustering algorithm** (discussed in the next chapter, `Chapter 11`, *Hierarchical Clustering*) to segment the data.

In a streaming process, when a sample x_i is fed into the model, it's placed into the root CF and propagated along the tree until it reaches a terminal node (leaf) whose distance between its centroid and the sample is the minimum. To control the structure of the tree, a threshold, T, is employed. The usage is very technical; however, it's helpful to know that to guarantee a balanced structure, whenever the addition of the new sample yields a subcluster whose radius is greater than T and the total number of $CFs > B$, then a new block is allocated. At that point, BIRCH will scan the sub-clusters to find the most dissimilar ones and then split them into two parts (one of them will be moved into the new empty space). In this way, all sub-clusters will remain very compact and therefore it's easier to merge them before the final global clustering.

Let's consider the previously defined dataset and cluster it using the `Birch` class with `branching_factor=100` and `threshold=0.15`:

```
from sklearn.cluster import Birch

birch = Birch(n_clusters=5, threshold=0.15, branching_factor=100)

X_batch = []
Y_preds = []

for i in range(0, nb_samples, batch_size):
    birch.partial_fit(X[i:i + batch_size])
    X_batch.append(X[:i + batch_size])
    Y_preds.append(birch.predict(X[:i + batch_size]))

print(adjusted_rand_score(birch.predict(X), Y))
```

Advanced Clustering

The following is the output:

```
0.902818524742
```

In this case, the performance is slightly worse than the mini-batch K-means, and also very sensitive to small changes in the hyperparameters. The final result is shown in the following output:

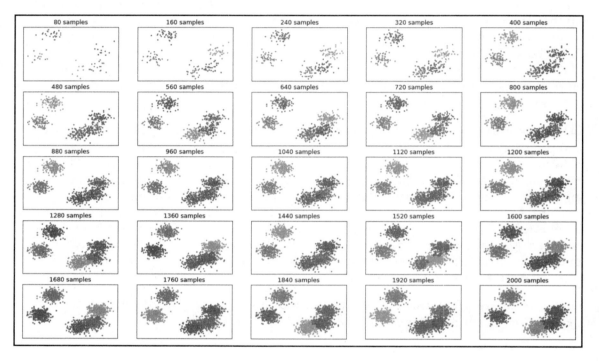

Partial clustering obtained using BIRCH

If we compare this result to the previous one, we can immediately notice some changes during the training process (due to the merging policy explained previously) and a few imprecisions in the block containing the three close blobs. In general, the performance of this algorithm is very similar to the mini-batch K-means, except when the number of sub-clusters is very large. In this case, in fact, BIRCH is much more flexible in the merging process and yields better results. However, finding the right branching factor and threshold is not always an immediate operation; hence, when the ground truth is known, I recommend performing a grid search. Alternatively, it's possible to employ methods such as **silhouette plots** or **cluster instability**, to get better insight into the dynamic and final behaviors.

Biclustering

In some specific scenarios, the dataset can be structured like a matrix, where the rows represent a category and the columns represent another category. For example, let's suppose we have a set of feature vectors representing the preference (or rating) that a user expressed for a group of items. In this example, we can randomly create such a matrix, forcing 50% of ratings to be null (this is realistic considering that a user never rates all possible items):

```
import numpy as np

nb_users = 100
nb_products = 150
max_rating = 10

up_matrix = np.random.randint(0, max_rating + 1, size=(nb_users, nb_products))
mask_matrix = np.random.randint(0, 2, size=(nb_users, nb_products))
up_matrix *= mask_matrix
```

In this case, we are assuming that *0* means that no rating has been provided, while a value bounded between 1 and 10 is an actual rating. The resulting matrix is shown in the following screenshot:

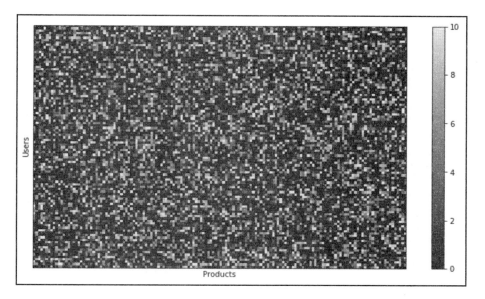

User-product matrix for the biclustering example

Given such a matrix, it's possible to assume that there's an underlying checkerboard structure. In other words, our assumption authorizes us to suppose that it is possible to rearrange rows and columns to determine coherent blocks inside the matrix. This is the main concept of **Biclustering**, which is extremely helpful whenever it's necessary to segment a set of feature vectors according to their common properties (for example, for marketing purposes or in recommendation engines). In our example, our goal is to find a matrix where each block is associated with a group of users and products (for example, all the users that rated 8 the 4th product).

The algorithm we are going to employ is called **Spectral Biclustering**, and it's based on some specific steps. The first one is a row-column normalization, whose goal is to find a set of values so that all the row sums and column sums become equal. At this point, the normalized matrix A_n is factorized using **Singular Value Decomposition (SVD)**:

$$A_n = U \Sigma V^T \text{ where } A_n \in \mathbb{R}^{m \times n},\ U \in \mathbb{R}^{m \times m},\ V \in \mathbb{R}^{n \times n},\ \text{and } \Sigma \in diag(\mathbb{R}^{m \times n})$$

Matrices U and V contain the left and right singular vectors of A_n (that is, the eigenvectors of $A_n A_n^T$ and $A n^T A_n$ respectively). The first singular vectors (u_1 and v_1) are discarded and the remaining ones are selected. The actual Biclustering is performed by rearranging the singular vectors according to a precise criterion. Let's suppose we consider a generic piece-wise constant vector:

$$\bar{p} = (0, 0, \ldots 0, 2, 2, \ldots 2, \ldots, 9, 9, \ldots, 9, 10, 10, \ldots, 10)^T$$

The singular vectors are ranked proportionally based on their similarity to a piece-wise constant vector (which represents the building block of our checkerboard structure, like the sample one shown previously). This operation is normally carried out using the K-means algorithm and measuring the difference between the original vector and the clustered one. Once the top k vectors have been determined, the dataset is projected onto them, obtaining the final representation. At this point, K-means is applied and the clusters (for rows and columns) are computed.

In our example, we are interested in segmenting according to the ranking; let's apply the algorithm, setting `n_clusters=10`:

```
import numpy as np

from sklearn.cluster.bicluster import SpectralBiclustering

sbc = SpectralBiclustering(n_clusters=10, random_state=1000)
sbc.fit(up_matrix)
```

```
up_clustered = np.outer(np.sort(sbc.row_labels_) + 1,
np.sort(sbc.column_labels_) + 1)
```

The final `up_clustered` matrix is obtained as the outer product of the sorted rows and columns labels (available through the `row_labels_` and `column_labels_` instance variables):

$$A_c = \bar{r} \otimes \bar{c} = \begin{pmatrix} r_1 \\ \vdots \\ r_m \end{pmatrix} \begin{pmatrix} c_1 & \cdots & c_n \end{pmatrix} = \begin{pmatrix} r_1 c_1 & \cdots & r_1 c_n \\ \vdots & \ddots & \vdots \\ r_m c_1 & \cdots & r_m c_n \end{pmatrix}$$

The result is shown in the following screenshot:

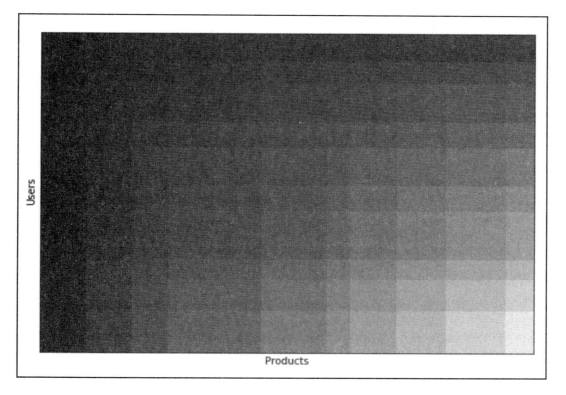

User-product matrix biclustered with Spectral Biclustering

Advanced Clustering

As you can see, the final structure is a checkerboard, where every block represents a rating. Let's now suppose that we want to find out all the rows (users) assigned to cluster 6 (rating 7); we can do this using the `rows_` instance variable, which is a Boolean array containing (n_clusters, n_rows) values. All `True` values represent a specific row belonging to a cluster:

```
import numpy as np

print(np.where(sbc.rows_[6, :] == True))
```

The following is the output:

```
(array([12, 15, 18, 39, 48, 74, 75, 81, 90], dtype=int64),)
```

The result shows that we need to look at specific rows, excluding the remaining ones. In the same way, we can do this with the columns (products):

```
print(np.where(sbc.columns_[6, :] == True))
```

The following is the output:

```
(array([  4,  12,  29,  46,  48,  76,  86,  88,  96, 110, 111, 126, 147, 148], dtype=int64),)
```

In this way, we have a double segmentation that can be employed to find similar users, with respect to ratings, and similar products, with respect to users that expressed the same rating (from a visual viewpoint, a segment is a homogeneous rectangle in the final matrix). We are going to discuss similar approaches in Chapter 12, *Introducing Recommendation Systems*. In that case, our goal will be to find optimal suggestions based on different similarity criteria.

Summary

In this chapter, we discussed two algorithms that can easily solve non-convex clustering problems. The first one is called DBSCAN and is a simple algorithm that analyzes the differences between points surrounded by other samples and boundary samples. In this way, it can easily determine high-density areas (which become clusters) and low-density spaces between them. There are no assumptions about the shape or the number of clusters, so it's necessary to tune up the other parameters to generate the right number of clusters.

Spectral Clustering is a family of algorithms based on a measure of affinity among samples. It uses a classic method (such as K-means) on subspaces generated by the Laplacian of the affinity matrix. In this way, it's possible to exploit the power of many kernel functions to cluster non-convex datasets.

We have also discussed two online algorithms: mini-batch K-means and BIRCH. The former is the direct counterpart of the standard K-means, with the dynamic reassignment of the centroids when the existing balance is altered by new samples. We have shown how, generally, this algorithm is able to achieve almost the same performance as K-means, with minimal memory consumption. BIRCH is an alternative solution that is based on a particular tree, itself based on tuples containing a summary of each sub-cluster.

The last topic we discussed is the biclustering algorithm, which can be employed whenever the dataset has an underlying checkerboard structure. The task of biclustering is to rearrange rows and columns to reveal all the homogeneous regions.

In `Chapter 11`, *Hierarchical Clustering*, we're going to discuss another approach, called Hierarchical Clustering. It allows us to segment data by splitting and merging clusters until a final configuration is reached.

11
Hierarchical Clustering

In this chapter, we're going to discuss a particular clustering technique called **Hierarchical Clustering**. Instead of working with relationships that exist in the whole dataset, this approach starts with a single entity containing all the elements (**divisive**) or N separate elements (agglomerative), and proceeds by splitting or merging the clusters according to some specific criteria, which we're going to analyze and compare.

In particular, we are going to discuss the following:

- Hierarchical strategies
- Agglomerative Clustering (metrics, linkages, and dendrograms)
- Connectivity constraints for Agglomerative Clustering

Hierarchical strategies

Hierarchical Clustering is based on the general concept of finding a hierarchy of partial clusters, built using either a bottom-up or a top-down approach. More formally, they are split into two categories:

- **Agglomerative Clustering**: The process starts from the bottom (each initial cluster is made up of a single element) and proceeds by merging the clusters until a stop criterion is reached. In general, the target has a sufficiently small number of clusters at the end of the process.
- **Divisive Clustering**: In this case, the initial state is a single cluster with all samples, and the process proceeds by splitting the intermediate cluster until all the elements are separated. At this point, the process continues with an aggregation criterion based on dissimilarity between elements. A famous approach is called **Divisive Analysis** (**DIANA**); however, that algorithm is beyond the scope of this book.

Hierarchical Clustering

scikit-learn implements only Agglomerative Clustering. However, this is not a real limitation because the complexity of divisive clustering is higher and the performance of Agglomerative Clustering is quite similar to that achieved by the divisive approach.

Agglomerative Clustering

Let's consider the following dataset:

$$X = \{\bar{x}_1, \bar{x}_2, \ldots, \bar{x}_n\} \quad where \quad \bar{x}_i \in \mathbb{R}^m$$

We define **affinity**, a metric function of two arguments with the same dimensionality, m. The most common metrics (also supported by scikit-learn) are the following:

- **Euclidean** or *L2 (Minkowski distance with p=2)*:

$$d_{Euclidean}(\bar{x}_1, \bar{x}_2) = \|\bar{x}_1 - \bar{x}_2\|_2 = \sqrt{\sum_j (\bar{x}_1^{(j)} - \bar{x}_2^{(j)})^2}$$

- **Manhattan** (also known as **city block**) or *L1 (Minkowski distance with p=1)*:

$$d_{Manhattan}(\bar{x}_1, \bar{x}_2) = \|\bar{x}_1 - \bar{x}_2\|_1 = \sum_j \left| \bar{x}_1^{(j)} - \bar{x}_2^{(j)} \right|$$

- **Cosine distance**:

$$d_{Cosine}(\bar{x}_1, \bar{x}_2) = 1 - \frac{\bar{x}_1 \cdot \bar{x}_2}{\|\bar{x}_1\|_2 \|\bar{x}_2\|_2} = 1 - cos(\alpha_{12})$$

The Euclidean distance is normally a good choice, but sometimes it's useful to have a metric whose difference from the Euclidean one gets larger and larger. As discussed in `Chapter 9`, *Clustering Fundamentals*, the Manhattan metric has this property. In the following graph, there's a plot representing the distances from the origin of points belonging to the line $y = x$:

Chapter 11

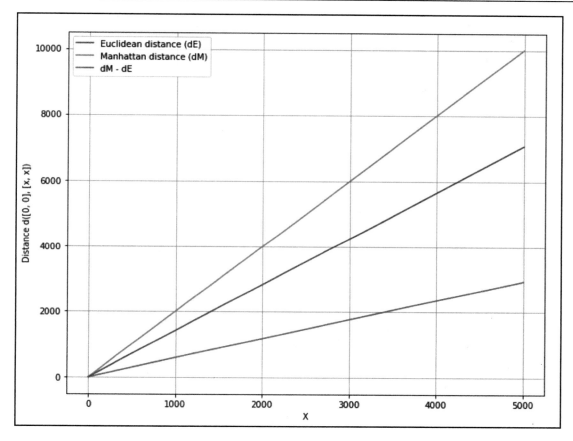

Distances of the point (x, x) from (0, 0) using the Euclidean and Manhattan metrics

The cosine distance is instead useful when we need a distance proportional to the angle between two vectors. If the direction is the same, the distance is null, while it is the maximum when the angle is equal to 180° (meaning opposite directions). This distance can be employed when the clustering must not consider the *L2* norm of each point. For example, a dataset could contain bidimensional points with different scales, and we need to group them into clusters corresponding to circular sectors. Alternatively, we could be interested in their position according to the four quadrants because we have assigned a specific meaning (invariant to the distance between a point and the origin) to each of them.

Hierarchical Clustering

Once a metric has been chosen (let's simply call it $d(x,y)$), the next step is to define a strategy (called **linkage**) to aggregate different clusters. There are many possible methods, but scikit-learn supports the three most common ones:

- **Complete linkage**: For each pair of clusters, the algorithm computes and merges them to minimize the maximum distance between the clusters (in other words, the distance of the furthest elements):

$$\forall C_i, C_j \Rightarrow L_{ij} = max\{d(\bar{x}_a, \bar{x}_b) \ \forall \ \bar{x}_a \in C_i \text{ and } \bar{x}_b \in C_j\}$$

- **Average linkage**: It's similar to complete linkage, but in this case the algorithm uses the average distance between the pairs of clusters:

$$\forall C_i, C_j \Rightarrow L_{ij} = \frac{1}{|C_i||C_j|} \sum_{\bar{x}_a \in C_i} \sum_{\bar{x}_b \in C_j} d(\bar{x}_a, \bar{x}_b)$$

- **Ward's linkage**: In this method, all clusters are considered, and the algorithm computes the sum of squared distances within the clusters and merges them to minimize it. From a statistical viewpoint, the process of agglomeration leads to a reduction in the variance of each resulting cluster. The measure is as follows:

$$\forall C_i, C_j \Rightarrow L_{ij} = \sum_{\bar{x}_a \in C_i} \sum_{\bar{x}_b \in C_j} \|\bar{x}_a - \bar{x}_b\|_2^2$$

Ward's linkage supports only the Euclidean distance.

Dendrograms

To better understand the agglomeration process, it's useful to introduce a graphical method called a **dendrogram**, which shows in a static way how the aggregations are performed, starting from the bottom (where all samples are separated) to the top (where the linkage is complete). Unfortunately, scikit-learn doesn't support it. However, SciPy provides some useful built-in functions.

Let's start by creating a dummy dataset:

```
from sklearn.datasets import make_blobs

nb_samples = 25
X, Y = make_blobs(n_samples=nb_samples, n_features=2, centers=3,
cluster_std=1.5)
```

To avoid excessive complexity in the resulting plot, the number of samples has been kept very low. In the following graph, there's a representation of the dataset:

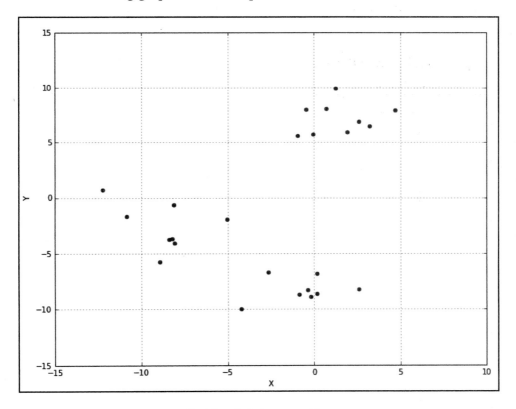

Simple dataset for dendrogram computations

Now, we can compute the dendrogram. The first step is computing a distance matrix:

```
from scipy.spatial.distance import pdist

Xdist = pdist(X, metric='euclidean')
```

Hierarchical Clustering

We have chosen a `euclidean` metric, which is the most suitable in this case. At this point, it's necessary to decide which `linkage` we want. Let's take `ward`; however, all known methods are supported:

```
from scipy.cluster.hierarchy import linkage

Xl = linkage(Xdist, method='ward')
```

Now, it's possible to create and visualize `dendrogram`:

```
from scipy.cluster.hierarchy import dendrogram

Xd = dendrogram(Xl)
```

The resulting plot is shown in the following screenshot:

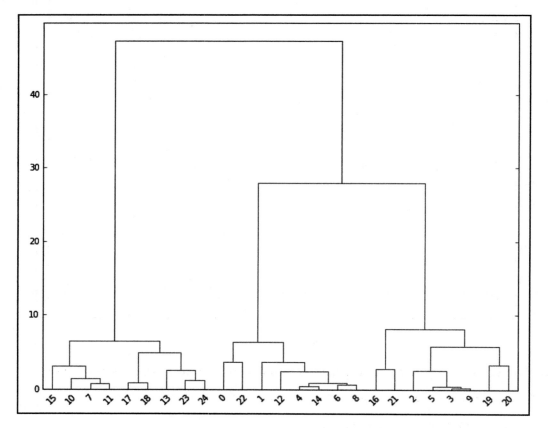

Dendrogram describing our dummy dataset

In the x axis, there are the samples (numbered progressively), while the y axis represents the distance (that is, the dissimilarity). Every arch connects two clusters that are merged together by the algorithm. For example, **23** and **24** are single elements merged together. Element **13** is then aggregated to the resulting cluster, and so the process continues. In general, the agglomerative procedure starts from the bottom (corresponding to a null dissimilarity) and proceeds along the y axis, merging the sub-clusters until the desired number of clusters has been reached.

For example, if we decide to cut the graph at a distance slightly below **30** (where the sub-cluster on the right part is merged into a single cluster), we get two separate clusters: the first one from **15** to **24**, and the other one from **0** to **20**. Instead, if we cut the graph at **10**, we obtain three clusters: **15** to **14**, **0** to **8**, and **16** to **20**. Looking at the previous dataset plot, all the points with *Y < 10* are considered to be part of the first cluster, while the others belong to the second cluster. If we increase the distance, the linkage becomes very aggressive (particularly in this example, which has only a few samples), and with values greater than **27**, only one cluster is generated (even if the internal variance is quite high!).

Agglomerative Clustering in scikit-learn

Let's consider a more complex dummy dataset with 8 centers:

```
from sklearn.datasets import make_blobs

nb_samples = 3000
X, Y = make_blobs(n_samples=nb_samples, n_features=2, centers=8, 
cluster_std=2.0)
```

Hierarchical Clustering

A graphical representation is shown in the following graph:

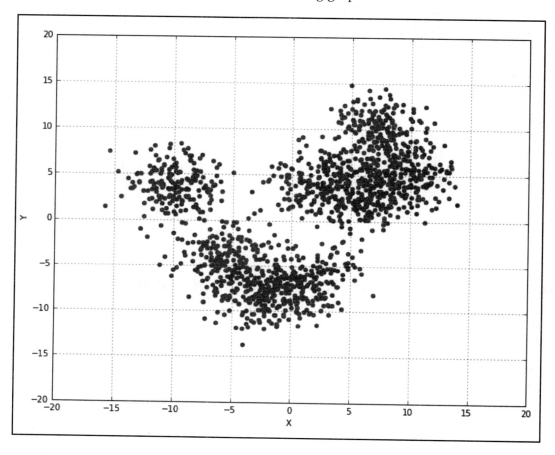

Dataset created for Agglomerative Clustering examples

We can now perform Agglomerative Clustering with different linkages (always keeping the Euclidean distance) and compare the results. Let's start with a complete linkage (AgglomerativeClustering uses the fit_predict(); method to train the model and transform the original dataset), also computing silhouette_score and adjusted_rand_score:

```
from sklearn.cluster import AgglomerativeClustering
from sklearn.metrics import silhouette_score, adjusted_rand_score

ac = AgglomerativeClustering(n_clusters=8, linkage='complete')
Y_pred = ac.fit_predict(X)
```

```
print('Silhouette score (Complete): %.3f' % silhouette_score(X, Y_pred))
Silhouette score (Complete): 0.352    #This is the output

print('Adjusted Rand score (Complete): %.3f' % adjusted_rand_score(Y,
Y_pred))
Adjusted Rand score (Complete): 0.481    #This is the output
```

A plot of the result (using both different markers and colors) is shown in the following graph:

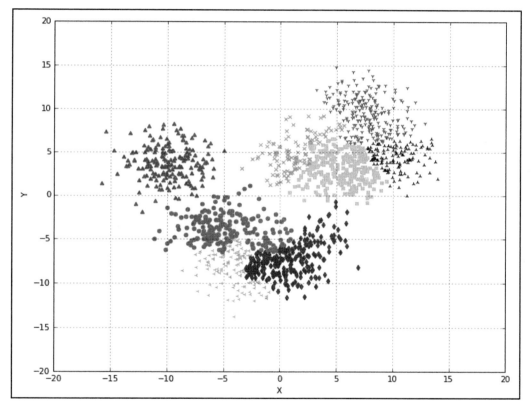

Clustering with complete linkage

The result is almost unacceptable. This approach penalizes the inter-variance and merges clusters that in most cases should be different (however, their final structure shows no overlaps, as also indicated by `silhouette_score`). In the previous plot, the three clusters in the middle are quite fuzzy, and the probability of wrong placement is very high, considering the variance of the cluster represented by dots.

Hierarchical Clustering

Let's now consider the `'average'` linkage:

```
ac = AgglomerativeClustering(n_clusters=8, linkage='average')
Y_pred = ac.fit_predict(X)

print('Silhouette score (Average): %.3f' % silhouette_score(X, Y_pred))
Silhouette score (Average): 0.320    #This is the output

print('Adjusted Rand score (Average): %.3f' % adjusted_rand_score(Y, Y_pred))
Adjusted Rand score (Average): 0.476 #This is the output
```

The result is shown in the following graph:

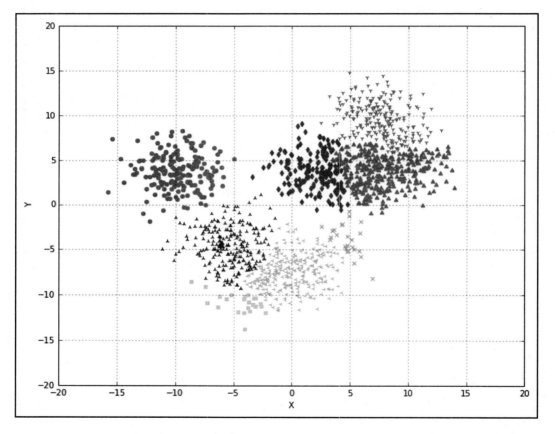

Clustering with average linkage

In this case, the clusters seem visually better defined, even if some of them could have become really small, and this is also indicated by a worsening of both Silhouette and Adjusted Rand scores. It can also be useful to try other metrics (in particular *L1*) and compare the results. The last method, which is often the best (it's the default one), is Ward's linkage; it can only be used with a Euclidean metric (also the default one):

```
ac = AgglomerativeClustering(n_clusters=8)
Y_pred = ac.fit_predict(X)

print('Adjusted Rand score (Ward): %.3f' % adjusted_rand_score(Y, Y_pred))
Adjusted Rand score (Ward): 0.318    #This is the output

print('Adjusted Rand score (Ward): %.3f' % adjusted_rand_score(Y, Y_pred))
Adjusted Rand score (Ward): 0.531    #This is the output
```

The resulting plot is shown in the following graph:

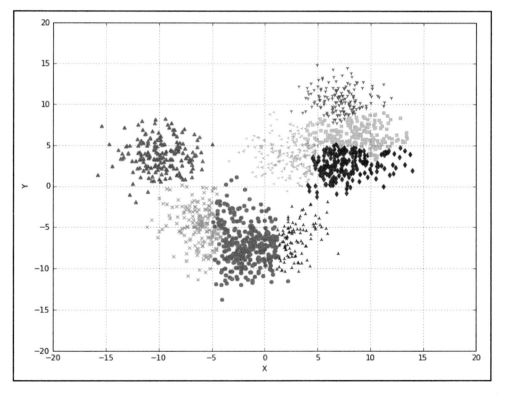

Clustering with Ward's linkage

Hierarchical Clustering

In this case, the Silhouette score is only slightly worse than average linkage, but there's an improvement in the Adjusted Rand Index, which indicates a superior accuracy in the assignments. As the ground truth is known, this score should be considered more reliable than the Silhouette one. Conversely, whenever there are no pieces of information about the true labels, the Silhouette score provides a good way to measure cohesion and homogeneity. As with Ward's linkage, it is impossible to modify the metric, so a valid alternative could be the average linkage, which can be used with every affinity and a grid search.

Connectivity constraints

scikit-learn also allows specifying a connectivity matrix, which can be used as a constraint when finding the clusters to merge. In this way, clusters that are far from one another (non-adjacent in the connectivity matrix) are skipped. A very common method for creating such a matrix involves using the **k-Nearest Neighbors (k-NN)** graph function (implemented as `kneighbors_graph()`), which is based on the number of neighbors a sample has (according to a specific metric). In the following example, we consider a **circular dummy dataset**, without the ground truth:

```
from sklearn.datasets import make_circles

nb_samples = 3000
X, Y = make_circles(n_samples=nb_samples, noise=0.05)
```

A graphical representation is shown as following:

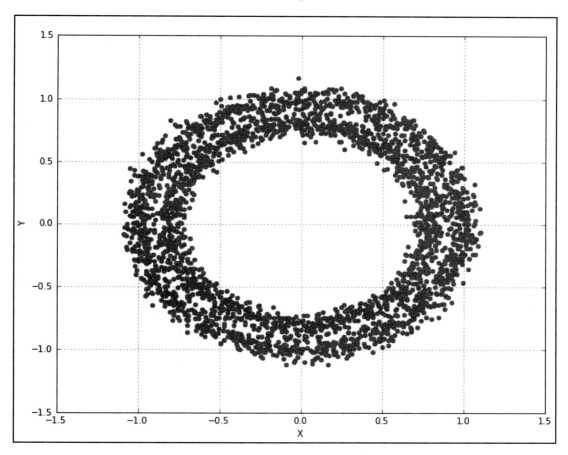

Circle dataset for connectivity constraint examples

We start with unstructured Agglomerative Clustering based on `'average'` linkage, impose 20 clusters, and compute the Silhouette score:

```
from sklearn.metrics import silhouette_score

ac = AgglomerativeClustering(n_clusters=20, linkage='average')
Y_pred = ac.fit_predict(X)

print('Silhouette score: %.3f' % silhouette_score(X, Y_pred))
Silhoette score: 0.310    #This is the output
```

Hierarchical Clustering

In this case, we have used the `fit_predict()` method because the `AgglomerativeClustering` class, after being trained, exposes the labels (cluster number) through the `labels_` instance variable, and it's easier to use this variable when the number of clusters is very high. `silhouette_score` is 0.310, which indicates no overlaps but also imperfectly balanced clusters. However, as we don't know the ground truth and we have imposed 20 clusters, we can use this value as an initial benchmark (of course, I invite the reader to repeat the exercise to test other solutions and improve the results).

A graphical plot of the result is shown in the following graph:

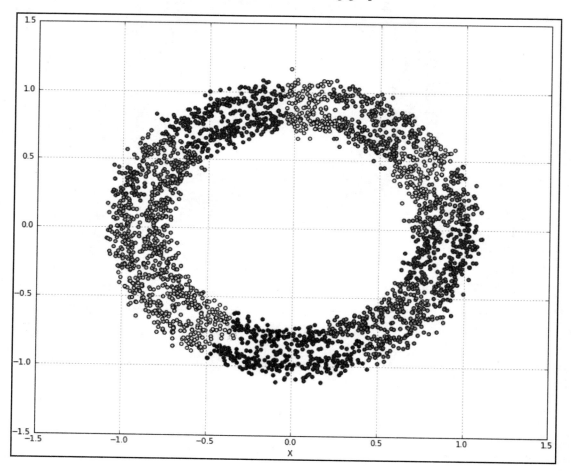

Clustering result with average linkage

Now we can try to impose a constraint with different values for k, also computing the silhouette_score:

```
from sklearn.neighbors import kneighbors_graph
from sklearn.metrics import silhouette_score

acc = []
k = [50, 100, 200, 500]

for i in k:
    kng = kneighbors_graph(X, i)
    ac1 = AgglomerativeClustering(n_clusters=20, connectivity=kng,
linkage='average')
    Y_pred = ac1.fit_predict(X)
    print('Silhouette score (k=%d): %.3f' % (i, silhouette_score(X,
Y_pred)))
    acc.append(ac1)
```

The output is as follows:

```
Silhouette score (k=50): -0.751
Silhouette score (k=100): -0.221
Silhouette score (k=200): 0.277
Silhouette score (k=500): 0.310
```

The resulting plots are shown in the following graph:

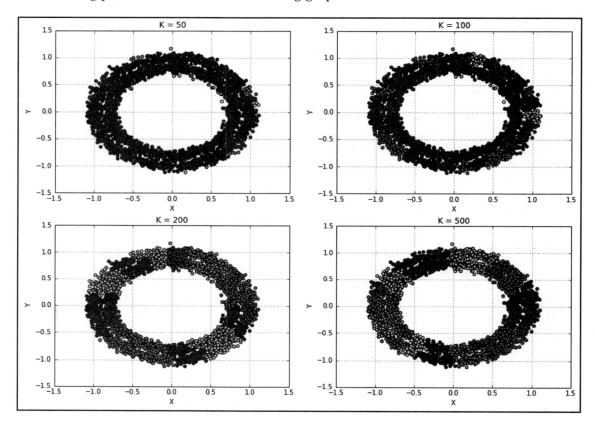

Clustering results for different k values

As you can see, imposing a constraint (in this case, based on k-NN) allows us to control how the Agglomeration creates new clusters and can be a powerful tool for tuning the models, or for filtering those elements whose distance in the originals space is very large could be taken into account during the merging phase (this is particularly useful when clustering images). Considering the previous example, k=50 yields a very large cluster with only a few outliers. The structure starts appearing with k=100, but the clustering is still very unbalanced, with a dominant group and some small, isolated clusters. The result appears to be rather good for k=200, even if the homogeneity isn't perfect yet. For k=500, the Silhouette score is positive but still quite low; however, it indicates that there are no overlaps but some potential wrong assignments. Moreover, it corresponds to the score obtained without the constraints; hence, we can assume that this level of k weakens the constraints too much. At this point, a choice must be made, considering all the possible pieces of information about the dataset. Excluding the first two scenarios, another analysis can be performed in the range *(200, 500)*, to try to determine an optimal solution that matches the constraints that we know (for example, some samples belong to a single cluster even if the final structure is unbalanced). As a supplementary exercise, I suggest analyzing the range *(200, 500)*, choosing some potential ground-truth constraints, and checking which value can yield the optimal solution.

Summary

In this chapter, we presented Hierarchical Clustering, focusing our attention on the Agglomerative version, which is the only one supported by scikit-learn. We discussed its philosophy, which is rather different to the one adopted by many other methods. In Agglomerative Clustering, the process begins by considering each sample as a single cluster and proceeds by merging the blocks until the number of desired clusters is reached. To perform this task, two elements are needed: a metric function (also called affinity) and a linkage criterion. The former is used to determine the distance between the elements, while the latter is a target function that is used to determine which clusters must be merged.

We also saw how to visualize this process through dendrograms, using SciPy. This technique is quite useful when it's necessary to maintain complete control of the process and the final number of clusters is initially unknown (it's easier to decide where to cut the graph). We showed how to use scikit-learn to perform Agglomerative Clustering with different metrics and linkages, and, at the end of the chapter, we also introduced the connectivity constraints that are useful when it's necessary to force the process, to avoid merging clusters that are too far apart.

In `Chapter 12`, *Introducing Recommendation Systems,* we're going to introduce the recommendation systems that are employed daily by many different systems to automatically suggest items to a user, according to their similarity to other users and their preferences.

12
Introducing Recommendation Systems

Imagine an online shop with thousands of items. If you're not a registered user, you'll probably see a home page with some highlights, but if you've already bought some items, it would be interesting if the website showed products that you would probably buy, instead of a random selection. This is the purpose of a recommender system, and, in this chapter, we're going to discuss the most common techniques to create such a system.

The basic concepts are users, items, and ratings (or implicit feedback about the products, such as the fact you have bought them). Every model must work with known data (as in a supervised scenario) to be able to suggest the most suitable items or to predict ratings for all the items not evaluated yet.

We're going to discuss two different kinds of strategies:

- User- or content-based
- Collaborative filtering

The first approach is based on the information we have about users or products, and its target is to associate a new user with an existing group of peers to suggest all the items positively rated by the other members, or to cluster the products according to their features and propose a subset of items similar to the one taken into account. The second approach, which is a little bit more sophisticated, works with explicit ratings, and its purpose is to predict this value for every item and every user. Even if collaborative filtering requires mode computational power, the great availability of cheap resources allows using these algorithms with every large datasets, yielding almost real time recommendations. The model can also be retrained or updated every day.

Naive user-based systems

In this first scenario, we assume that we have a set of users represented by m-dimensional feature vectors:

$$U = \{\bar{u}_1, \bar{u}_2, \ldots, \bar{u}_m\} \;\; where \;\; \bar{u}_i \in \mathbb{R}^m$$

Typical features are age, gender, interests, and so on. All of them must be encoded using one of the techniques discussed in the previous chapters (for example, they can be binarized, normalized in a fixed range, or transformed into one-hot vectors). However, in general, it's useful to avoid different variances that can negatively impact the computation of the distance between neighbors.

We have a set of k items:

$$I = \{i_1, i_2, \ldots, i_k\}$$

Let's also assume that there is a relation that associates each user with a subset of items (bought or positively reviewed), items for which an explicit action or feedback has been performed:

$$g(\bar{u}) \rightarrow \{i_1, i_2, \ldots, i_p\} \;\; where \;\; p \in (1, k)$$

In a user-based system, the users are periodically clustered (normally using a **k-Nearest Neighbors** (**k-NN**) approach), and therefore considering a generic user u (also a new sample), we can immediately determine a ball with radius R containing all the users who are similar (and therefore neighbors) to our sample:

$$B_R(\bar{u}) = \{\bar{u}_i \;\; with \;\; d(\bar{u}, \bar{u}_i) \leqslant R\}$$

At this point, we can create the set of suggested items using the relation previously introduced:

$$I_{Suggested}(\bar{u}) = \left\{ \bigcup_i g(\bar{u}_i) \;\; where \;\; \bar{u}_i \in B_R(\bar{u}) \right\}$$

In other words, the set contains all the unique products positively rated or bought by the neighborhood. I've used the adjective *naive* because there's a similar alternative that we're going to discuss in the section dedicated to collaborative filtering.

Implementing a user-based system with scikit-learn

For our purposes, we need to create a dummy dataset of users and products (but it's very easy to employ real datasets):

```
import numpy as np

nb_users = 1000
users = np.zeros(shape=(nb_users, 4))

for i in range(nb_users):
    users[i, 0] = np.random.randint(0, 4)
    users[i, 1] = np.random.randint(0, 2)
    users[i, 2] = np.random.randint(0, 5)
    users[i, 3] = np.random.randint(0, 5)
```

We assume that we have 1000 users with 4 features, represented by integer numbers bounded between 0 and 4 or 5. It doesn't matter what they mean; their role is to characterize a user and to allow for the clustering of the set. Of course, the features must be consistent with the task. In other words, we are assuming that the distances can be **grounded**. Hence, if $d(A, B) < d(B, C)$, it means that the actual user A is more similar to B than C.

For the products, we also need to create the association:

```
nb_product = 20
user_products = np.random.randint(0, nb_product, size=(nb_users, 5))
```

We assume that we have 20 different items (from 1 to 20; 0 means that a user didn't buy anything) and an association matrix where each user is linked to a number of products bounded between 0 and 5 (maximum), for example:

$$M_{U \times I} = \begin{pmatrix} i_1 & i_2 & 0 & 0 & 0 \\ i_{15} & i_3 & i_{12} & 0 & 0 \\ \vdots & \vdots & \vdots & \vdots & \vdots \\ i_4 & i_8 & i_{11} & i_2 & i_5 \\ i_8 & 0 & 0 & 0 & 0 \end{pmatrix}$$

At this point, we need to cluster the users using the NearestNeighbors implementation provided by scikit-learn:

```
from sklearn.neighbors import NearestNeighbors

nn = NearestNeighbors(n_neighbors=20, radius=2.0)
nn.fit(users)
```

We have selected to have 20 neighbors and a Euclidean radius equal to 2.0. This parameter is used when we want to query the model to know which items are contained in the ball whose center is a sample and with a fixed radius. In our case, we are going to query the model to get all the neighbors of a test user:

```
test_user = np.array([2, 0, 3, 2])
d, neighbors = nn.kneighbors(test_user.reshape(1, -1))

print(neighbors)
array([[933,  67, 901, 208,  23, 720, 121, 156, 167,  60, 337, 549,  93,
        563, 326, 944, 163, 436, 174,  22]], dtype=int64)
```

Now, we need to build the recommendation list using the association matrix (for didactic purposes, this step is not well optimized):

```
suggested_products = []

for n in neighbors:
    for products in user_products[n]:
        for product in products:
            if product != 0 and product not in suggested_products:
                suggested_products.append(product)

print(suggested_products)
[14, 5, 13, 4, 8, 9, 16, 18, 10, 7, 1, 19, 12, 11, 6, 17, 15, 3, 2]
```

For each neighbor, we retrieve the products they bought and perform a union, avoiding the inclusion of items with zero value (meaning no product) and duplicate elements. The result is a list (not sorted) of suggestions that can be obtained almost in real time for many different systems. In some cases, when the number of users or items is too large, it's possible to limit the list to a fixed number of elements and to reduce the number of neighbors. This approach is also naive because it doesn't consider the actual distance (or similarity) between users to weigh the suggestions. It's possible to consider the distance as a weighing factor, but it's simpler to adopt the collaborative filtering approach, which provides a more robust solution.

Content-based systems

This is probably the simplest method, and it's based only on products modeled as m-dimensional feature vectors:

$$I = \{\bar{i}_1, \bar{i}_2, \ldots, \bar{i}_n\} \quad where \quad \bar{i}_j \in \mathbb{R}^m$$

Just like users, features can also be categorical (indeed, for products it's easier), for example, the genre of a book or a movie, and they can be used together with numerical values (such as price, length, number of positive reviews, and so on) after encoding them.

Then, a clustering strategy is adopted, even if the most used strategy is k-NN, as it allows us to control the size of each neighborhood to determine, given a sample product, the quality and the number of suggestions.

Using scikit-learn, first of all we create a dummy product dataset:

```
nb_items = 1000
items = np.zeros(shape=(nb_items, 4))

for i in range(nb_items):
    items[i, 0] = np.random.randint(0, 100)
    items[i, 1] = np.random.randint(0, 100)
    items[i, 2] = np.random.randint(0, 100)
    items[i, 3] = np.random.randint(0, 100)
```

In this case, we have 1000 samples with 4 integer features bounded between 0 and 100. Then we proceed, as in the previous example, to cluster them:

```
nn = NearestNeighbors(n_neighbors=10, radius=5.0)
nn.fit(items)
```

At this point, it's possible to query our model with the `radius_neighbors()` method, which allows us to restrict our research only to a limited subset. The default radius (set through the `radius` parameter) is 5.0, but we can change it dynamically:

```
test_product = np.array([15, 60, 28, 73])
d, suggestions = nn.radius_neighbors(test_product.reshape(1, -1),
radius=20)

print(suggestions)
[array([657, 784, 839, 342, 446, 196], dtype=int64)]

d, suggestions = nn.radius_neighbors(test_product.reshape(1, -1),
```

```
radius=30)

print(suggestions)
[ array([844, 340, 657, 943, 461, 799, 715, 863, 979, 784, 54, 148, 806,
    465, 585, 710, 839, 695, 342, 881, 864, 446, 196, 73, 663, 580, 216],
    dtype=int64)]
```

Of course, when trying these examples, the number of suggestions can be different, as we are using random datasets, so I suggest trying different values for the radius (in particular when using different metrics).

When clustering with k-NN, it's important to consider the metric adopted for determining the distance between the samples. The default for scikit-learn is the **Minkowski distance** $d_p(a, b)$, which, as discussed in Chapter 11, *Hierarchical Clustering*, is a generalization of the Euclidean distance. Parameter p controls the type of distance, and the default value is 2 so that the resulting metric is a classical Euclidean distance. Other distances are offered by SciPy (in the scipy.spatial.distance package) and include, for example, the **Hamming** and **Jaccard distances**. The former is defined as the disagreement proportion between two vectors (if they are binary, this is the normalized number of different bits), for example:

```
from scipy.spatial.distance import hamming

a = np.array([0, 1, 0, 0, 1, 0, 1, 1, 0, 0])
b = np.array([1, 1, 0, 0, 0, 1, 1, 1, 1, 0])
d = hamming(a, b)

print(d)
0.40000000000000002
```

It means there's a disagreement proportion of 40%, or considering that both vectors are binary, there are four different bits (out of 10). This measure can be useful when it's necessary to emphasize the presence/absence of a particular feature.

The Jaccard distance is defined as follows:

$$d_{Jaccard} = 1 - J(A, B) = 1 - \frac{|A \cap B|}{|A \cup B|}$$

It's particularly useful to measure the dissimilarity between two different sets (A and B) of items. If our feature vectors are binary, it's immediate to apply this distance considering the standard boolean operators OR and AND. Using the previous test values, we get this:

```
from scipy.spatial.distance import jaccard
```

```
d = jaccard(a, b)
print(d)
0.5714285714285714
```

This measure is bounded between *zero* (equal vectors) and *one* (total dissimilarity).

As for the Hamming distance, it can be very useful when it's necessary to compare items where their representation is made up of binary states (such as *present/absent*, *yes/no*, and so forth). If you want to adopt a different metric for k-NN, it's possible to specify it directly using the metric parameter:

```
nn = NearestNeighbors(n_neighbors=10, radius=5.0, metric='hamming')
nn.fit(items)

nn = NearestNeighbors(n_neighbors=10, radius=5.0, metric='jaccard')
nn.fit(items)
```

Model-free (or memory-based) collaborative filtering

As with the user-based approach, let's consider two sets of elements: users and items. However, in this case, we don't assume that they have explicit features. Instead, we try to model a user-item matrix based on the preferences of each user (rows) for each item (columns), for example:

$$M_{U \times I} = \begin{pmatrix} 0 & 2 & 4 & \cdots & 1 \\ 2 & 3 & 0 & \cdots & 4 \\ \vdots & \vdots & \vdots & \vdots & \vdots \\ 4 & 1 & 2 & \cdots & 5 \\ 2 & 5 & 0 & \cdots & 0 \end{pmatrix}$$

In this case, the ratings are bounded between *1* and *5* (*0* means no rating), and our goal is to cluster the users according to their rating vector (which is, an internal representation based on a particular kind of feature). This allows us to produce recommendations even when there are no explicit pieces of information about the user. However, it has a drawback, called **cold startup**, which means that when a new user has no ratings, it's impossible to find the right neighborhood, because they can belong to virtually any cluster.

Introducing Recommendation Systems

Once the clustering is done, it's easy to check which products (not rated yet) have the highest rating for a given user and therefore are more likely to be bought. It's possible to implement a solution in scikit-learn as we've done before, but I'd like to introduce a small framework called **Crab** (see the box at the end of this section) that simplifies this process.

To build the model, we first need to define user_item_matrix as a Python dictionary with this structure:

```
{ user_1: { item1: rating, item2: rating, ... }, ..., user_n: ... }
```

A missing value in an internal user dictionary means no rating. In our example, we consider 5 users with 5 items:

```
from scikits.crab.models import MatrixPreferenceDataModel

user_item_matrix = {
    1: {1: 2, 2: 5, 3: 3},
    2: {1: 5, 4: 2},
    3: {2: 3, 4: 5, 3: 2},
    4: {3: 5, 5: 1},
    5: {1: 3, 2: 3, 4: 1, 5: 3}
}

model = MatrixPreferenceDataModel(user_item_matrix)
```

Once user_item_matrix has been defined, we need to pick a metric and, therefore, a distance function $d(u_i, u_j)$, to build a similarity matrix:

$$S = \begin{pmatrix} d(\bar{u}_1, \bar{u}_1) & \cdots & d(\bar{u}_n, \bar{u}_1) \\ \vdots & \ddots & \vdots \\ d(\bar{u}_1, \bar{u}_n) & \cdots & d(\bar{u}_n, \bar{u}_n) \end{pmatrix}$$

Using Crab, we do this in the following way (using a Euclidean metric):

```
from scikits.crab.similarities import UserSimilarity
from scikits.crab.metrics import euclidean_distances

similarity_matrix = UserSimilarity(model, euclidean_distances)
```

There are many metrics, such as Pearson and Jaccard, so I suggest visiting the website (http://muricoca.github.io/crab) for further information. At this point, it's possible to build the recommendation system (based on the k-NN clustering method) and test it:

```
from scikits.crab.recommenders.knn import UserBasedRecommender

recommender = UserBasedRecommender(model, similarity_matrix,
with_preference=True)

print(recommender.recommend(2))
[(2, 3.6180339887498949), (5, 3.0), (3, 2.5527864045000417)]
```

So, the `recommender` suggests the following predicted rating for user 2:

- **Item 2**: 3.6 (which can be rounded to 4.0)
- **Item 5**: 3
- **Item 3**: 2.5 (which can be rounded to 3.0)

When running the code, you may see some warnings (Crab is under continuous development). However, they don't affect the functionality. If you want to avoid them, you can use the `catch_warnings()` context manager:

```
import warnings

with warnings.catch_warnings():
    warnings.simplefilter("ignore")
    print(recommender.recommend(2))
```

It's possible to suggest all the items, or limit the list to the highest ratings (so, for example, avoiding item 3). This approach is quite similar to the user-based model. However, it's faster (very big matrices can be processed in parallel) and it doesn't take care of details that can produce misleading results. Only the ratings are considered as useful features for defining a user. Like model-based collaborative filtering, the cold startup problem can be addressed in two ways:

- Asking the user to rate some items (this approach is often adopted because it's easy to show some movie/book covers and ask the user to select what they like and what they don't).
- Placing the user in an average neighborhood by randomly assigning some mean ratings. In this approach, it's possible to start using the recommendation system immediately. However, it's necessary to accept a certain degree of error at the beginning and to correct the dummy ratings when the real ones are produced.

Crab is an open source framework for building collaborative filtering systems. It's still under development and therefore doesn't implement all possible features. However, it's very easy to use and is quite powerful for many tasks. The home page, with installation instructions and documentation, is http://muricoca.github.io/crab/index.html. Crab depends on scikits.learn, which still has some issues with Python 3. Therefore, I recommend using Python 2.7 for this example. It's possible to install both packages using pip: pip install -U scikits.learn and pip install -U crab.

Model-based collaborative filtering

This is currently one of the most advanced approaches and is an extension of what was already seen in the previous section. The starting point is always a rating-based user-item matrix:

$$M_{U \times I} = \begin{pmatrix} r_{11} & \cdots & r_{1n} \\ \vdots & \ddots & \vdots \\ r_{m1} & \cdots & r_{mn} \end{pmatrix}$$

However, in this case, we assume the presence of **latent factors** for both the users and the items. In other words, we define a generic user as follows:

$$\bar{p}_i = (p_{i1}, p_{i2}, \ldots, p_{ik}) \ \ where \ \ p_{ij} \in \mathbb{R}$$

A generic item is defined as follows:

$$\bar{q}_j = (q_{j1}, q_{j2}, \ldots, q_{jk}) \ \ where \ \ q_{ji} \in \mathbb{R}$$

We don't know the value of each vector component (for this reason they are called latent), but we assume that a ranking is obtained as follows:

$$r_{ij} = \bar{p}_i \cdot \bar{q}_j^T$$

So, we can say that a ranking is obtained from a latent space of rank k, where k is the number of latent variables we want to consider in our model. In general, there are rules to determine the right value for k, so the best approach is to check different values and test the model with a subset of known ratings. However, there's still a big problem to solve: finding the latent variables. There are several strategies, but before discussing them, it's important to understand the dimensionality of our problem. If we have 1,000 users and 500 products, M has 500,000 elements. If we decide to have a rank equal to 10, it means that we need to find 5,000,000 variables constrained by the known ratings. As you can imagine, this problem can easily become impossible to solve with standard approaches, and parallel solutions must be employed.

Singular value decomposition strategy

The first approach is based on the **Singular Value Decomposition** (**SVD**) of the user-item matrix. This technique allows the transforming of a matrix through a low-rank factorization and can also be used in an incremental way, as described in *Incremental Singular Value Decomposition Algorithms for Highly Scalable Recommender Systems, Sarwar B, Karypis G, Konstan J, Riedl J, 2002*. In particular, if the user-item matrix has m rows and n columns:

$$M_{U \times I} = U \Sigma V^T \text{ where } U \in \mathbb{R}^{m \times m}, \Sigma \in \mathbb{R}^{m \times n}, \text{ and } V \in \mathbb{R}^{n \times n}$$

We have assumed that we have real matrices (which is often true in our case), but in general they are complex. U and V are unitary, while Σ is a rectangular diagonal matrix. The columns of U contain the left singular vectors, the rows of transposed V contain the right singular vectors, while the diagonal matrix Σ contains the singular values. Selecting k latent factors means taking the first k singular values and therefore the corresponding k left and right singular vectors:

$$M_k = U_k \Sigma_k V_k^T$$

This technique has the advantage of minimizing the Frobenius norm of the difference between M and M_k for any value of k, and therefore it's an optimal choice to approximate the full decomposition. Before moving to the prediction stage, let's create an example using SciPy. The first thing to do is to create a dummy user-item matrix:

```
M = np.random.randint(0, 6, size=(20, 10))

print(M)
array([[0, 4, 5, 0, 1, 4, 3, 3, 1, 3],
```

```
                [1, 4, 2, 5, 3, 3, 3, 4, 3, 1],
                [1, 1, 2, 2, 1, 5, 1, 4, 2, 5],
                [0, 4, 1, 2, 2, 5, 1, 1, 5, 5],
                [2, 5, 3, 1, 1, 2, 2, 4, 1, 1],
                [1, 4, 3, 3, 0, 0, 2, 3, 3, 5],
                [3, 5, 2, 1, 5, 3, 4, 1, 0, 2],
                [5, 2, 2, 0, 1, 0, 4, 4, 1, 0],
...
```

We're assuming that we have 20 users and 10 products. The ratings are bounded between 1 and 5, and 0 means no rating. Now, we can decompose M:

```
from scipy.linalg import svd

import numpy as np

U, s, V = svd(M, full_matrices=True)
S = np.diag(s)

print(U.shape)
(20L, 20L)

print(S.shape)
(10L, 10L)

print(V.shape)
(10L, 10L)
```

Now, let's consider only the first eight singular values, which will have eight latent factors for both the users and items:

```
Uk = U[:, 0:8]
Sk = S[0:8, 0:8]
Vk = V[0:8, :]
```

It's important to bear in mind that in SciPy's SVD implementation, V is already transposed. When the full matrices are not needed, it's possible to set the `full_matrices=False` parameter. In this case, the two matrices containing the singular vectors will be cut using the minimum dimension $k = min(m, n)$. Hence, the dimensions become the following:

$$U \in \mathbb{R}^{m \times k}, \Sigma \in \mathbb{R}^{k \times k}, \text{ and } V \in \mathbb{R}^{k \times n}$$

If the number of desired components is less than or equal to k, this solution allows a faster computation and it's highly recommended.

According to *Incremental Singular Value Decomposition Algorithms for Highly Scalable Recommender Systems, Sarwar B, Karypis G, Konstan J, Riedl J, 2002*, we can easily get a prediction considering the cosine similarity (which is proportional to the dot product) between customers and products. The two latent factor matrices are the following:

$$\begin{cases} S_U = U_k \cdot \sqrt{\Sigma_k}^T \\ S_I = \sqrt{\Sigma_k} \cdot V_k^T \end{cases}$$

To take into account the loss of precision, it's useful to also consider the average rating per user (which corresponds to the mean row value of the user-item matrix), so that the resultant rating prediction for user i and item j becomes the following:

$$\tilde{r}_{ij} = E[r_i] + S_U(i) \cdot S_I(j)$$

Here, $S_U(i)$ and $S_I(j)$ are the user and product vectors respectively. Continuing with our example, let's determine the rating prediction for user 5 and item 2:

```
Su = Uk.dot(np.sqrt(Sk).T)
Si = np.sqrt(Sk).dot(Vk).T
Er = np.mean(M, axis=1)

r5_2 = Er[5] + Su[5].dot(Si[2])

print(r5_2)
2.38848720112
```

This approach is of medium complexity. In particular, SVD is $O(m^3)$ and an incremental strategy (as described in *Incremental Singular Value Decomposition Algorithms for Highly Scalable Recommender Systems, Sarwar B, Karypis G, Konstan J, Riedl J, 2002*) must be employed when new users or items are added; however, it can be effective when the number of elements is not too big. In all other cases, the next strategy (together with a parallel architecture) can be adopted.

Alternating least squares strategy

The problem of finding the latent factors can be easily expressed as a least squares optimization problem by defining the following loss function:

$$L = \sum_{(i,j)} \left(r_{ij} - \bar{p}_i \cdot \bar{q}_j^T \right)^2 + \alpha \left(\|\bar{p}_i\|^2 + \|\bar{q}_j\|^2 \right)$$

L is limited only to known samples (user, item). The second term works as a regularization factor, and the whole problem can easily be solved with any optimization method. However, there's an additional issue: we have two different sets of variables to determine (user and item factors). We can solve this problem with an approach called **Alternating Least Squares** (**ALS**), described in *Matrix Factorization Techniques for Recommender Systems, Koren Y, Bell R, Volinsky C, IEEE Computer Magazine, August 2009*. The algorithm is very easy to describe and can be summarized in two main iterating steps:

- p_i is fixed and q_j is optimized
- q_j is fixed and p_i is optimized

The algorithm stops when a predefined precision has been achieved. It can be easily implemented with parallel strategies to be able to process huge matrices in a short time. Moreover, considering the price of virtual clusters, it's also possible to retrain the model periodically, to immediately (with an acceptable delay) include new products and users.

ALS with Apache Spark MLlib

Apache Spark is beyond the scope of this book, so if you want to know more about this powerful framework, I suggest you read the online documentation or one of the many books available. In *Machine Learning with Spark, Pentreath N, Packt Publishing*, there's an interesting introduction to the **Machine Learning Library** (**MLlib**) and how to implement most of the algorithms discussed in this book.

Spark is a parallel computational engine that is now part of the Hadoop project (even if it doesn't use its code), which can run in local mode or on very large clusters (with thousands of nodes) to execute complex tasks using huge amounts of data. It's mainly based on Scala, though there are interfaces for Java, Python, and R. In this example, we're going to use PySpark, which is the built-in shell for running Spark with Python code.

After launching PySpark in local mode, we get a standard Python prompt and we can start working, just like with any other standard Python environment:

```
# Linux
>>> ./pyspark

# Mac OS X
>>> pyspark

# Windows
>>> pyspark
```

```
Python 3.X.X |Anaconda X.0.0 (64-bit)| (default, Apr 26 2018, 11:07:13)
Type "help", "copyright", "credits" or "license" for more information.
Anaconda is brought to you by Continuum Analytics.
Please check out: http://continuum.io/thanks and https://anaconda.org
Using Spark's default log4j profile: org/apache/spark/log4j-
defaults.properties
Setting default log level to "WARN".
To adjust logging level use sc.setLogLevewl(newLevel).
Welcome to
      ____              __
     / __/__  ___ _____/ /__
    _\ \/ _ \/ _ `/ __/  '_/
   /__ / .__/\_,_/_/ /_/\_\   version 2.0.2
      /_/

Using Python version 3.X.X (default,Apr 26 2018, 11:07:13)
SparkSession available as 'spark'.
>>>
```

Spark MLlib implements the ALS algorithm through a very simple mechanism. The `Rating` class is a wrapper for the tuple (user, product, rating), so we can easily define a dummy dataset (which must be considered only as an example, because it's very limited):

```
from pyspark.mllib.recommendation import Rating

import numpy as np

nb_users = 200
nb_products = 100

ratings = []

for _ in range(10):
    for i in range(nb_users):
        rating = Rating(user=i,
                        product=np.random.randint(1, nb_products),
                        rating=np.random.randint(0, 5))
        ratings.append(rating)

ratings = sc.parallelize(ratings)
```

Introducing Recommendation Systems

We assumed that we have 200 users and 100 products, and we have populated a list of `ratings` by iterating the main loop that assigns a rating to a random product 10 times. We're not controlling repetitions or other uncommon situations. The last `sc.parallelize()` command is a way to ask Spark to transform our list into a structure called a **Resilient Distributed Dataset (RDD)**, which will be used for the remaining operations. There are no actual limits to the size of these structures, because they are distributed across different executors (if in clustered mode) and can work with petabyte datasets just as we work with kilobyte ones.

At this point, we can train an ALS model (which is formally `MatrixFactorizationModel`) and use it to make some predictions:

```
from pyspark.mllib.recommendation import ALS

model = ALS.train(ratings, rank=5, iterations=10)
```

We want 5 latent factors and 10 optimization iterations. As discussed before, it's not very easy to determine the right rank for each model, so after a training phase, there should always be a validation phase with known data. The **Mean Squared Error (MSE)** is a good measure to understand how the model is working. We can do it using the same training dataset. The first thing to do is to remove the ratings (because we only need the tuple made up of `user` and `product`):

```
test = ratings.map(lambda rating: (rating.user, rating.product))
```

If you're not familiar with the MapReduce paradigm, you only need to know that `map()` applies the same function (in this case, a `lambda`) to all elements. Now, we can massively predict the ratings:

```
predictions = model.predictAll(test)
```

However, to compute the error, we also need to add `user` and `product` to have tuples that can be compared:

```
full_predictions = predictions.map(lambda pred: ((pred.user, pred.product),
pred.rating))
```

The result is a sequence of rows with a structure `((user, item), rating)` just like a standard dictionary entry `(key, value)`. This is useful because, using Spark, we can join two RDDs by using their keys. We do the same thing for the original dataset also, and then we proceed by joining the training values with the predictions:

```
split_ratings = ratings.map(lambda rating: ((rating.user, rating.product),
rating.rating))
joined_predictions = split_ratings.join(full_predictions)
```

Now, for each key (`user`, `product`) we have two values: `target` and `prediction`. Therefore, we can compute the MSE:

```
mse = joined_predictions.map(lambda x: (x[1][0] - x[1][1]) ** 2).mean()
```

The first map transforms each row into the squared difference between the `target` and `prediction`, while the `mean()` function computes the average value. At this point, let's check our error and produce a `prediction`:

```
print('MSE: %.3f' % mse)
MSE: 0.580

prediction = model.predict(10, 20)
print('Prediction: %3.f' % prediction)
Prediction: 2.810
```

So, our error is quite low but it can be improved by changing the rank or the number of iterations. The `prediction` for the rating of the product 20 by user 10 is about 2.810 (which can be rounded to 3). If you run the code, these values can be different, as we're using a random user-item matrix. Moreover, if you don't want to use the shell and run the code directly, you need to declare `SparkContext` explicitly at the beginning of your file:

```
from pyspark import SparkContext, SparkConf

conf = SparkConf().setAppName('ALS').setMaster('local[*]')
sc = SparkContext(conf=conf)
```

We have created a configuration through the `SparkConf` class and specified both an application name and a master (in local mode with all cores available). This is enough to run our code. However, if you need further information, visit the page mentioned in the information box at the end of the chapter. To run the application (since Spark 2.0), you must execute the following command:

```
# Linux, Mac OSx
./spark-submit als_spark.py

# Windows
spark-submit als_spark.py
```

When running a script using `spark-submit`, you will see hundreds of log lines that inform you about all the operations that are being performed. Among them, at the end of the computation, you'll also see the print function messages (`stdout`).

Of course, this is only an introduction to Spark ALS, but I hope it was useful to understand how easy this process can be and, at the same time, how the dimensional limitations can be effectively addressed.

If you don't know how to set up the environment and launch PySpark, I suggest reading the online quick-start guide (https://spark.apache.org/docs/2.1.0/quick-start.html), which can be useful even if you don't know all the details and configuration parameters.

Summary

In this chapter, we discussed the main techniques for building a recommender system. In a user-based scenario, we assume that we have enough pieces of information about the users to be able to cluster them, and we implicitly assume that similar users would like the same products. In this way, it's quick to determine the neighborhood of every new user and to suggest products positively rated by their peers. In a similar way, a content-based scenario is based on the clustering of products according to their peculiar features. In this case, the assumption is weaker, because it's probable that a user who bought an item or rated it positively will do the same with similar products.

Then, we introduced collaborative filtering, which is a technique based on explicit ratings, used to predict all missing values for all users and products. In the memory-based variant, we don't train a model but we try to work directly with a user-product matrix, looking for the k-NN of a test user, and computing the ranking through an average. This approach is very similar to the user-based scenario and has the same limitations; in particular, it's very difficult to manage large matrices. On the other hand, the model-based approach is more complex, but after training the model, it can predict the ratings in real time. Moreover, there are parallel frameworks such as Spark, that can be employed to process a huge amount of data using a cluster of cheap servers.

In Chapter 13, *Introducing Natural Language Processing*, we're going to introduce some **Natural Language Processing** (**NLP**) techniques, which are very important when automatically classifying texts or working with machine translation systems.

13
Introducing Natural Language Processing

Natural Language Processing (**NLP**) is a set of machine learning techniques that allow working with text documents, considering their internal structure, and the distribution of words. In this chapter, we're going to discuss all common methods to collect texts, split them into atoms, and transform them into numerical vectors. In particular, we'll compare different methods to tokenize documents (separate each word), to filter them, to apply special transformations to avoid inflected or conjugated forms, and finally to build a common vocabulary. Using the vocabulary, it will be possible to apply different vectorization approaches, to build feature vectors that can easily be used for classification or clustering purposes. To show you how to implement the whole pipeline, at the end of the chapter, we're going to set up a simple classifier for news lines.

In particular, we're going to discuss these topics:

- **Natural Language Toolkit** (**NLTK**) corpora and how to access them to train the models
- The **Bag-of-Words** strategy (tokenization, stopword removal, stemming, and vectorization)
- **Part-of-Speech** (**POS**) Tagging and **Named Entity Recognition** (**NER**)

NLTK and built-in corpora

NLTK is a very powerful Python framework that implements most NLP algorithms and will be adopted in this chapter together with scikit-learn. Moreover, NLTK provides some built-in corpora that can be used to test algorithms. Before starting to work with NLTK, it's normally necessary to download all the additional elements (corpora, dictionaries, and so on) using a specific graphical interface. This can be done in the following way:

```
import nltk

nltk.download()
```

This command will launch the user interface, as shown in the following screenshot:

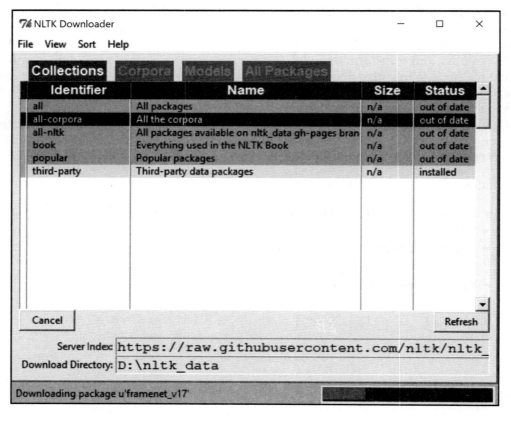

NLTK downloader window

It's possible to select every single feature or download all elements (I suggest this option if you have enough free space) to immediately exploit all NLTK functionalities. Alternatively, it's possible to install all dependencies using the following command:

```
python -m nltk.downloader all
```

 NLTK can be installed using `pip` (`pip install -U nltk`) or with one of the binary distributions available at `http://www.nltk.org`. On the same website, there's complete documentation that can be useful for going deeper into each topic.

Corpora examples

A subset of the Gutenberg project is provided and can be freely accessed in this way:

```
from nltk.corpus import gutenberg

print(gutenberg.fileids())

[u'austen-emma.txt', u'austen-persuasion.txt', u'austen-sense.txt', u'bible-kjv.txt', u'blake-poems.txt', u'bryant-stories.txt', u'burgess-busterbrown.txt', u'carroll-alice.txt', u'chesterton- ...
```

A single document can be accessed as a raw version, or it can be split into sentences or words:

```
print(gutenberg.raw('milton-paradise.txt'))

[Paradise Lost by John Milton 1667]

Book I

Of Man's first disobedience, and the fruit
Of that forbidden tree whose mortal taste...

print(gutenberg.sents('milton-paradise.txt')[0:2])

[[u'[', u'Paradise', u'Lost', u'by', u'John', u'Milton', u'1667', u']'], [u'Book', u'I']]

print(gutenberg.words('milton-paradise.txt')[0:20])

[u'[', u'Paradise', u'Lost', u'by', u'John', u'Milton', u'1667', u']', u'Book', u'I', u'Of', u'Man', u"'", u's', u'first', u'disobedience', u',', u'and', u'the', u'fruit']
```

As we're going to discuss, in many cases, it can be useful to have the raw text so as to split it into words using a custom strategy. In many other situations, accessing sentences directly allows working with the original structural subdivision. Other corpora include web texts, Reuters news lines, the Brown Corpus, and many more. For example, the Brown Corpus is a famous collection of documents divided by genre:

```
from nltk.corpus import brown

print(brown.categories())

[u'adventure', u'belles_lettres', u'editorial', u'fiction', u'government',
u'hobbies', u'humor', u'learned', u'lore', u'mystery', u'news',
u'religion', u'reviews', u'romance', u'science_fiction']

print(brown.sents(categories='editorial')[0:100])

[[u'Assembly', u'session', u'brought', u'much', u'good'], [u'The',
u'General', u'Assembly', u',', u'which', u'adjourns', u'today', u',',
u'has', u'performed', u'in', u'an', u'atmosphere', u'of', u'crisis',
u'and', u'struggle', u'from', u'the', u'day', u'it', u'convened', u'.'],
...]
```

Further information about corpora can be found at http://www.nltk.org/book/ch02.html.

The Bag-of-Words strategy

In NLP, a very common pipeline can be subdivided into the following steps:

1. Collecting a document into a corpus
2. Tokenizing, stopword (articles, prepositions, and so on) removal, and stemming (reduction to radix-form)
3. Building a common vocabulary
4. Vectorizing the documents
5. Classifying or clustering the documents

The pipeline is called Bag-of-Words and will be discussed in this chapter. A fundamental assumption is that the order of every single word in a sentence is not important. In fact, when defining a feature vector, as we're going to see, the measures taken into account are always related to frequencies, and therefore they are insensitive to the local positioning of all elements. From some viewpoints, this is a limitation because in a natural language the internal order of a sentence is necessary to preserve the meaning; however, there are many models that can work efficiently with texts without the complication of local sorting. When it's absolutely necessary to consider small sequences, it will be done by adopting groups of tokens (called **n-grams**) but considering them as a single atomic element during the vectorization step.

In the following diagram, there's a schematic representation of this process (without the fifth step) for a sample document (sentence):

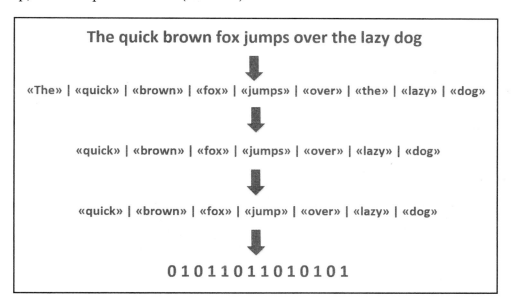

Sequence of steps describing the Bag-of-Words strategy

Introducing Natural Language Processing

There are many different methods used to carry out each step, and some of them are context-specific. However, the goal is always the same: maximizing the information of a document and reducing the size of the common vocabulary by removing terms that are too frequent or derived from the same radix (such as verbs). The information content of a document is in fact determined by the presence of specific terms (or a group of terms), whose frequency in the corpus is limited. In the example shown in the previous diagram, <<fox>> and <<dog>> are important terms, while <<the>> is useless (often called a **stopword**). Moreover, <<jumps>> can be converted to the standard form <<jump>>, which expresses a specific action when present in different forms (such as *jumping* or *jumped*). The last step is transforming into a numerical vector because our algorithms work with numbers, and it's important to limit the length of the vectors so as to improve the learning speed and the memory consumption. In the following sections, we're going to discuss each step in detail, and, at the end, we're going to build a sample classifier for news lines.

Tokenizing

The first step in processing a piece of text or a corpus is splitting it into atoms (sentences, words, or parts of words), normally defined as **tokens**. Such a process is quite simple; however, there can be different strategies to solve particular problems.

Sentence tokenizing

In many cases, it's useful to split large texts into sentences, which are normally delimited by a full stop or another equivalent mark. As every language has its own orthographic rules, NLTK offers the `sent_tokenize()` method which accepts a language (the default is English) and splits the text according to the specific rules. In the following example, we show the usage of this function with different languages:

```
from nltk.tokenize import sent_tokenize

generic_text = 'Lorem ipsum dolor sit amet, amet minim temporibus in sit. Vel ne impedit consequat intellegebat.'

print(sent_tokenize(generic_text))

['Lorem ipsum dolor sit amet, amet minim temporibus in sit.',
 'Vel ne impedit consequat intellegebat.']

english_text = 'Where is the closest train station? I need to reach London'

print(sent_tokenize(english_text, language='english'))
```

```
['Where is the closest train station?', 'I need to reach London']

spanish_text = u'¿Dónde está la estación más cercana? Inmediatamente me
tengo que ir a Barcelona.'

for sentence in sent_tokenize(spanish_text, language='spanish'):
    print(sentence)

¿Dónde está la estación más cercana?
Inmediatamente me tengo que ir a Barcelona.
```

Word tokenizing

The simplest way to tokenize a sentence into words is provided by the `TreebankWordTokenizer` class; however, this has some limitations:

```
from nltk.tokenize import TreebankWordTokenizer

simple_text = 'This is a simple text.'

tbwt = TreebankWordTokenizer()

print(tbwt.tokenize(simple_text))

['This', 'is', 'a', 'simple', 'text', '.']

complex_text = 'This isn\'t a simple text'

print(tbwt.tokenize(complex_text))

['This', 'is', "n't", 'a', 'simple', 'text']
```

As you can see, in the first case, the sentence has been correctly split into words, keeping the punctuation separate (this is not a real issue because it can be removed in a second step). However, in the complex example, the contraction `isn't` has been split into `is` and `n't`. Unfortunately, without a further processing step, it's not so easy to convert a token with a contraction into its normal form (in this instance, `not`); therefore, another strategy must be employed. A good way to solve the problem of separate punctuation is provided by the `RegexpTokenizer` class, which offers a flexible way to split words according to a regular expression:

```
from nltk.tokenize import RegexpTokenizer

complex_text = 'This isn\'t a simple text.'
```

```
ret = RegexpTokenizer('[a-zA-Z0-9\'\.]+')

print(ret.tokenize(complex_text))

['This', "isn't", 'a', 'simple', 'text.']
```

Most of the common problems can be easily solved using this class, so I suggest you learn how to write simple regular expressions that can match specific patterns. For example, we can remove all numbers, commas, and other punctuation marks from a sentence:

```
complex_text = 'This isn\'t a simple text. Count 1, 2, 3 and then go!'

ret = RegexpTokenizer('[a-zA-Z\']+')

print(ret.tokenize(complex_text))

['This', "isn't", 'a', 'simple', 'text', 'Count', 'and', 'the', 'go']
```

Even if there are other classes provided by NLTK, they can always be implemented using a customized `RegexpTokenizer`, which is powerful enough to solve almost every particular problem, so I'd prefer not to go deeper into this subject in this chapter.

Stopword removal

Stopwords are part of normal speech (articles, conjunctions, and so on), but their occurrence frequency is very high and they don't provide any useful semantic information. For these reasons, it's a good practice to filter sentences and corpora by removing them all. NLTK provides lists of stopwords for the most common languages, and their usage is immediate:

```
from nltk.corpus import stopwords

sw = set(stopwords.words('english'))
```

A subset of English stopwords is shown in the following snippet:

```
print(sw)

{u'a',
 u'about',
 u'above',
 u'after',
 u'again',
 u'against',
 u'ain'...
```

To filter a sentence, it's possible to adopt a functional approach:

```
complex_text = 'This isn\'t a simple text. Count 1, 2, 3 and then go!'

ret = RegexpTokenizer('[a-zA-Z\']+')
tokens = ret.tokenize(complex_text)
clean_tokens = [t for t in tokens if t not in sw]
print(clean_tokens)
['This', "isn't", 'simple', 'text', 'Count', 'go']
```

Language detection

Stopwords, as with other important features, are strictly related to a specific language, so it's often necessary to detect the language before moving on to any other step. A simple, free, and reliable solution is provided by the `langdetect` library, which has been ported from Google's language detection system. Its usage is immediate:

```
from langdetect import detect

print(detect('This is English'))
en

print(detect('Dies ist Deutsch'))
de
```

The function returns the ISO 639-1 codes (https://en.wikipedia.org/wiki/List_of_ISO_639-1_codes), which can be used as keys in a dictionary to get the complete language name. Where the text is more complex, the detection can be more difficult, and it's useful to know whether there are any ambiguities. It's possible to get the probabilities for the expected languages through the `detect_langs()` method:

```
from langdetect import detect_langs

print(detect_langs('I really love you mon doux amour!'))
[fr:0.714281321163, en:0.285716747181]
```

 `langdetect` can be installed using `pip` (`pip install --upgrade langdetect`). Further information is available at `https://pypi.python.org/pypi/langdetect`.

Stemming

Stemming is a process that is used to transform particular words (such as verbs or plurals) into their radical form so as to preserve the semantics without increasing the number of unique tokens. For example, if we consider the three expressions *I run, he runs,* and *running,* they can be reduced into a useful (though grammatically incorrect) form: *I run, he run, run.* In this way, we have a single token that defines the same concept (*run*), which, for clustering or classification purposes, can be used without any precision loss. There are many stemmer implementations provided by NLTK. The most common (and flexible) is `SnowballStemmer`, based on a multilingual algorithm:

```
from nltk.stem.snowball import SnowballStemmer

ess = SnowballStemmer('english', ignore_stopwords=True)
print(ess.stem('flies'))
fli

fss = SnowballStemmer('french', ignore_stopwords=True)
print(fss.stem('courais'))
cour
```

The `ignore_stopwords` parameter informs the stemmer not to process the stopwords. Other implementations are `PorterStemmer` and `LancasterStemmer`. Very often, the result is the same, but in some cases, a stemmer can implement more selective rules, for example:

```
from nltk.stem.snowball import PorterStemmer
from nltk.stem.lancaster import LancasterStemmer

print(ess.stem('teeth'))
teeth

ps = PorterStemmer()
print(ps.stem('teeth'))
teeth

ls = LancasterStemmer()
print(ls.stem('teeth'))
tee
```

As you can see, Snowball and Porter algorithms keep the word unchanged, while Lancaster extracts a radix (which is meaningless). On the other hand, the latter algorithm implements many specific English rules, which can really reduce the number of unique tokens:

```
print(ps.stem('teen'))
teen
```

```
print(ps.stem('teenager'))
teenag

print(ls.stem('teen'))
teen

print(ls.stem('teenager'))
teen
```

Unfortunately, both Porter and Lancaster stemmers are available in NLTK only in English, so the default choice is often Snowball, which is available in many languages and can be used in conjunction with an appropriate stopword set.

Vectorizing

This is the last step of the Bag-of-Words pipeline and it is necessary for transforming text tokens into numerical vectors. The most common techniques are based on a count or frequency computation, and they are both available in scikit-learn with sparse matrix representations (this is a choice that can save a lot of space considering that many tokens appear only a few times, while the vectors must have the same length).

Count vectorizing

The algorithm is very simple and it's based on representing a token considering how many times it appears in a document. Of course, the whole corpus must be processed to determine how many unique tokens are present and their frequencies. Let's see an example of the `CountVectorizer` class on a simple corpus:

```
from sklearn.feature_extraction.text import CountVectorizer

corpus = [
        'This is a simple test corpus',
        'A corpus is a set of text documents',
        'We want to analyze the corpus and the documents',
        'Documents can be automatically tokenized'
]

cv = CountVectorizer()
vectorized_corpus = cv.fit_transform(corpus)
print(vectorized_corpus.todense())
[[0 0 0 0 0 1 0 1 0 0 1 1 0 0 1 0 0 0]
 [0 0 0 0 0 1 1 1 1 1 0 0 1 0 0 0 0 0]
 [1 1 0 0 0 1 1 0 0 0 0 0 2 0 1 0 1 1]
 [0 0 1 1 1 0 1 0 0 0 0 0 0 0 0 1 0 0]]
```

As you can see, each document has been transformed into a fixed-length vector, where 0 means that the corresponding token is not present, while a positive number represents the occurrences. If we need to exclude all tokens whose document frequency is less than a predefined value, we can set it through the `min_df` parameter (the default value is *1*). Sometimes, it can be useful to avoid terms that are very common; however, the next strategy will manage this problem in a more reliable and complete way.

The vocabulary can be accessed through the `vocabulary_` instance variable:

```
print(cv.vocabulary_)
{u'and': 1, u'be': 3, u'we': 18, u'set': 9, u'simple': 10, u'text': 12,
u'is': 7, u'tokenized': 16, u'want': 17, u'the': 13, u'documents': 6,
u'this': 14, u'of': 8, u'to': 15, u'can': 4, u'test': 11, u'corpus': 5,
u'analyze': 0, u'automatically': 2}
```

Given a generic vector, it's possible to retrieve the corresponding list of tokens with an inverse transformation:

```
vector = [0, 0, 0, 0, 0, 1, 0, 1, 0, 0, 1, 1, 0, 0, 1, 0, 0, 1, 1]
print(cv.inverse_transform(vector))
[array([u'corpus', u'is', u'simple', u'test', u'this', u'want', u'we'],
       dtype='<U13')]
```

Both this and the following method can also use an external tokenizer (through the `tokenizer` parameter); it can be customized using the techniques discussed in previous sections:

```
ret = RegexpTokenizer('[a-zA-Z0-9\']+')
sw = set(stopwords.words('english'))
ess = SnowballStemmer('english', ignore_stopwords=True)

def tokenizer(sentence):
    tokens = ret.tokenize(sentence)
    return [ess.stem(t) for t in tokens if t not in sw]

cv = CountVectorizer(tokenizer=tokenizer)
vectorized_corpus = cv.fit_transform(corpus)
print(vectorized_corpus.todense())
[[0 0 1 0 0 1 1 0 0 0]
 [0 0 1 1 1 0 0 1 0 0]
 [1 0 1 1 0 0 0 0 0 1]
 [0 1 0 1 0 0 0 1 0]]
```

With our tokenizer (using stopwords and stemming), the vocabulary is shorter and so are the vectors.

N-grams

So far, we have considered only single tokens (also called **unigrams**), but, in many contexts, it's useful to consider short sequences of words (**bigrams** or **trigrams**) as atoms for our classifiers, just like all the other tokens. For example, if we are analyzing the sentiment of some texts, it could be a good idea to consider bigrams such as *pretty good, very bad,* and so on. From a semantic viewpoint, in fact, it's important to consider not just the adverbs but the whole compound form. It's possible to inform our vectorizers about the range of n-grams we want to consider. For example, if we need unigrams and bigrams, we can use this snippet:

```
cv = CountVectorizer(tokenizer=tokenizer, ngram_range=(1, 2))
vectorized_corpus = cv.fit_transform(corpus)
print(vectorized_corpus.todense())
[[0 0 0 0 0 1 0 1 0 0 1 1 0 0 1 0 0 0 0]
 [0 0 0 0 0 1 1 1 1 0 0 1 0 0 0 0 0 0]
 [1 1 0 0 0 1 1 0 0 0 0 0 2 0 1 0 1 1]
 [0 0 1 1 1 0 1 0 0 0 0 0 0 0 0 1 0 0]]

print(cv.vocabulary_)
{u'and': 1, u'be': 3, u'we': 18, u'set': 9, u'simple': 10, u'text': 12,
u'is': 7, u'tokenized': 16, u'want': 17, u'the': 13, u'documents': 6,
u'this': 14, u'of': 8, u'to': 15, u'can': 4, u'test': 11, u'corpus': 5,
u'analyze': 0, u'automatically': 2}
```

As you can see, the vocabulary now contains the bigrams, and the vectors include their relative frequencies.

TF-IDF vectorizing

The most common limitation of count vectorizing is that the algorithm doesn't consider the whole corpus while considering the frequency of each token. The goal of vectorizing is normally preparing the data for a classifier; therefore, it's necessary to avoid features that are present very often, because their information decreases when the number of global occurrences increases. For example, in a corpus about a sport, the word `match` could be present in a huge number of documents; therefore, it's almost useless as a classification feature. To address this issue, we need a different approach. If we have a corpus C with n documents, we define **Term Frequency** (**TF**), the number of times a token occurs in a document, as the following:

$$t_f(t, d) \ \forall \, d \in C \ and \ \forall \, t \in d$$

We define **Inverse Document Frequency (IDF)** as the following measure:

$$idf(t, C) = \log \frac{n}{1 + count(D, t)} \quad \text{where} \quad count(D, t) = \sum_t 1(t \in D)$$

In other words, *idf(t, C)* measures how much information is provided by every single term. In fact, if *count(D, t) = n*, it means that a token is always present and *idf(t, C)* comes close to 0, and vice versa. The term *1* in the denominator is a correction factor, which avoids a null *idf* when the count *(D, t) = n*. So, instead of considering only the TF, we weigh each token by defining a new measure:

$$t_f \cdot idf(t, d, C) = t_f(t, d) \cdot idf(t, C)$$

scikit-learn provides the `TfIdfVectorizer` class, which we can apply to the same toy corpus used in the previous paragraph:

```
from sklearn.feature_extraction.text import TfidfVectorizer

tfidfv = TfidfVectorizer()
vectorized_corpus = tfidfv.fit_transform(corpus)
print(vectorized_corpus.todense())
[[ 0.          0.          0.          0.          0.          0.31799276
   0.          0.39278432  0.          0.          0.49819711  0.49819711
   0.          0.          0.49819711  0.          0.          0.
   0.          ]
 ...
```

Let's now check the vocabulary to make a comparison with simple count vectorizing:

```
print(tfidfv.vocabulary_)
{u'and': 1, u'be': 3, u'we': 18, u'set': 9, u'simple': 10, u'text': 12,
u'is': 7, u'tokenized': 16, u'want': 17, u'the': 13, u'documents': 6,
u'this': 14, u'of': 8, u'to': 15, u'can': 4, u'test': 11, u'corpus': 5,
u'analyze': 0, u'automatically': 2}
```

The term documents is the sixth feature in both vectorizers and appears in the last three documents. As you can see, its weight is about 0.3, while the term the is present twice only in the third document, and its weight is about 0.64. The general rule is this: if a term is representative of a document, its weight becomes close to 1.0, while it decreases if finding it in a sample document doesn't allow us to easily determine its category.

Also, in this case, it's possible to use an external tokenizer and specify the desired n-gram range. Moreover, it's possible to normalize the vectors (through the `norm` parameter) and decide whether to include or exclude the addend 1 to the denominator of `idf` (through the `smooth_idf` parameter). It's also possible to define the range of accepted document frequencies using the `min_df` and `max_df` parameters so as to exclude tokens whose occurrences are below or beyond a minimum/maximum threshold. They accept both integers (number of occurrences) or floats in the range of *[0.0, 1.0]* (proportion of documents). In the next example, we use some of these parameters:

```
tfidfv = TfidfVectorizer(tokenizer=tokenizer, ngram_range=(1, 2), norm='l2')
vectorized_corpus = tfidfv.fit_transform(corpus)
print(vectorized_corpus.todense())
[[ 0.          0.          0.          0.          0.30403549  0.
   0.
   0.          0.          0.          0.          0.47633035  0.47633035
   0.47633035  0.47633035  0.          0.          0.          0.
   0.          ]
 ...

print(tfidfv.vocabulary_)
{u'analyz corpus': 1, u'set': 9, u'simpl test': 12, u'want analyz': 19,
u'automat': 2, u'want': 18, u'test corpus': 14, u'set text': 10, u'corpus
set': 6, u'automat token': 3, u'corpus document': 5, u'text document': 16,
u'token': 17, u'document automat': 8, u'text': 15, u'test': 13, u'corpus':
4, u'document': 7, u'simpl': 11, u'analyz': 0}
```

In particular, normalizing vectors is always a good choice if they must be used as input for a classifier, as we'll see in the next chapter, *Chapter 14*, *Topic Modeling and Sentiment Analysis in NLP*.

Part-of-Speech

In some cases, it can be helpful to detect the single syntactical components of a text to perform specific analyses. For example, given a sentence, we can be interested in finding the verb that represents the intent of an action. Alternatively, we could need to extract other attributes such as locations, names, and temporal dependencies. Even though this topic is quite complex and beyond the scope of this book, we wanted to provide some examples that can be immediately applied to more complex scenarios.

The first step of this process is called **POS Tagging** and consists of adding a syntactic identifier to each token. NLTK has a built-in model based on the **Penn Treebank** POS corpus, which provides a large number of standard tags for the English language (for a complete list, please check out `https://www.ling.upenn.edu/courses/Fall_2003/ling001/penn_treebank_pos.html`). To better understand the dynamics, let's consider a couple of simple sentences tokenized using the standard `work_tokenize()` function (of course, there are no limitations in the tokenizing strategy) and processed using the `pos_tag()` function:

```
from nltk import word_tokenize, pos_tag

sentence_1 = 'My friend John lives in Paris'

tokens_1 = word_tokenize(sentence_1)
tags_1 = pos_tag(tokens_1)

print(tags_1)
[('My', 'PRP$'), ('friend', 'NN'), ('John', 'NNP'), ('lives', 'VBZ'),
('in', 'IN'), ('Paris', 'NNP')]
```

As you can see, each token is transformed into a tuple containing the POS tag too. In the aforementioned link, the reader can find a description of each tag; however, it's easy to intuitively understand some of them. For example, `PRP$` identifies a possessive pronoun, `NNP` is a proper noun, and `VBZ` is a verb conjugated in the third-person singular. If we need to identify the action, we can skip all tokens but `'lives'` (which is a verb) and focus on the intent. A very simple chatbot could parse the following request:

```
sentence_2 = 'Search a hotel in Rome'

print(tags_2)
[('Search', 'VB'), ('a', 'DT'), ('hotel', 'NN'), ('in', 'IN'), ('Rome',
'NNP')]
```

The verb is `'Search'`, which can be associated with a specific action. The preposition `'in'` followed by `'Rome'` provides an extra piece of information to complete the request. However, as `'Rome'` is a proper noun, we cannot be sure that it matches the intent (for example, for a searching action, we expect a location). For this reason, we need to introduce a more sophisticated tool that can help in the disambiguation process.

Named Entity Recognition

Named Entity Recognition (**NER**) is a method for extracting specific pieces of semantic information from tokens. Consider the last example: we are interested in understanding whether Rome is a city, a lake, or a company. The action performed by our simple chatbot is, in fact, strictly related to the contextual element indicated by the generic proper noun (for example, if Rome is a city, the request can be easily fulfilled, while if it were a company, the chatbot should ask the exact location). Let's consider another example (the first steps are the same as the previous ones):

```
from nltk import word_tokenize, pos_tag, ne_chunk, tree2conlltags

sentence_2 = 'Search a hotel in Cambridge near the MIT'

print(tree2conlltags(ne_chunk(tags_2)))
[('Search', 'VB', 'O'), ('a', 'DT', 'O'), ('hotel', 'NN', 'O'), ('in',
'IN', 'O'), ('Cambridge', 'NNP', 'B-GPE'), ('near', 'IN', 'O'), ('the',
'DT', 'O'), ('MIT', 'NNP', 'B-ORGANIZATION')]
```

The `ne_chunk()` function performs the NER, but it outputs a tree showing the relations. As we want to obtain a list of tuples, we need to employ the `tree2colltags()` function. Now each tuple has an additional element, indicating the recognized entity. Let's focus on `'Cambridge'` and `'MIT'`. The tag `'B-GPE'` indicates a **Geo-political entity** (**GPE**) (which can be a city or a state, but, in many cases, it's enough to restrict our research), while `'B-ORGANIZATION'` clearly indicates a generic company (either public or private). With these two additional pieces of information, we can use the intent `'Search'` and perform a lookup for `MIT` in `Cambridge`. Once we get a valid address (for example, using the Google Maps API), we can look for the target (a hotel) close to the desired destination.

As this is an introduction to NLP, we cannot analyze more complex scenarios; however, I invite the reader to checkout the book *Hands-On Natural Language Processing with Python, Arumugam R, Shanmugamani R, Packt Publishing, 2018,* for further information and examples. I also suggest reading the standard NLTK documentation, where it's possible to find detailed descriptions of the linguistic algorithms and methods employed in every specific function.

A sample text classifier based on the Reuters corpus

We are going to build a sample text classifier based on the NLTK Reuters corpus. This one is made up of thousands of news lines divided into 90 categories:

```
from nltk.corpus import reuters

print(reuters.categories())
[u'acq', u'alum', u'barley', u'bop', u'carcass', u'castor-oil', u'cocoa',
u'coconut', u'coconut-oil', u'coffee', u'copper', u'copra-cake', u'corn',
...
```

To simplify the process, we'll take only two categories, which have a similar number of documents:

```
import numpy as np

Xr = np.array(reuters.sents(categories=['rubber']))
Xc = np.array(reuters.sents(categories=['cotton']))
Xw = np.concatenate((Xr, Xc))
```

As each document is already split into tokens and we want to apply our custom tokenizer (with stopword removal and stemming), we need to rebuild the full sentences:

```
X = []

for document in Xw:
    X.append(' '.join(document).strip().lower())
```

Now we need to prepare the label vector, by assigning 0 to `rubber` and 1 to `cotton`:

```
Yr = np.zeros(shape=Xr.shape)
Yc = np.ones(shape=Xc.shape)
Y = np.concatenate((Yr, Yc))
```

At this point, we can vectorize our corpus:

```
tfidfv = TfidfVectorizer(tokenizer=tokenizer, ngram_range=(1, 2),
norm='l2')
Xv = tfidfv.fit_transform(X)
```

Now the dataset is ready, and we can proceed by splitting it into train and test subsets and finally train our classifier. I've decided to adopt a random forest, because it's particularly effective for this kind of task, but the reader can try different classifiers and compare the results:

```
from sklearn.model_selection import train_test_split
from sklearn.ensemble import RandomForestClassifier

X_train, X_test, Y_train, Y_test = train_test_split(Xv, Y, test_size=0.25, random_state=1000)

rf = RandomForestClassifier(n_estimators=25)
rf.fit(X_train, Y_train)
score = rf.score(X_test, Y_test)
print('Score: %.3f' % score)
Score: 0.874
```

The score is about 88%, which is quite a good result, but let's try a prediction with a fake news line:

```
test_newsline = ['Trading tobacco is reducing the amount of requests for cotton and this has a negative impact on our economy']

yvt = tfidfv.transform(test_newsline)
category = rf.predict(yvt)
print('Predicted category: %d' % int(category[0]))
Predicted category: 1
```

The classification result is correct; however, by adopting some techniques that we're going to discuss in the next chapter, `Chapter 14`, *Topic Modeling and Sentiment Analysis in NLP*, it's also possible to get better performance in more complex real-life problems.

Summary

In this chapter, we discussed all the basic NLP techniques, starting with the definition of a corpus up to the final transformation into feature vectors. We analyzed different tokenizing methods to address particular problems or situations of splitting a document into words. Then, we introduced some filtering techniques that are necessary to remove all useless elements (also called stopwords) and to convert the inflected forms into standard tokens.

These steps are important to increase the information content by removing frequently used terms. When the documents have been successfully cleaned, it is possible to vectorize them using a simple approach such as the one implemented by the count-vectorizer, or a more complex one that takes into account the global distribution of terms, such as TF-IDF. The latter was introduced to complete the work done by the stemming phase; in fact, its purpose is to define vectors where each component will be close to 1 when the amount of information is high, and vice versa. Normally, a word that is present in many documents isn't a good marker for a classifier; therefore, if not already removed by the previous steps, TF-IDF will automatically reduce its weight. At the end of the chapter, we built a simple text classifier that implemented the whole Bag-of-Words pipeline and used a random forest to classify news lines.

In the next chapter, `Chapter 14`, *Topic Modeling and Sentiment Analysis in NLP*, we're going to complete this introduction with a brief discussion of advanced techniques, such as topic modeling, latent semantic analysis, and sentiment analysis.

14
Topic Modeling and Sentiment Analysis in NLP

In this chapter, we're going to introduce some well-known modeling methods, and discuss some applications. **Topic modeling** is a very important part of **Natural Language Processing** (**NLP**) and its purpose is to extract semantic pieces of information out of a corpus of documents. We're going to discuss **Latent Semantic Analysis** (**LSA**), one of the most famous methods; it's based on the same philosophy already discussed for model-based recommendation systems. We'll also discuss its probabilistic variant, **Probabilistic Latent Semantic Analysis** (**PLSA**), which is aimed at building a latent factor probability model without any assumption of prior distributions. On the other hand, the **Latent Dirichlet Allocation** (**LDA**) is a similar approach that assumes a prior Dirichlet distribution for latent variables. In the last section, we're going to discuss the basics of **Word2vec** and sentiment analysis with a concrete example based on a freely available Twitter dataset.

In particular, we are going to discuss the following topics:

- Topic modeling
- Word2vec with Gensim
- Sentiment analysis

Topic modeling

The main goal of topic modeling in NLP is to analyze a corpus, to identify common topics among documents. In this context, even if we talk about semantics, this concept has a particular meaning, driven by a very important assumption. A topic derives from the usage of particular terms in the same document, and it is confirmed by the multiplicity of different documents where the first condition is true.

In other words, we don't consider human-oriented semantics but a statistical modeling that works with meaningful documents (this guarantees that the usage of terms is aimed to express a particular concept, and, therefore, there's a human semantic purpose behind them). For this reason, the starting point of all our methods is an **occurrence matrix**, normally defined as a **document-term matrix** (we have already discussed count vectorizing and TF-IDF in Chapter 12, *Introducing Natural Language Processing*):

$$M_{dw} = \begin{pmatrix} f(d_1, w_1) & \cdots & f(d_1, w_n) \\ \vdots & \ddots & \vdots \\ f(d_m, w_1) & \cdots & f(d_m, w_n) \end{pmatrix} \text{ where } f(d_i, w_j) \text{ is a frequency measure}$$

In many papers, this matrix is transposed (it's a term-document one); however, scikit-learn produces document-term matrices, and, to avoid confusion, we are going to consider this structure.

Latent Semantic Analysis

The idea behind LSA is factorizing M_{dw} so as to extract a set of latent variables (this means that we can assume their existence, but they cannot be observed directly) that work as connectors between the document and terms. As discussed in Chapter 11, *Introducing Recommendation Systems*, a very common decomposition method is **Singular Value Decomposition (SVD)**:

$$M_{U \times I} = U \Sigma V^T \text{ where } U \in \mathbb{R}^{m \times m}, \Sigma \in \mathbb{R}^{m \times n}, \text{ and } V \in \mathbb{R}^{n \times n}$$

However, we're not interested in a full decomposition; we are interested only in the subspace defined by the top *k* singular values:

$$M_k = U_k \Sigma_k V_k^T$$

This approximation has the reputation of being the best one, considering the Frobenius norm, so it guarantees a very high level of accuracy. When applying it to a document-term matrix, we obtain the following decomposition:

$$M_{dwk} = \begin{pmatrix} g(d_1, t_1) & \cdots & g(d_1, t_k) \\ \vdots & \ddots & \vdots \\ g(d_m, t_1) & \cdots & g(d_m, t_k) \end{pmatrix} \cdot \begin{pmatrix} h(t_1, w_1) & \cdots & h(d_1, w_n) \\ \vdots & \ddots & \vdots \\ h(t_k, w_1) & \cdots & h(t_k, w_n) \end{pmatrix}$$

Here is a more compact version:

$$M_{dwk} = M_{dtk} \cdot M_{twk}$$

Here, the first matrix defines a relationship among documents and *k* latent variables, and the second a relationship among *k* latent variables and words. Considering the structure of the original matrix and what is explained at the beginning of this chapter, we can consider the latent variables as **topics** that define a subspace where the documents are projected. A generic document can now be defined as follows:

$$d_i = \sum_{j=1}^{k} g(d_i, t_j)$$

Furthermore, each topic becomes a linear combination of words. As the weight of many words is close to zero, we can decide to take only the top *r* words to define a topic; therefore, we get the following:

$$t_i \approx \sum_{j=1}^{r} h_{ji} w_j$$

Here, each h_{ji} is obtained after sorting the columns of M_{twk}. To better understand the process, let's show a complete example based on a subset of the Brown Corpus (500 documents from the 'news' category):

```
from nltk.corpus import brown

sentences = brown.sents(categories=['news'])[0:500]
corpus = []
```

Topic Modeling and Sentiment Analysis in NLP

```
for s in sentences:
    corpus.append(' '.join(s))
```

After defining the corpus, we need to tokenize and vectorize using a TF-IDF approach:

```
from sklearn.feature_extraction.text import TfidfVectorizer

vectorizer = TfidfVectorizer(strip_accents='unicode', stop_words='english',
norm='l2', sublinear_tf=True)
Xc = vectorizer.fit_transform(corpus).todense()
```

Now it's possible to apply an SVD to the Xc matrix (remember that, in SciPy, the V matrix is already transposed):

```
from scipy.linalg import svd

U, s, V = svd(Xc, full_matrices=False)
```

As the corpus is not very small, it's useful to set the `full_matrices=False` parameter to save computational time. We assume we have two topics, so we can extract our sub-matrices:

```
import numpy as np

rank = 2

Uk = U[:, 0:rank]
sk = np.diag(s)[0:rank, 0:rank]
Vk = V[0:rank, :]
```

If we want to analyze the top ten words per topic, we need to consider this:

$$M_{twk} = V_k$$

Therefore, we can obtain the most significant words per topic after sorting the matrix using the `get_feature_names()` method provided by the vectorizers:

```
Mtwks = np.argsort(Vk, axis=1)[::-1]

for t in range(rank):
    print('\nTopic ' + str(t))
    for i in range(10):
        print(vectorizer.get_feature_names()[Mtwks[t, i]])

Topic 0
said
mr
```

```
city
hawksley
president
...

Topic 1
plainfield
wasn
copy
released
absence
...
```

In this case, we're considering only non-negative values in the `Vk` matrix; however, as a topic is a mixture of words, the negative components should also be taken into account. In this case, we need to sort the absolute values of `Vk`:

```
Mtwks = np.argsort(np.abs(Vk), axis=1)[::-1]
```

If we want to analyze how a document is represented in this subspace, we must use this:

$$M_{dtk} = U_k \Sigma_k$$

Let's consider, for example, the first document of our `corpus`:

```
print(corpus[0])
The Fulton County Grand Jury said Friday an investigation of Atlanta's
recent primary election produced `` no evidence '' that any irregularities
took place .

Mdtk = Uk.dot(sk)

print('d0 = %.2f*t1 + %.2f*t2' % (Mdtk[0][0], Mdtk[0][1]))
d0 = 0.15*t1 + -0.12*t2
```

As we are working in a bidimensional space, it's interesting to plot all the points corresponding to each document:

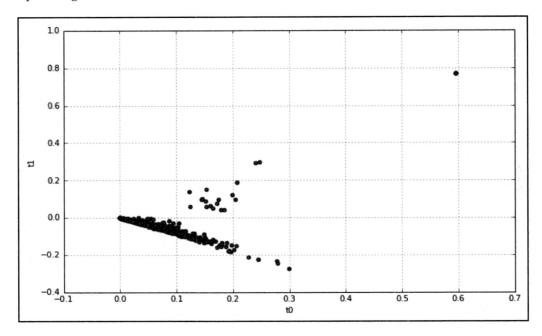

Document distribution as a function of two topics

In the previous graph, we can see that many documents are correlated, with a small group of outliers. This is probably due to the fact that our choice of two topics is restrictive. If we repeat the same experiment using two Brown Corpus categories (`'news'` and `'fiction'`), we observe a different behavior:

```
sentences = brown.sents(categories=['news', 'fiction'])
corpus = []

for s in sentences:
   corpus.append(' '.join(s))
```

I won't repeat the remaining calculations, because they are similar. (The only difference is that our corpus is now a little bigger, and this leads to a longer computational time. For this reason, we're going to discuss an alternative, that is much faster.) Plotting the points corresponding to the documents, we now get the following:

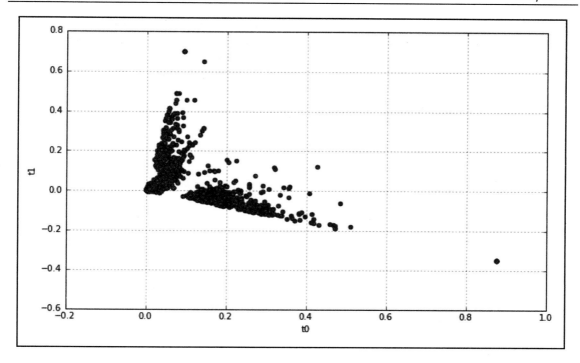

Document distribution as a function of two topics

Now it's easier to distinguish two groups, which are almost *orthogonal* (meaning that many documents belong to only one category). I suggest repeating this experiment with different corpora and ranks. Unfortunately, it's impossible to plot more than three dimensions, but it's always possible to check whether the subspace describes the underlying semantics correctly using only numerical computations.

As anticipated, the standard SciPy SVD implementation can be really slow when the occurrence matrix is huge; however, scikit-learn provides a truncated SVD implementation, TruncatedSVD, that works only with the subspace. The result is much faster (it can directly manage sparse matrices too). Let's repeat the previous experiments (with a complete corpus) using this class:

```
from sklearn.decomposition import TruncatedSVD

tsvd = TruncatedSVD(n_components=rank)
Xt = tsvd.fit_transform(Xc)
```

Through the `n_components` parameter, it's possible to set the desired rank, discarding the remaining parts of the matrices. After fitting the model, we get the document-topic matrix M_{dtk} directly as the output of the `fit_transform()` method, while the topic-word matrix M_{twk} can be accessed using the `components_instance` variable:

```
Mtws = np.argsort(tsvd.components_, axis=1)[::-1]

for t in range(rank):
    print('\nTopic ' + str(t))
    for i in range(10):
        print(vectorizer.get_feature_names()[Mwts[t, i]])

Topic 0
said
rector
hans
aloud
liston
...

Topic 1
bong
varnessa
schoolboy
kaboom
keeeerist
...
```

The reader can verify how much faster this process can be; therefore, I suggest using a standard SVD implementation only when it's needed to have access to the full matrices. Unfortunately, as is also written in the documentation, this method is very sensitive to the algorithm and the random state. It also suffers from a phenomenon called **sign indeterminacy**, which means that the signs of all components can change if a different random seed is used. As we have already done in other chapters, I suggest to declare a fixed random seed:

```
import numpy as np

np.random.seed(1000)
```

Do this with a fixed seed at the beginning of every file (even Jupyter notebooks) to be sure that it's possible to repeat the calculations and to always obtain the same result.

Moreover, I invite the reader to repeat this experiment using the **Non-Negative Matrix Factorization (NNMF)** algorithm, as described in `Chapter 3`, *Feature Selection and Feature Engineering*.

Probabilistic Latent Semantic Analysis

The previous model was based on a deterministic approach, but it's also possible to define a probabilistic model PLSA over the space determined by documents and words. In this case, we're not making any assumption about Apriori probabilities (this will be done in the next approach), and we're going to determine the parameters that maximize the log-likelihood of our model. In particular, consider the plate notation (if you want to know more about this technique, read `https://en.wikipedia.org/wiki/Plate_notation`) shown in the following diagram:

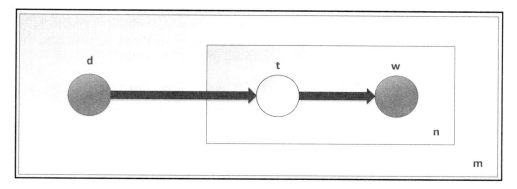

Plate diagram for PLSA

We assume we have a corpus of **m** documents, and each of them is composed of **n** words (both elements are observed and therefore represented as gray circles); however, we also assume the presence of a limited set of *k* common latent factors (topics) that link a document with a group of words (as they are not observed, the circle is white). As already written, we cannot observe them directly, but we're allowed to assume their existence.

The joint probability to find a document with a particular word is as follows:

$$p(d, w) = p(w|d)p(d)$$

Therefore, after introducing the latent factors, the conditional probability to find a word in a specific document can be written as follows:

$$p(w|d) = \sum_{i=1}^{k} p(w|t_i)p(t_i|d)$$

The initial joint probability P(d, w) can be also expressed using the latent factors:

$$p(d, w) = \sum_{i=1}^{k} p(t_i)p(w|t_i)p(d|t_i)$$

This includes the prior probability p(t). As we don't want to work with it, it's preferable to use the expression p(w|d). To determine the two conditional probability distributions, a common approach is the **Expectation-Maximization** (**EM**) algorithm, which was introduced in Chapter 2, *Important Elements in Machine Learning*. A full description can be found in *Unsupervised Learning by Probabilistic Latent Semantic Analysis, Hofmann T, Machine Learning 42, 177-196, 2001, Kluwer Academic Publishers*. In this context, considering the complexity, we show only the final results, without any formal proof.

The log-likelihood can be written as follows:

$$L = \sum_d \sum_w M_{dw}(d, w) \cdot \log p(d, w)$$

If we expand all the terms, the previous formula becomes this:

$$L = \sum_d \sum_w \left(M_{dw}(d, w) \cdot \log p(d) + M_{dq}(d, w) \cdot \log \sum_k p(t_k|d)p(w|t_k) \right)$$

M_{dw} is an occurrence matrix (normally obtained with a count vectorizer), and $M_{dw}(d, w)$ is the frequency of the word w in document d. For simplicity, we are going to approximate it by excluding the first term (which doesn't depend on t_k):

$$L \approx \sum_d \sum_w M_{dq}(d, w) \cdot \log \sum_k p(t_k|d)p(w|t_k)$$

Moreover, it's useful to introduce the conditional probability p(t|d,w), which is the probability of a topic, given a document and a word. The EM algorithm maximizes the expected complete log-likelihood under the posterior probability P(t|d,w) (which represents a proxy of the actual log-likelihood):

$$E[L_c] = \sum_d \sum_w M_{dw}(d, w) \sum_k p(t|d, w) \cdot \log p(t_k|d)p(w|t_k)$$

The E phase of the algorithm can be expressed as follows:

$$p(t|d,w) = \frac{p(t|d)p(w|t)}{\sum_k p(t_k|d)p(w|t_k)}$$

It must be extended to all topics, words, and documents, and it must be normalized with the sum per topic to always have consistent probabilities.

The M phase is split into two computations:

$$\begin{cases} p(w|t) = \frac{\sum_d M_{dw}(d,w) \cdot p(t|d,w)}{\sum_w \sum_d M_{dw}(d,w) \cdot p(t|d,w)} \\ p(t|d) = \frac{\sum_w M_{dw}(d,w) \cdot p(t|d,w)}{\sum_w M_{dw}(d,w)} \end{cases}$$

Also, in this case, the calculations must be extended to all topics, words, and documents. But in the first case, we sum by document and normalize by summing by word and document, while in the second, we sum by word and normalize by the length of the document.

The algorithm must be iterated until the log-likelihood stops increasing its magnitude. Unfortunately, scikit-learn doesn't provide a PLSA implementation (maybe because the next strategy, LDA, is considered much more powerful and efficient), so we need to write some code from scratch. Let's start by defining a small subset of the Brown Corpus, taking 10 sentences from the `'editorial'` category and 10 from the `'fiction'` one:

```
sentences_1 = brown.sents(categories=['editorial'])[0:10]
sentences_2 = brown.sents(categories=['fiction'])[0:10]
corpus = []

for s in sentences_1 + sentences_2:
    corpus.append(' '.join(s))
```

Now we can vectorize using the `CountVectorizer` class:

```
import numpy as np

from sklearn.feature_extraction.text import CountVectorizer

cv = CountVectorizer(strip_accents='unicode', stop_words='english')
Xc = np.array(cv.fit_transform(corpus).todense())
```

At this point, we can define the `rank` (we choose 2 for simplicity), two constants that will be used later, and the matrices to hold the probabilities $p(t|d)$, $p(w|t)$, and $p(t|d,w)$:

```python
import numpy as np

rank = 2
alpha_1 = 1000.0
alpha_2 = 10.0

Ptd = np.random.uniform(0.0, 1.0, size=(len(corpus), rank))
Pwt = np.random.uniform(0.0, 1.0, size=(rank, len(cv.vocabulary_)))
Ptdw = np.zeros(shape=(len(cv.vocabulary_), len(corpus), rank))

for d in range(len(corpus)):
    nf = np.sum(Ptd[d, :])
    for t in range(rank):
        Ptd[d, t] /= nf

for t in range(rank):
    nf = np.sum(Pwt[t, :])
    for w in range(len(cv.vocabulary_)):
        Pwt[t, w] /= nf
```

The two matrices $p(t|d)$, $p(w|t)$ must be normalized so as to be coherent with the algorithm; the other one is initialized to zero. Now we can define the `log_likelihood` function:

```python
def log_likelihood():
    value = 0.0

    for d in range(len(corpus)):
        for w in range(len(cv.vocabulary_)):
            real_topic_value = 0.0

            for t in range(rank):
                real_topic_value += Ptd[d, t] * Pwt[t, w]

            if real_topic_value > 0.0:
                value += Xc[d, w] * np.log(real_topic_value)

    return value
```

We can also define the `expectation()` and `maximization()` functions:

```python
def expectation():
    global Ptd, Pwt, Ptdw

    for d in range(len(corpus)):
        for w in range(len(cv.vocabulary_)):
```

```
        nf = 0.0

        for t in range(rank):
            Ptdw[w, d, t] = Ptd[d, t] * Pwt[t, w]
            nf += Ptdw[w, d, t]

        Ptdw[w, d, :] = (Ptdw[w, d, :] / nf) if nf != 0.0 else 0.0
```

In the preceding function, when the normalization factor is 0, the probability *p(t|w, d)* is set to 0.0 for each topic:

```
def maximization():
    global Ptd, Pwt, Ptdw

    for t in range(rank):
        nf = 0.0

        for d in range(len(corpus)):
            ps = 0.0

            for w in range(len(cv.vocabulary_)):
                ps += Xc[d, w] * Ptdw[w, d, t]

            Pwt[t, w] = ps
            nf += Pwt[t, w]

        Pwt[:, w] /= nf if nf != 0.0 else alpha_1

    for d in range(len(corpus)):
        for t in range(rank):
            ps = 0.0
            nf = 0.0

            for w in range(len(cv.vocabulary_)):
                ps += Xc[d, w] * Ptdw[w, d, t]
                nf += Xc[d, w]

            Ptd[d, t] = ps / (nf if nf != 0.0 else alpha_2)
```

The constants `alpha_1` and `alpha_2` are used when a normalization factor becomes 0. In that case, it can be useful to assign the probability a small value; therefore, we divided the numerator for those constants. I suggest trying with different values so as to tune up the algorithm for different tasks.

At this point, we can try our algorithm with a limited number of iterations:

```
print('Initial Log-Likelihood: %f' % log_likelihood())

for i in range(50):
    expectation()
    maximization()
    print('Step %d - Log-Likelihood: %f' % (i, log_likelihood()))

Initial Log-Likelihood: -1242.878549
Step 0 - Log-Likelihood: -1240.160748
Step 1 - Log-Likelihood: -1237.584194
Step 2 - Log-Likelihood: -1236.009227
Step 3 - Log-Likelihood: -1234.993974
Step 4 - Log-Likelihood: -1234.318545
...
```

It's possible to verify the convergence after the 30th step. At this point, we can check the top five words per topic considering the *p(w|t)* conditional distribution sorted in descending mode per topic weight:

```
Pwts = np.argsort(Pwt, axis=1)[::-1]

for t in range(rank):
    print('\nTopic ' + str(t))
    for i in range(5):
        print(cv.get_feature_names()[Pwts[t, i]])

Topic 0
years
questions
south
reform
social

Topic 1
convened
maintenance
penal
year
legislators
```

Latent Dirichlet Allocation

In the previous method, we didn't make any assumptions about the topic prior to distribution, and this can result in a limitation, because the algorithm isn't driven by any real-world intuition. LDA, instead, is based on the idea that a topic is characterized by a small ensemble of important words, and normally a document doesn't cover many topics. For this reason, the main assumption is that the prior topic distribution is a **symmetric Dirichlet** one ($\alpha_1 = \alpha_2 = ... = \alpha_K = \alpha$). The probability density function is defined as follows:

$$f(\bar{x}; \bar{\alpha}) = \frac{\Gamma(\sum_k \alpha_k)}{\prod_k \Gamma(\alpha_k)} \prod_k x_k^{\alpha_k - 1} = \frac{\Gamma(\alpha K)}{\Gamma(\alpha)^K} \prod_k x_k^{\alpha - 1}$$

If the concentration parameter α is less than 1.0, the distribution will be sparse, as desired. This allows us to model topic-document and topic-word distributions, which will always be concentrated on a few values. In this way we can avoid the following:

- The topic mixture assigned to a document becoming flat (many topics with similar weight)
- The structure of a topic considering the word ensemble becoming similar to a background (in fact, only a limited number of words must be important; otherwise, the semantic boundaries fade out)

Using the plate notation, we can represent the relationship among documents, topics, and words, as shown in the following diagram:

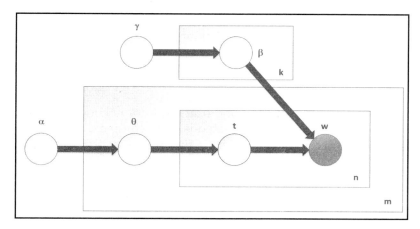

Plate diagram for a Latent Dirichlet Allocation

In the previous diagram, **α** is the Dirichlet parameter for the topic-document distribution, while **γ** has the same role for the topic-word distribution. **θ**, instead, is the topic distribution for a specific document, while **β** is the topic distribution for a specific word.

If we have a corpus of **m** documents and a vocabulary of **n** words (each document has n_i words) and we assume to have **k** different topics, the generative algorithm can be described with the following steps:

1. For each document, we draw a sample (a topic mixture) from the topic-document distribution:

$$\theta_i \sim Dir(\alpha) \; \forall \; i \in (1, m)$$

2. For each topic, we draw a sample from the topic-word distribution:

$$\beta_i \sim Dir(\gamma) \; \forall \; i \in (1, k)$$

Both parameters must be estimated. At this point, considering the occurrence matrix M_{dw} and the notation z_{mn} to define the topic assigned to the *n-th* word in the *m-th* document, we can iterate over documents (index *d*) and words (index *w*):

- A topic for document *d* and word *w* is chosen according to a **categorical distribution**, considering that the Dirichlet distribution is its conjugate prior:

$$z_{dw} \sim Categorical(\theta_d)$$

- A word is chosen according to the following:

$$w_{wj} \sim Categorical(\beta_{z_{dw}})$$

In both cases, a *categorical* distribution is a one-trial multinomial one. A complete description of how the parameters, are estimated is quite complex, and it's beyond the scope of this book; however, the main problem is finding the distribution of latent variables:

$$p(\bar{z}, \theta, \beta | \bar{w}, \alpha, \gamma) = \frac{p(\bar{z}, \theta, \beta | \alpha, \gamma)}{p(\bar{w} | \alpha, \gamma)}$$

The reader can find a lot more information in *Latent Dirichlet Allocation, Blei D, Ng A, Jordan M, Journal of Machine Learning Research, 3, (2003) 993-1022*. However, a very important difference between LDA and PLSA is about the generative ability of LDA, which allows working with unseen documents. In fact, the PLSA training process finds the optimal parameters *p(t|d)* only for the corpus, while LDA adopts random variables. It's possible to understand this concept by defining the probability of theta (a topic mixture) as joint with a set of topics and a set of words, and conditioned to the following model parameters:

$$p(\theta, \bar{z}, \bar{w} | \alpha, \gamma) = p(\theta | \alpha) \prod_i p(z_i | \theta) p(w_i | z_i, \gamma)$$

As shown in the previously mentioned paper, the probability of a document (a set of words) conditioned to the model parameters, can be obtained by integration (all the mathematical proofs are omitted due to their complexity, but they can be found in the aforementioned paper):

$$p(\bar{w} | \alpha, \gamma) = \int p(\theta | \alpha) \left(\prod_i \sum_{z_i} p(z_i | \theta) p(w_i | z_i, \gamma) \right) d\theta$$

This expression shows the difference between PLSA and LDA. Once learned, *p(t|d)* PLSA cannot generalize, while LDA, sampling from the random variables, can always find a suitable topic mixture for an unseen document.

scikit-learn provides a full LDA implementation through the `LatentDirichletAllocation` class. We're going to use it with a bigger dataset (4,000 documents) built from a subset of the Brown Corpus:

```
sentences_1 = brown.sents(categories=['reviews'])[0:1000]
sentences_2 = brown.sents(categories=['government'])[0:1000]
sentences_3 = brown.sents(categories=['fiction'])[0:1000]
sentences_4 = brown.sents(categories=['news'])[0:1000]
corpus = []

for s in sentences_1 + sentences_2 + sentences_3 + sentences_4:
    corpus.append(' '.join(s))
```

Now we can vectorize, define, and train our LDA model by assuming that we have eight main topics:

```
from sklearn.decomposition import LatentDirichletAllocation

cv = CountVectorizer(strip_accents='unicode', stop_words='english',
analyzer='word', token_pattern='[a-z]+')
Xc = cv.fit_transform(corpus)

lda = LatentDirichletAllocation(n_topics=8, learning_method='online',
max_iter=25)
Xl = lda.fit_transform(Xc)
```

In `CountVectorizer`, we added a regular expression to filter the tokens through the `token_pattern` parameter. This is useful, as we are not using a full tokenizer, and, in the corpus, there are also many numbers that we want to filter out. The `LatentDirichletAllocation` class allows us to specify the learning method (through `learning_method`), which can be either `'batch'` or `'online'`. We have chosen `'online'` because it's faster; however, both methods adopt variational Bayes to learn the parameters. The former adopts the whole dataset, while the latter works with mini-batches. The `'online'` option will be removed in the 0.20 release; therefore, you can see a deprecation warning when using it now. Both theta and beta Dirichlet parameters can be specified through `doc_topic_prior` (theta) and `topic_word_prior` (beta). The default value (adopted by us too) is 1.0 / n_topics. It's important to keep both values small, and, in particular, less than 1.0, to encourage sparseness. The maximum number of iterations (`max_iter`) and other learning-related parameters can be applied by reading the built-in documentation or visiting http://scikit-learn.org/stable/modules/generated/sklearn.decomposition.LatentDirichletAllocation.html.

Now we can test our model by extracting the top five keywords per topic. Just like `TruncatedSVD`, the topic-word distribution results are stored in the `components_` instance variable:

```
Mwts_lda = np.argsort(lda.components_, axis=1)[::-1]

for t in range(8):
    print('\nTopic ' + str(t))
    for i in range(5):
        print(cv.get_feature_names()[Mwts_lda[t, i]])

Topic 0
code
cadenza
unlocks
ophthalmic
```

```
quo

Topic 1
countless
harnick
leni
addle
chivalry
```

...

There are some repetitions, probably due to the composition of some topics, and the reader can try different prior parameters to observe the changes. It's possible to do an experiment to check whether the model works correctly.

Let's consider two documents:

```
print(corpus[0])
It is not news that Nathan Milstein is a wizard of the violin.

print(corpus[2500])
The children had nowhere to go and no place to play , not even sidewalks.
```

They are quite different and so are their topic distributions:

```
print(Xl[0])
[ 0.85412134 0.02083335 0.02083335 0.02083335 0.02083335 0.02083677
  0.02087515 0.02083335]

print(Xl[2500])
[ 0.22499749 0.02500001 0.22500135 0.02500221 0.025 0.02500219
  0.02500001 0.42499674]
```

We have a dominant topic ($0.85 t_0$) for the first document and a mixture ($0.22 t_0 + 0.22 t_2 + 0.42 t_7$) for the second one. Now let's consider the concatenation of both documents:

```
test_doc = corpus[0] + ' ' + corpus[2500]
y_test = lda.transform(cv.transform([test_doc]))

print(y_test)
[[ 0.61242771 0.01250001 0.11251451 0.0125011 0.01250001 0.01250278
   0.01251778 0.21253611]]
```

In the resultant document, as expected, the mixture has changed: $0.61t_0 + 0.11t_2 + 0.21t_7$. In other words, the algorithm introduced the previously dominant topic 5 (which is now stronger) by weakening both topic 2 and topic 7. This is reasonable, because the length of the first document is less than the second one, and therefore topic 5 cannot completely cancel the other topics out.

Introducing Word2vec with Gensim

One of the most common problems in NLP and topic modeling is represented by the semantic-free structure of the Bag-of-Words strategy. In fact, as discussed in the previous chapter, `Chapter 13`, *Introducing Natural Language Processing*, this strategy is based on frequency counts and doesn't take into account the positions and the similarity of the tokens. The problem can be partially mitigated by employing n-grams; however, it's still impossible to detect the contextual similarity of words. For example, let's suppose that a corpus contains the sentences *John lives in Paris* and *Mark lives in Rome*. If we perform a **Part-of-Speech** (**POS**) and **Named Entity Recognition** (**NER**) tagging, we can discover that *John* and *Mark* are proper nouns and *Paris* and *Rome* are cities. Hence, we can deduce that the two sentences share the same structure; *Paris* and *Rome* must share a similarity (at least, limited to a specific context).

The technique we're going to present (called **Word2vec**) has been designed to overcome this problem, providing us with feature vectors ($w_i \in \Re^m$ where m is the dimensionality) representing every word, so that their distance is proportional to their dissimilarity. Let's consider the following diagram:

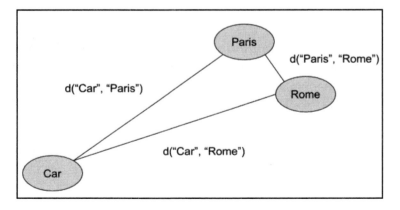

Feature vectors representing three words and their relationships

In this case, we are analyzing three words: *Paris*, *Rome*, and *Car* (the blocks represent the corresponding feature vectors). If we add another sentence to our corpus (for example, *A car is a vehicle*), we expect the following:

$$\begin{cases} d(Paris, Rome) \approx 0 \\ d(Car, Paris) \gg d(Paris, Rome) \text{ and } d(Car, Rome) \gg d(Paris, Rome) \end{cases}$$

In other words, the contextual semantics must force the cities to be rather closer than the term *car*, which appears in a completely different context. Of course, this is not an extremely accurate way to determine the semantics, but it can improve dramatically the performances of many models. Clearly, the price is the extra space needed to store the feature vectors (which can be quite long, for example, $m > 200$), but, on the other hand, can reduce the computational cost when determining the similarity of terms. A full description of the algorithm (which is indeed not very complex) requires some deep learning knowledge, and, as this is an introductory book, we've preferred to show directly some examples using the Gensim framework (see the info box at the end of the section for further information). The reader who's interested in the details can read the original paper *Distributed Representations of Words and Phrases and their Compositionality, Mikolov T, Sutskever I, Chen K, Corrado G., Dean J., arXiv:1310.4546 [cs.CL]* and the website `https://code.google.com/archive/p/word2vec/`.

We can now test a Word2vec model using the **Natural Language Toolkit** (**NLTK**) built-in Brown Corpus. The first step is loading the dataset, setting all words as lowercase (otherwise, the two versions can be considered different tokens) and removing the stopwords:

```
from nltk.corpus import brown
from nltk.corpus import stopwords

brown_corpus = brown.sents()

sw = set(stopwords.words('english'))

corpus = []

for sent in brown_corpus:
    c_sent = [w.strip().lower() for w in sent if w.strip().lower() not in sw]
    corpus.append(c_sent)
```

At this point, we can instantiate the Gensim `Word2Vec` class, setting `size=300` (feature vector dimensionality), and `window=10` (this parameter is strictly related to the underlying algorithm; however, it determines the number of surrounding words to consider when predicting the target). For our purposes, we can say that small windows restrict the context, while large ones can create wrong semantic relationships. I suggest to start with the default value (`window=5`) and to increase it only if the final accuracy is lower than expected. The `min_count` parameter determines the minimum number of occurrences of a token to take it into account, while `workers` is helpful to exploit the multiprocessing features of modern CPUs (we are going to use all available cores):

```
import multiprocessing

from gensim.models import Word2Vec

model = Word2Vec(corpus, size=300, window=10, min_count=1,
workers=multiprocessing.cpu_count())
wv = model.wv
del model
```

Once the model has been trained, it's a good idea to copy the word vectors and delete the instance to save memory. We can now check the features of this algorithm. Let's start analyzing the structure of a vector:

```
print(wv['committee'])

[  3.43790025e-01   1.70713723e-01   4.63349819e-02  -3.11405450e-01
   2.85413533e-01   4.22946483e-01   7.18410164e-02   6.64607957e-02
   4.88715507e-02  -2.26669595e-01  -1.02209471e-01  -3.95602554e-01
   4.93697792e-01   3.61298062e-02  -1.56762660e-01   1.78436086e-01
   1.88913181e-01  -2.47268111e-01  -3.87201369e-01  -2.34532371e-01
   5.29331207e-01  -5.41749746e-02  -1.57853425e-01  -2.29428243e-02
...
```

As you can see, each word contained in the dictionary is now represented as a 300-dimensional real vector. These values can be fed into complex models (for example, **Deep Neural Networks (DNN)** or **Support Vector Machines (SVMs)**), but Gensim offers also some very helpful utility functions. Let's suppose, for example, that we have found a word that has not been analyzed and we want to check the terms that are strictly related to it. We can obtain the list of the most similar tokens using the `most_similar()` function:

```
print(wv.most_similar('house'))

[('door', 0.9965364933013916), ('room', 0.9964739084243774), ('turned', 0.9958587884902954), ('left', 0.9955481886863708), ('walked', 0.9954644441604614), ('plunged', 0.9951649904251099), ('corridor', 0.9951382875442505), ('side', 0.9950708150863647), ('open', 0.9949836134910583), ('deduce', 0.993992805480957)]
```

In this case, we haven't found synonyms, but a set of terms that share the same contextual semantics of the query word (`'house'`). In general, the result of such queries depends on the corpus and can become extremely accurate when it is built to fulfill specific requirements. For example, a business chatbot can be trained using FAQs, so it is able to immediately associate a question to a potential matching answer. Considering our example, a furniture company can have a knowledge base with many questions containing the word `'door'` associated to `'house'`; hence, when the latter term is detected, an automated system can show some suggestions based on the terms `'door'` and `'room'`. Another problem is determining the similarity between words:

```
print(wv.similarity('committee', 'president'))
0.967507
```

The `similarity()` function computes the inverse distance of the corresponding feature vectors. In this case, for example, we know that, given the Brown Corpus, the words `'committee'` and `'president'` are strictly related because they appeared in the same context in many sentences. It's important to remember that Word2vec works with a local semantic model; therefore, the similarity could sometimes appear to be *weird*. I invite the reader to repeat the exercise with different corpora (possibly very targeted ones) and visualize a subset of the word vectors using **t-Distributed Stochastic Neighbor Embedding (t-SNE)**.

 Gensim (https://radimrehurek.com/gensim/) is an optimized Python framework for advanced NLP, topic modeling, and word embedding. It can be installed using the standard `pip install -U gensim` command. For further details and a complete documentation, I suggest checking the official website.

Sentiment analysis

One the most widespread applications of NLP is **sentiment analysis** of short texts (tweets, posts, comments, reviews, and so on). From a marketing viewpoint, it's very important to understand the semantics of these pieces of information, in terms of the sentiment expressed. As you can understand, this task can be very easy when the comment is precise and contains only a set of positive/negative words, but it becomes more complex when in the same sentence there are different propositions that can conflict with each other. For example, *I loved that hotel. It was a wonderful experience* is clearly a positive comment, while *The hotel is good; however, the restaurant was bad, and, even if the waiters were kind, I had to fight with a receptionist to have another pillow*. In this case, the situation is more difficult to manage, because there are both positive and negative elements, resulting in a neutral review. For this reason, many applications aren't based on a binary decision but admit intermediate levels (at least one to express the neutrality).

These kind of problems are normally supervised (as we're going to do), but there are also cheaper and more complex solutions. The simplest way to evaluate the sentiment is to look for particular keywords. This dictionary-based approach is fast and, together with a good stemmer, can immediately mark positive and negative documents. On the flip side, it doesn't consider the relationship between terms and cannot learn how to weight the different components. For example, *Lovely day; bad mood* will result in a neutral *(+1, -1)*, while with a supervised approach, it's possible to make the model learn that *mood* is very important and *bad mood* will normally drive to a negative sentiment. Other approaches (much more complex) are based on topic modeling (you can now understand how to apply LSA or LDA to determine the underlying topics in terms of positivity or negativity); however, they need further steps to use topic-word and topic-document distributions. It can be helpful in the real semantics of a comment, where, for example, a positive adjective is normally used together with other similar components (such as verbs). For example, *Lovely hotel; I'm surely coming back*. In this case (if the number of samples is big enough), a topic can emerge from the combination of words, such as *lovely* or *amazing,* and (positive) verbs, such as *returning* or *coming back*.

An alternative is to consider the topic distribution of positive and negative documents and work with a supervised approach in the topic subspace. Other approaches include deep learning techniques (such as Word2vec or fastText) and are based on the idea of generating a vectorial space where similar words are close to one another, to easily manage synonyms. For example, as explained in the previous section, if the training set contains the sentence *Lovely hotel,* but it doesn't contain *Wonderful hotel,* a Word2vec model can learn from other examples that *lovely* and *wonderful* are very close. Therefore, the new document *Wonderful hotel* is immediately classified using the knowledge provided by the first comment.

Let's now consider our example, which is based on a subset of the *Twitter Sentiment Analysis Training Corpus* dataset. To speed up the process, we have limited the experiment to 100,000 tweets. After downloading the file (see the box at the end of this paragraph), it's necessary to parse it (using the UTF-8 encoding):

```
dataset = 'dataset.csv'

corpus = []
labels = []

with open(dataset, 'r', encoding='utf-8') as df:
    for i, line in enumerate(df):
        if i == 0:
            continue

        parts = line.strip().split(',')
        labels.append(float(parts[1].strip()))
        corpus.append(parts[3].strip())
```

The `dataset` variable must contain the full path to the CSV file. This procedure reads all the lines, skipping the first one (which is the header), and stores each tweet as a new list entry in the `corpus` variable, and the corresponding sentiment (which is binary, 0 or 1) in the `labels` variable. At this point, we proceed as usual, tokenizing, vectorizing, and preparing the training and test sets:

```
from nltk.tokenize import RegexpTokenizer
from nltk.corpus import stopwords
from nltk.stem.lancaster import LancasterStemmer

from sklearn.feature_extraction.text import TfidfVectorizer
from sklearn.model_selection import train_test_split

rt = RegexpTokenizer('[a-zA-Z0-9\.]+')
ls = LancasterStemmer()
sw = set(stopwords.words('english'))
```

```
def tokenizer(sentence):
    tokens = rt.tokenize(sentence)
    return [ls.stem(t.lower()) for t in tokens if t not in sw]

tfv = TfidfVectorizer(tokenizer=tokenizer, sublinear_tf=True,
ngram_range=(1, 2), norm='l2')
X = tfv.fit_transform(corpus[0:100000])
Y = np.array(labels[0:100000])

X_train, X_test, Y_train, Y_test = train_test_split(X, Y, test_size=0.1,
random_state=1000)
```

We have chosen to include dots, together with letters and numbers in the `RegexpTokenizer` instance, because they are useful for expressing particular emotions. Moreover, the `ngram_range` has been set to `(1, 2)`, so we include bigrams (the reader can try with trigrams too). At this point, we can train a random forest:

```
from sklearn.ensemble import RandomForestClassifier

import multiprocessing

rf = RandomForestClassifier(n_estimators=20,
n_jobs=multiprocessing.cpu_count())
rf.fit(X_train, Y_train)
```

Now we can produce some metrics to evaluate the model:

```
from sklearn.metrics import precision_score, recall_score

print('Precision: %.3f' % precision_score(Y_test, rf.predict(X_test)))
Precision: 0.720

print('Recall: %.3f' % recall_score(Y_test, rf.predict(X_test)))
Recall: 0.784
```

The performances are not excellent (it's possible to achieve better accuracies using Word2vec). However, they are acceptable for many tasks. In particular, a 78% recall means that the number of false negatives is about 20%, and it can be useful when using sentiment analysis for an automatic processing task (in many cases, the risk threshold to auto-publish a negative review is quite a bit lower, and, therefore, a better solution must be employed). The performances can be also confirmed by the corresponding ROC curve:

Chapter 14

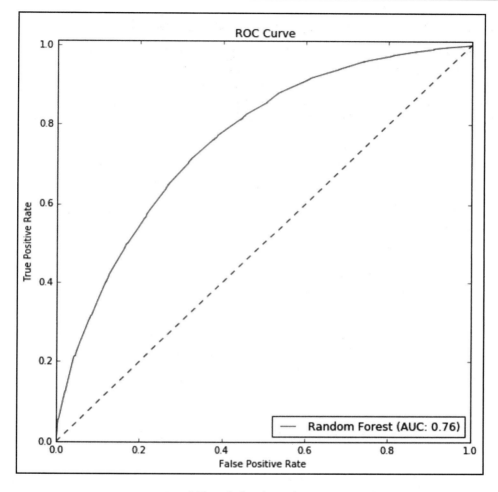

ROC curve for the sentiment analyzer

 The *Twitter Sentiment Analysis Training Corpus* dataset (as a CSV file) used in the example can be downloaded from http://thinknook.com/wp-content/uploads/2012/09/Sentiment-Analysis-Dataset.zip. Considering the amount of data, the training process can be very long (even taking hours on slower machines).

VADER sentiment analysis with NLTK

For the English language, NLTK provides an already trained model called **Valence Aware Dictionary and sEntiment Reasoner** (**VADER**) which works in a slightly different way and adopts a rule engine together with a lexicon to infer the sentiment intensity of a piece of text. More information and details can be found in *VADER: A Parsimonious Rule-based Model for Sentiment Analysis of Social Media Text, Hutto C J, Gilbert E, AAAI, 2014*.

The NLTK version uses the `SentimentIntensityAnalyzer` class and can immediately be used to have a polarity sentiment measure made up of four components:

- Positive factor
- Negative factor
- Neutral factor
- Compound factor

The first three don't need any explanation, while the last one is a particular measure (a normalized overall score), which is computed as follows:

$$Compound = \frac{\sum_i Sentiment(w_i)}{\sqrt{(\sum_i Sentiment(w_i))^2 + \alpha}}$$

Here, $Sentiment(w_i)$ is the score valence of the word w_i and α is a normalization coefficient that should approximate the maximum expected value (the default value set in NLTK is 15). The usage of this class is immediate, as the following snippet can confirm:

```
from nltk.sentiment.vader import SentimentIntensityAnalyzer

text = 'This is a very interesting and quite powerful sentiment analyzer'

vader = SentimentIntensityAnalyzer()
print(vader.polarity_scores(text))
{'neg': 0.0, 'neu': 0.535, 'pos': 0.465, 'compound': 0.7258}
```

The NLTK VADER implementation uses the Twython library for some functionalities. Even though it's not necessary, to avoid a warning, it's possible to install it using `pip` (`pip install twython`).

Summary

In this chapter, we introduced topic modeling. We discussed latent semantic analysis based on truncated SVD, PLSA (which aims to build a model without assumptions about latent factor prior probabilities), and LDA, which outperformed the previous method and is based on the assumption that the latent factor has a sparse prior Dirichlet distribution. This means that a document normally covers only a limited number of topics and a topic is characterized by only a few important words.

In the last section, we discussed the basics of Word2vec and the sentiment analysis of documents, which is aimed at determining whether a piece of text expresses a positive or negative feeling. To show a feasible solution, we built a classifier based on an NLP pipeline and a random forest with average performances that can be used in many real-life situations.

In the next chapter, `Chapter 15`, *Introducing Neural Networks,* we're going to briefly introduce deep learning, together with the TensorFlow framework. As this topic alone requires a dedicated book, our goal is to define the main concepts, with some practical examples. If the reader wants to have further information, at the end of the chapter, a complete reference list will be provided.

15
Introducing Neural Networks

In this chapter, I'm going to briefly introduce deep learning with some examples based on Keras. This topic is quite complex and needs dedicated books; however, my goal is to allow the reader to understand some basic concepts that can be helpful before starting a complete course. In the first section, I'm presenting the structure of artificial neural networks and how they can be transformed in a complex computational graph with several different layers. In the second one, I'm going to introduce the basic concepts of Keras, and we'll see an example based on a very famous test dataset.

In particular, we are going to discuss the following:

- Structure of an artificial neuron
- **Multi-layer Perceptrons** (**MLP**)
- The back propagation algorithm
- How to build and train an MLP with Keras
- How to interface Keras with scikit-learn

Deep learning at a glance

Deep learning has become very famous over the last few decades, thanks to hundreds of applications that are changing the way we interact with many electronic (and non-electronic) systems. Speech, text, and image recognition; autonomous vehicles; and intelligent bots (just to name a few) are common applications normally based on deep learning models that have outperformed any previous classical approach. However, to better understand what a deep architecture is, we need to step back and talk about standard **artificial neural networks** (**ANNs**).

Artificial neural networks

An ANN or simply a neural network is a directed or recurrent computational structure that connects an input layer to an output one. Normally, all operations are differentiable, and the overall vectorial function can be easily written as follows:

$$\bar{y} = f(\bar{x})$$

The function transforms a vector into another one by applying the same operator element-wise. Therefore, we assume the following:

$$\bar{x} = (x_1, x_2, \ldots, x_n) \quad and \quad \bar{y} = (y_1, y_2, \ldots, y_m)$$

The adjective *neural* comes from two important elements: the internal structure of a basic computational unit and the interconnections among them. Let's start with the former. In the following diagram, there's a schematic representation of an artificial neuron:

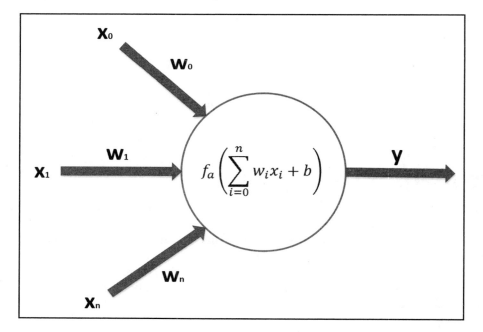

Generic structure of an artificial neuron

A neuron core is connected with **n** input channels, each of them characterized by a synaptic weight w_i. The input is split into its components, and they are multiplied by the corresponding weight and calculated. An optional bias can be added to this sum (it works like another weight connected to a unitary input). The resultant sum is filtered by an activation function f_a (for example, a sigmoid, if you recall how a logistic regression works), and the output is therefore produced. In Chapter 5, *Linear Classification Algorithms*, we also discussed perceptrons (the first ANNs), which correspond exactly to this architecture with a binary-step activation function. On the other hand, even a logistic regression can be represented as a single neuron neural network, where $f_a(x)$ is a sigmoid. The main problem with this architecture is that it's intrinsically linear because the output is always a function of the dot product between the input vector and the weight one. You already know all the limitations that such a system has; therefore, it's necessary to step forward and create the first MLP. In the following diagram, there's a schematic representation of an MLP with an n-dimensional input, p hidden neurons, and a k-dimensional output:

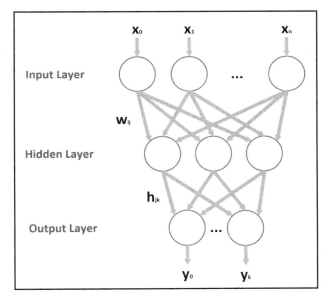

Structure of MLP with a single hidden layer

There are three layers (even though the number can be larger): the **input layer**, which receives the input vectors; a **hidden layer**; and the **output layer**, which is responsible for producing the output. As you can see, every neuron is connected to all the neurons belonging the next layer, and now we have two weight matrices, $W = (w_{ij})$ and $H = (h_{jk})$, using the convention that the first index is referred to the previous layer and the second to the following one.

Therefore, the net input to each hidden neuron and the corresponding output is as follows:

$$\begin{cases} z_j^{Input} = w_{0j}x_0 + w_{1j}x_1 + \ldots + w_{nj}x_n = \sum_i w_{ij}x_i \\ x_j^{Output} = f_a^{Hidden}\left(z_j^{Input} + b_j^{Hidden}\right) \end{cases}$$

In the same way, we can compute the network output:

$$\begin{cases} y_k^{Input} = h_{0k}z_0^{Output} + h_{1k}z_1^{Output} + \ldots + h_{pk}z_p^{Output} = \sum_j h_{jk}z_j^{Output} \\ y_k^{Output} = f_a^{Output}\left(y_k^{Input} + b_k^{Output}\right) \end{cases}$$

As you can see, the network has become highly non-linear, and this feature allows us to model complex scenarios that were impossible to manage with linear methods. But how can we determine the values for all synaptic weights and biases? The most famous algorithm is called **back propagation** and it works in a very simple way (the only important assumption is that both $f_a(x)$ must be differentiable).

First of all, we need to define an error (loss/cost) function. For many classification tasks, it can be the total (or mean) squared error:

$$L = \frac{1}{2}\sum_i \left\|\bar{y}_i^{Predicted} - \bar{y}_i^{Target}\right\|^2$$

Here, we have assumed to have N input samples. Expanding it, we obtain the following:

$$L = \frac{1}{2}\sum_i \sum_k \left[f_a^{Output}\left(\sum_j h_{jk}z_j^{Target}\right) - y_k^{Target}\right]^2 = \frac{1}{2}\sum_i\sum_k \delta_k^2$$

This function depends on all variables (weights and biases), but we can start from the bottom and consider first only h_{jk} (for simplicity, I'm not considering the biases as normal weights). Therefore, we can compute the gradients and update the weights:

$$\frac{\partial L}{\partial h_{jk}} = \sum_i \delta_k \frac{\partial f_a^{Output}}{\partial y_k^{Input}} \frac{\partial y_k^{Input}}{\partial h_{jk}} = \sum_i \delta_k z_j^{Output} \frac{\partial f_a^{Output}}{\partial y_k^{Input}} = \sum_i \alpha_j z_j^{Output}$$

In the same way, employing the chain rule of derivatives, we can derive the gradient with respect to w_{ij}:

$$\frac{\partial L}{\partial w_{ij}} = \sum_i \sum_k \delta_k \frac{\partial f_a^{Output}}{\partial y_k^{Input}} \frac{y_k^{Input}}{z_j^{Output}} \frac{\partial z_j^{Output}}{\partial z_j^{Input}} \frac{\partial z_j^{Input}}{\partial w_{ij}} =$$

$$= \sum_i \sum_k \delta_k h_{jk} x_i \frac{\partial f_a^{Output}}{\partial y_k^{Input}} \frac{\partial z_j^{Output}}{\partial z_j^{Input}} = \sum_i \sum_k \alpha_k h_{jk} x_i \frac{\partial z_j^{Output}}{\partial z_j^{Input}}$$

As you can see, the term α (which is proportional to the error δ) is back propagated from the output layer to the hidden one. If there are many hidden layers, this procedure should be repeated recursively until the first layer. The algorithm adopts the gradient descent method; therefore, it updates the weights iteratively until convergence:

$$\begin{cases} h_{jk}^{(t+1)} = h_{jk}^{(t)} - \eta \frac{\partial L}{\partial h_{jk}} \\ w_{ij}^{(t+1)} = w_{ij}^{(t)} - \eta \frac{\partial L}{\partial w_{ij}} \end{cases}$$

Here, the parameter η is the learning rate, which must be chosen to avoid too many changes in the presence of outliers. In fact, a large η forces the weights to move in the direction of the negative gradient with *big jumps* that can slow down the convergence toward the global minimum. Conversely, very small values avoid the problem of wrong corrections but require many more iterations. The optimal value depends on every single problem and must be selected with a grid search starting from a set of common choices (for example, 0.001, 0.01, and 0.1) and *zooming* into the range where the accuracy and convergence speed reach their optimal values.

In many real problems, the **stochastic gradient descent** (**SGD**) method is adopted (check out Chapter 5, *Linear Classification Algorithms*, for further information), which works with batches of input samples instead of considering the entire dataset. Moreover, many optimizations can be employed to speed up the convergence, but they are beyond the scope of this book. In *Mastering Machine Learning Algorithms, Bonaccorso G, Packt Publishing, 2018*, the reader can find all the details about the majority of them. For our purposes, it's important to know that we can build a complex network and, after defining a global loss function, optimize all the weights with a standard procedure (for example, SGD).

MLPs with Keras

Keras (https://keras.io) is a high-level deep learning framework that works seamlessly with low-level deep learning backends such as TensorFlow, Theano, and **Microsoft Cognitive Toolkit** (**CNTK**). In Keras, a model is like a sequence of layers where each output is fed into the following computational block until the final layer is reached and the cost function can be evaluated and differentiated.

The generic structure of a model is as follows:

```
from keras.models import Sequential

model = Sequential()

model.add(...)
model.add(...)
...
model.add(...)
```

The Sequential class defines a generic empty sequential model that already implements all the methods needed to add layers, compile the model according to the underlying framework (that is, transforming the high-level description into a set of commands compatible with the underlying backend), to fit and evaluate the model and to predict the output, given an input.

All the most common layers are already implemented (some of them will be explained in the next chapter, Chapter 16, *Advanced Deep Learning Models*), including the following:

- Dense (standard MLP layer), dropout, and flattening layers
- Convolutional (1D, 2D, and 3D) layers
- Pooling layers
- Zero padding layers
- **Recurrent Neural Network (RNN)** layers

A model can be compiled using several loss functions (such as **mean squared error** (**MSE**) or cross-entropy) and all the most diffused SGD optimization algorithms (such as **RMSProp** or **ADAM**). For further details about the mathematical foundation of these methods, please refer to *Mastering Machine Learning Algorithms, Bonaccorso G, Packt Publishing, 2018*.

Let's start this analysis with a concrete example of MLP based on a famous dataset created by nesting two spirals whose points (500 per spiral) belong to the same class:

```
import numpy as np

from sklearn.preprocessing import StandardScaler
from sklearn.utils import shuffle

nb_samples = 1000

X = np.zeros(shape=(nb_samples, 2), dtype=np.float32)
Y = np.zeros(shape=(nb_samples,), dtype=np.float32)

t = 15.0 * np.random.uniform(0.0, 1.0, size=(int(nb_samples / 2), 1))

X[0:int(nb_samples / 2), :] = t * np.hstack([-np.cos(t), np.sin(t)]) + \
                     np.random.uniform(0.0, 1.8,
size=(int(nb_samples / 2), 2))
Y[0:int(nb_samples / 2)] = 0

X[int(nb_samples / 2):, :] = t * np.hstack([np.cos(t), -np.sin(t)]) + \
                     np.random.uniform(0.0, 1.8,
size=(int(nb_samples / 2), 2))
Y[int(nb_samples / 2):] = 1

ss = StandardScaler()
X = ss.fit_transform(X)

X, Y = shuffle(X, Y, random_state=1000)
```

Introducing Neural Networks

Once generated, it's helpful to normalize the dataset using a `StandardScaler` class, to impose null mean and unit variance. A graphical representation of the dataset is shown in the following graph:

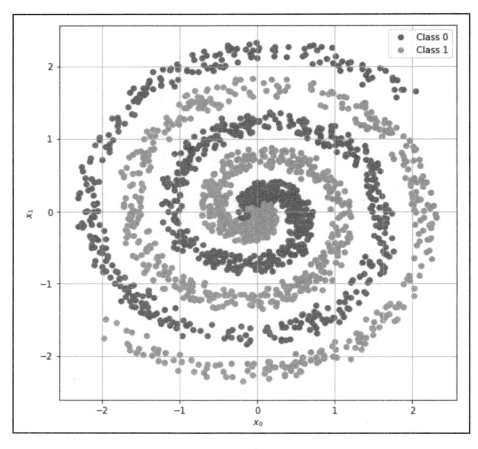

Graphical representation of the two-spirals dataset

As it's easy to understand, a linear separation is impossible, and even a more complex classifier needs a sufficient capacity to find a separation hypersurface, considering that the spirals are nested and the distance from the center becomes larger and larger. To better understand the behavior, let's try to employ a logistic regression:

```
import numpy as np

from sklearn.linear_model import LogisticRegression
from sklearn.model_selection import cross_val_score
```

```
lr = LogisticRegression(penalty='l2', C=0.01, random_state=1000)
print(np.mean(cross_val_score(lr, X, Y, cv=10)))
0.5694999999999999
```

The average cross-validation score is slightly higher than 0.5, which denotes a pure random guess (this value can be different because of the randomness; however, it can never overcome a maximum value, which is generally lower than 0.6). Such a result is not surprising considering that the separation is achieved with a line:

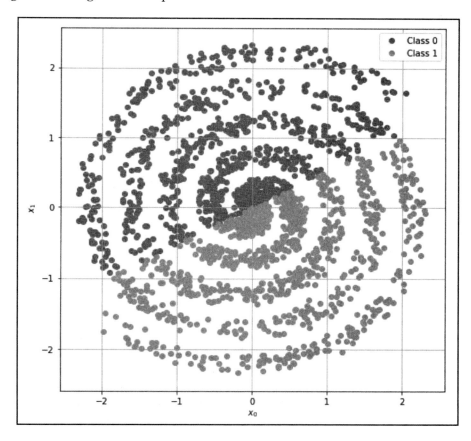

Classification result of a logistic regression

Introducing Neural Networks

It's obvious that there's no way to improve the performances of a linear model. Let's try to solve the problem using an MLP modeled with Keras. The first step consists in defining the structure:

```
from keras.models import Sequential
from keras.layers import Dense, Activation

model = Sequential()

model.add(Dense(64, input_dim=2))
model.add(Activation('relu'))

model.add(Dense(32))
model.add(Activation('relu'))

model.add(Dense(16))
model.add(Activation('relu'))

model.add(Dense(2))
model.add(Activation('softmax'))
```

After the declaration of a new `Sequential` model, we start adding the layers. In this case, we have chosen the configuration:

- Input layer with 64 neurons
- First hidden layer with 32 neurons
- Second hidden layer with 16 neurons
- Output layer with 2 neurons

All layers are instances of the `Dense` class, which represent a fully connected standard MLP layer. Keras requires the user to specify the input shape in the first layer (while it auto-detects it in the other ones). This can be done using the `input_shape` parameter (for example, `input_shape=(100, 100)` for 100 × 100 matrices) or, when the samples have a single axis, using the `input_dim` parameter. In both cases, the batch size must not be supplied and it is implicitly detected during the training process. As our dataset has a shape *(2000, 2)*, the only dimension of a sample is 2, and we have employed the second option.

The activation function (expressed as an additional layer based on an instance of the `Activation` class) of the first three layers is a very common one, called the **Rectified Linear Unit (ReLU)**, which is defined as follows:

$$f_{ReLU}(x) = max(0, x)$$

The main advantage of this function is that, even if it's non-linear, it has a constant gradient for $x > 0$. In this way, the convergence speed is not affected by the presence of a saturation, where the gradient becomes close to 0 (for example, sigmoid or a hyperbolic tangent). The output activation is instead based on a softmax function (which allows representing a probability distribution). With n values, this activation is defined as follows:

$$f_{Softmax} = \left(\frac{e^{z_1}}{\sum_{i=1}^{n} e^{z_i}}, \frac{e^{z_2}}{\sum_{i=1}^{n} e^{z_i}}, \ldots, \frac{e^{z_n}}{\sum_{i=1}^{n} e^{z_i}} \right)$$

The choice of such a function is very common in deep classifiers because it permits to manage multi-class problems using the cross-entropy loss (explained later). Once the model has been defined, it's necessary to compile it so that Keras can convert the description into a computational graph compatible with the underlying backend:

```
model.compile(optimizer='adam',
              loss='categorical_crossentropy',
              metrics=['accuracy'])
```

Keras will transform the high-level description into low-level operations, adding a 'categorical_crossentropy' loss (loss parameter) and the 'adam' optimizer (which is a very common choice in the majority of tasks). Moreover, it will apply an accuracy metric (metrics parameter) to dynamically evaluate the performance. Given a data-generating process, p_{data}, and a set of training samples drawn from it, the goal of the classifier is to find a distribution p_{mlp} whose Kullback–Leibler divergence with p_{data} is close to 0 (see Chapter 2, *Important Elements in Machine Learning*). The cross-entropy is defined as follows:

$$H(p_{data}, p_{mlp}) = -\sum_{\bar{x} \in X} p_{data}(\bar{x}) \log p_{mlp}(\bar{x}) = H(p_{data}) + D_{KL}(p_{data} || p_{mlp})$$

Hence, considering that $H(p_{data})$ is a constant, by minimizing the cross-entropy, we also minimize the Kullback–Leibler divergence and force the model to learn a distribution $p_{mlp} \approx p_{data}$.

Introducing Neural Networks

At this point, the model can be trained. We need only two preliminary operations:

- Splitting the dataset into train and test sets.
- Applying the one-hot encoding to the integer label. This operation is necessary because we are using a cross-entropy loss and the true labels are always in the form *(1, 0)* and *(0, 1)*, meaning that the probability is *1* respectively for the first and the second class.

The first operation can be performed using the scikit-learn `train_test_split()` function, while the second can be easily carried out using the Keras `to_categorical()` built-in function:

```
from sklearn.model_selection import train_test_split
from keras.utils import to_categorical

X_train, X_test, Y_train, Y_test = \
    train_test_split(X, to_categorical(Y), test_size=0.2,
random_state=1000)
```

We want to train with batches made up of `32` images and for a period of `100` epochs. The reader is free to change all these values, to compare the results. The process can be started using the `fit()` method, which also accepts a tuple containing the validation dataset:

```
model.fit(X_train, Y_train,
          epochs=100,
          batch_size=32,
          validation_data=(X_test, Y_test))
```

The output provided by Keras shows the progress in the learning phase:

```
Train on 1600 samples, validate on 400 samples
Epoch 1/100
1600/1600 [==============================] - 1s 687us/step - loss: 0.6839 - acc: 0.5406 - val_loss: 0.6707 - val_acc: 0.5925
Epoch 2/100
1600/1600 [==============================] - 0s 219us/step - loss: 0.6723 - acc: 0.5975 - val_loss: 0.6665 - val_acc: 0.6125
Epoch 3/100
1600/1600 [==============================] - 0s 273us/step - loss: 0.6668 - acc: 0.6138 - val_loss: 0.6612 - val_acc: 0.6150
Epoch 4/100
1600/1600 [==============================] - 0s 234us/step - loss: 0.6638 - acc: 0.6131 - val_loss: 0.6580 - val_acc: 0.6250
Epoch 5/100
1600/1600 [==============================] - 0s 244us/step - loss: 0.6596 - acc: 0.6238 - val_loss: 0.6523 - val_acc: 0.6300
...
```

```
Epoch 100/100
1600/1600 [==============================] - 0s 217us/step - loss: 0.0528 -
acc: 0.9838 - val_loss: 0.0569 - val_acc: 0.9800
```

At the end of the training process, the validation accuracy is about 98%, which means that the classifier is almost perfectly able to separate the two classes. For confirmation, let's predict the label for all points (as it is a probability, the final choice is made using the `argmax` operator):

```
Y_pred_mlp = np.argmax(model.predict(X), axis=1)
```

The resultant plot is shown in the following graph:

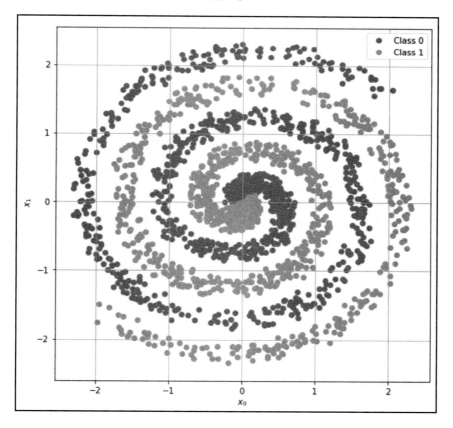

Classification result of a multilayer perceptron

Introducing Neural Networks

It's easy to see that the misclassifications regard only a few noisy points, while the majority of the samples are correctly assigned to the original class. To have further confirmation, we can plot the decision surface for both the logistic regression and the MLP. We can easily achieve this goal by creating a mesh grid bounded between -2.0 and 2.0 (both axes) and computing the predictions for all points:

```
import numpy as np

Xm = np.linspace(-2.0, 2.0, 1000)
Ym = np.linspace(-2.0, 2.0, 1000)
Xmg, Ymg = np.meshgrid(Xm, Ym)
X_eval = np.vstack([Xmg.ravel(), Ymg.ravel()]).T

Y_eval_lr = lr.predict(X_eval)
Y_eval_mlp = np.argmax(model.predict(X_eval), axis=1)
```

The resultant plots are shown in the following graph:

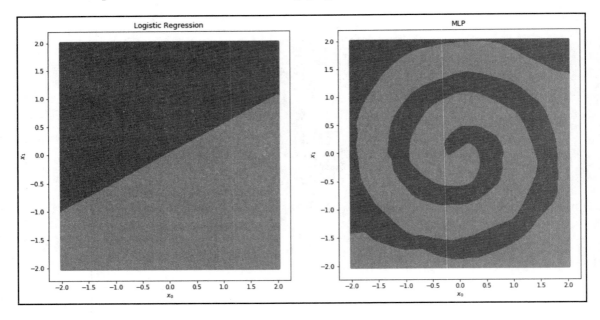

Decision surfaces for a logistic regression (left) and an MLP (right)

As expected, the logistic regression splits the plane into two sub-planes (always achieving an accuracy of about 50%), while the MLP has successfully learned the original distribution is also able to generalize with a minimum error. This simple example showed the power of deep architectures, using the most elementary building blocks (in the next chapter, Chapter 15, *Introducing Neural Networks*, we are going to analyze more complex examples). I invite the reader to repeat the exercise, using different activations (for example, hyperbolic tangent, larger or smaller batch sizes, and various numbers of layers).

> All the information needed to install Keras with different backends (with or without GPU support), and the official documentation can be found on the website: https://keras.io.

Interfacing Keras to scikit-learn

Even if Keras provides many support functions, in some cases, it's easier to perform specific operations with scikit-learn. For example, we could be interested in an automated cross-validation grid search; as we're going to discuss in Chapter 17, *Creating a Machine Learning Architecture*, it's helpful to create a complete processing pipeline that performs all the preliminary operations on the dataset and then train the model. To solve these problems, Keras provides two wrappers, KerasClassifier and KerasRegressor, that implement the standard scikit-learn interfaces (for classifiers and regressors) but work with Keras models.

To show how to use these wrappers, let's perform a grid search on the model previously defined. The first step is to define a function that must create and compile a Keras model:

```
from keras.models import Sequential
from keras.layers import Dense, Activation
from keras.optimizers import Adam

def build_model(lr=0.001):
 model = Sequential()

 model.add(Dense(64, input_dim=2))
 model.add(Activation('relu'))

 model.add(Dense(32))
 model.add(Activation('relu'))

 model.add(Dense(16))
 model.add(Activation('relu'))
```

```
model.add(Dense(2))
model.add(Activation('softmax'))

model.compile(optimizer=Adam(lr=lr),
              loss='categorical_crossentropy',
              metrics=['accuracy'])

return model
```

The `build_model()` function must accept all the parameters needed to build the model with a default value. In our case, for the optimizer, we have employed an instance of the `Adam` class, which allows setting a custom learning rate (a hyperparameter that we want to tune up). At this point, we can create a `KerasClassifier` instance:

```
from keras.wrappers.scikit_learn import KerasClassifier

skmodel = KerasClassifier(build_fn=build_model, epochs=100, batch_size=32, lr=0.001)
```

The first argument is the model-building function, while the other parameters are the default values needed to train the model. As the scikit-learn interface implements a `fit()` method that doesn't accept the number of epochs and the batch size, we need to supply their default values in the constructor. The wrapper will take care to convert a standard scikit-learn call to `fit()`, `score()`, and `predict()` functions into the corresponding Keras syntaxes.

The next step is setting up the grid search (in the same way we have done in the previous chapters):

```
from sklearn.model_selection import GridSearchCV
from keras.utils import to_categorical

parameters = {
 'lr': [0.001, 0.01, 0.1],
 'batch_size': [32, 64, 128]
}

gs = GridSearchCV(skmodel, parameters, cv=5)
gs.fit(X, to_categorical(Y, 2))
```

The grid search will check the optimal combination, considering three learning rate values and three batch sizes. As usual, at the end of the process (which can be quite long when the model is very complex), we can check the best score and parameter set:

```
print(gs.best_score_)
0.9815
```

```
print(gs.best_params_)
{'lr': 0.01, 'batch_size': 128}
```

Hence, the optimal learning rate is `0.01` with a batch size of `128` samples. In this case, the cross-validation accuracy is about `0.9815`, which is slightly better than the one obtained with our initial configuration. This is an example of a very common task, but the `skmodel` instance can be supplied to every scikit-learn function or class that requires a classifier/regressor. For example, it's possible to evaluate the cross-validation score using `cross_val_score()`, creating a learning curve using the `learning_curve()` function, or generating a random distribution of parameters using the `ParameterSampler` class. Moreover, in *Chapter 17*, *Creating a Machine Learning Architecture*, the reader will learn how to merge feature engineering methods, dimensionality reduction, and model training into a single block that can be optimized using a global grid search. I suggest testing these Keras functionalities with more hyperparameters, bearing in mind that the training time can become very long (hence, it's preferable to start with simple examples and, possibly, GPU support).

Summary

In this chapter, we have briefly discussed some basic deep learning concepts, and the reader should now understand what a computational sequential graph is and how it can be modeled using Keras. A deep architecture, in fact, can be seen as a sequence of layers connected to one another. They can have different characteristics and purposes, but the overall graph is always a directed structure that associates input values with a final output layer. Therefore, it's possible to derive a global loss function that will be optimized by a training algorithm.

We have presented Keras, which is a high-level framework that allows modeling and training complex deep-learning architectures. As an introductory example, we have shown the reader how to build an MLP that is able to solve the two spirals problem, and we have shown how to wrap a Keras model into a class that implements the standard scikit-learn classifier/regressor interface. In this way, it's possible to employ all scikit-learn functions and classes to perform grid search, cross-validations, and every other complex operation.

In the next chapter, we're going to discuss a few more advanced deep learning concepts, such as **convolutions** and **recurrent networks**, giving some examples based on Keras and TensorFlow.

16
Advanced Deep Learning Models

In this chapter, we're going to briefly discuss the most common deep learning layers, giving two examples based on Keras. The first one is a deep convolutional network employed to classify the MNIST dataset. The other one is an example of time-series processing using a recurrent network based on **Long Short-Term Memory** (**LSTM**) cells. We're also introducing the basic concepts of TensorFlow, giving some concrete examples based on algorithms already discussed in previous chapters.

In particular, we're going to discuss the following:

- Deep learning layers (convolutions, dropout, batch normalization, recurrent)
- An example of a deep convolutional network
- An example of a recurrent (LSTM-based) network
- A brief introduction to TensorFlow with examples of gradient computation, logistic regression, and convolution

Deep model layers

In this section, we're going to briefly discuss the most important layer types employed in deep learning architectures. Clearly, as this is an introductory book, we are not presenting all the mathematical details, but we are focusing the attention on the specific applications. Further details and theoretical foundations can be found in *Mastering Machine Learning Algorithms, Bonaccorso G, Packt Publishing, 2018*.

Fully connected layers

A **fully connected layer** (sometimes called a **dense layer**) is made up of n neurons, and each of them receives all the output values coming from the previous layer (such as the hidden layer in a **Multi-layer Perceptron (MLP)**). It can be characterized by a weight matrix, a bias vector, and an activation function:

$$\bar{y} = f\left(W\bar{x} + \bar{b}\right)$$

It's important to remember that an MLP must contain non-linear activations (for example, sigmoid, hyperbolic tangent, or ReLU). In fact, a network with n linear hidden layers is equivalent to a standard perceptron. In complex architectures, they are normally used as intermediate or output layers, in particular when it's necessary to represent a probability distribution. For example, a deep architecture could be employed for an image classification with m output classes. In this case, the *softmax* activation function allows having an output vector where each element is the probability of a class (and the sum of all outputs is always normalized to 1.0). In this case, the argument is considered as a *logit* or the logarithm of odds:

$$logit_i(p) = log\left(\frac{p}{1-p}\right) = W_i \bar{x} + b_i$$

W_i is the *i-th* row of W. The probability of a class y_i is obtained by applying the *softmax* function to each *logit*:

$$p(y_i) = Softmax\left(logit_i(p)\right) = \frac{e^{logit_i(p)}}{\sum_j e^{logit_j(p)}}$$

This type of output can easily be trained using a cross-entropy loss function, as already discussed in the previous chapter.

Chapter 16

Convolutional layers

Convolutional layers are normally applied to bidimensional inputs (even though they can be used for vectors and 3D matrices), and they became particularly famous thanks to their extraordinary performance in image classification tasks. They are based on the discrete convolution of a small kernel k with a bidimensional input (which can be the output of another convolutional layer):

$$(k * Y) = Z(i, j) = \sum_m \sum_n k(m, n) Y(i - m, j - n)$$

A layer is normally made up of n fixed-size kernels, whose values are weights that must be optimized using a back propagation algorithm. A convolutional architecture, in most cases, starts with layers with a few larger kernels (for example, 16 (8 x 8) matrices) and feeds their output to other layers with a higher number of smaller kernels (32 (5 x 5), 128 (4 x 4), and 256 (3 x 3)). In this way, the first layers should learn to capture more generic features (such as orientation), while the following ones will be trained to capture smaller and smaller elements (such as the position of the eyes, nose, and mouth in a face). The output of the last convolutional layer is normally flattened (transformed into a 1D vector) and used as input for one or more fully connected layers.

In the following diagram, there's a schematic representation of a convolution over a picture:

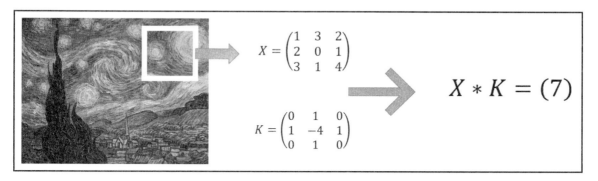

Example of a convolution with a 3 × 3 kernel

Each square set of 3 x 3 pixels is convoluted with a Laplacian kernel and transformed into a single value, which corresponds to the sum of upper, lower, left, and right pixels (considering the center) minus four times the central one. We're going to see a complete example using this kernel in the following section.

To reduce the complexity when the number of convolutions is very high, one or more **pooling layers** can be employed. Their task is to transform each group of input points (pixels in an image) into a single value using a predefined strategy. The most common pooling layers are as follows:

- **Max pooling**: Every bidimensional group of ($m \times n$) pixels is transformed into a single pixel whose value is the greatest in the group
- **Average pooling**: Every bidimensional group of ($m \times n$) pixels is transformed into a single pixel whose value is the average of the group

In this way, the dimensionality of the original matrix can be reduced with a loss of information, but that can often be discarded (in particular, in the first layers, where the granularity of the features is coarse). Moreover, pooling layers provide a medium robustness to small translations, increasing the generalization ability of the network.

Another important category of layers are the **zero-padding** ones. They work by adding null values (0) before and after the input (1D) or at the left, right, top, and bottom side of the 2D input.

Dropout layers

A dropout layer is used to prevent the overfitting of the network by randomly setting a fixed number of input elements to zero (generally, this is achieved by setting the probability of elements to nullify). This layer is adopted during the training phase, but it's normally deactivated during the test, validation, and production phases. Dropout networks can exploit higher learning rates, moving in different directions on the loss surface (setting to zero a few random input values in the hidden layers is equivalent to training different sub-models) and excluding all the error-surface areas that don't lead to a consistent optimization. Dropout is very useful in very big models, where it increases the overall performance and reduces the risk of freezing some weights and overfitting the model. A simple explanation of the behavior is based on the idea that, given the randomness of the dropout, every batch will be used to train a specific subnetwork, which has a lower capacity (and so it's less prone to overfitting). As the process is repeated many times, the overall network is forced to adapt its behavior to the overlap of the single components, which can become specialized on a specific subset of training samples. In this way, the negative effect of large training rates is mitigated, and the global network cannot easily overfit by increasing its variance.

Batch normalization layers

When the network is very deep, it's possible to observe a progressive modification in the mean and standard deviation of the batches throughout the network. This phenomenon is called **covariate shift** and is mainly responsible for a performance loss in terms of training speed. A batch normalization layer is responsible for correcting the statistical parameters of each batch and is normally inserted after a standard layer. Analogously to dropout, batch normalization operates only during the training phase, but, in this case, the model will apply a normalization computed over all samples during the prediction phase. It has been observed that, as a secondary effect, the batch normalization layers provide a regularization effect that prevents the model from overfitting (or at least it reduces the effect). For this reason, they are often used instead of dropout layers, so as to exploit the full capacity of the model while improving the convergence speed.

Recurrent Neural Networks

A recurrent layer is made up of particular neurons that present recurrent connections so as to bind the state at time t to its previous values (in general, only one). This category of computational cells is particularly useful when it's necessary to capture the temporal dynamics of an input sequence. In many situations, in fact, we expect an output value that must be correlated with the history of the corresponding inputs. But an MLP, as well as the other models that we've discussed, are stateless. Therefore, their output is determined only by the current input. **Recurrent Neural Networks** (**RNNs**) overcome this problem by providing an internal memory that can capture short-term and long-term dependencies.

The most common cells are LSTM and **Gated Recurrent Unit** (**GRU**), and they can both be trained using a standard back propagation approach. As this is only an introduction, I cannot go deeper (RNN mathematical complexity is non-trivial). However, it's useful to remember that whenever a temporal dimension must be included in a deep model, RNNs offer stable and powerful support. A full mathematical description of both LSTM and GRU is available in *Mastering Machine Learning Algorithms, Bonaccorso G, Packt Publishing, 2018*.

An example of a deep convolutional network with Keras

In this example, we are going to implement a complete deep convolutional network using Keras and the original MNIST handwritten digits dataset (available through a Keras utility function). It is made up of 60,000 grayscale 28 × 28 images for training and 10,000 for testing the model. An example is shown in the following screenshot:

Samples extracted from the original MNIST dataset

The first step consists in loading and normalizing the dataset so that each sample contains values bounded between 0 and 1:

```
from keras.datasets import mnist

(X_train, Y_train), (X_test, Y_test) = mnist.load_data()

width = height = X_train.shape[1]

X_train = X_train.reshape((X_train.shape[0], width, height,
1)).astype(np.float32) / 255.0
X_test = X_test.reshape((X_test.shape[0], width, height,
1)).astype(np.float32) / 255.0
```

The labels can be obtained using the `to_categorical()` function, so to train the model with a categorical cross-entropy loss:

```
from keras.utils import to_categorical

Y_train = to_categorical(Y_train, num_classes=10)
Y_test = to_categorical(Y_test, num_classes=10)
```

The network is based on the following configuration:

- Dropout with $p=0.25$
- 2D convolution with 16 (3 × 3) kernels and ReLU activation
- Dropout with $p=0.5$
- 2D convolution with 16 (3 × 3) kernels and ReLU activation
- Dropout with $p=0.5$
- 2D average pooling with areas of (2 × 2) pixels

- 2D convolution with 32 (3 × 3) kernels and ReLU activation
- 2D average pooling with areas of (2 × 2) pixels
- 2D convolution with 64 (3 × 3) kernels and ReLU activation
- Dropout with *p=0.5*
- 2D average pooling with areas of (2 × 2) pixels
- Flattening layer (needed to transform the multi-dimensional output into a 1-dimensional vector)
- Dense layer with 512 ReLU neurons
- Dropout with *p=0.5*
- Dense layer with ten softmax neurons

At this point, we can implement the model:

```
from keras.models import Sequential
from keras.layers import Dense, Activation, Dropout, Conv2D, 
AveragePooling2D, Flatten

model = Sequential()

model.add(Dropout(0.25, input_shape=(width, height, 1), seed=1000))

model.add(Conv2D(16, kernel_size=(3, 3), padding='same'))
model.add(Activation('relu'))
model.add(Dropout(0.5, seed=1000))

model.add(Conv2D(16, kernel_size=(3, 3), padding='same'))
model.add(Activation('relu'))
model.add(Dropout(0.5, seed=1000))

model.add(AveragePooling2D(pool_size=(2, 2), padding='same'))

model.add(Conv2D(32, kernel_size=(3, 3), padding='same'))
model.add(Activation('relu'))

model.add(AveragePooling2D(pool_size=(2, 2), padding='same'))

model.add(Conv2D(64, kernel_size=(3, 3), padding='same'))
model.add(Activation('relu'))
model.add(Dropout(0.5, seed=1000))

model.add(AveragePooling2D(pool_size=(2, 2), padding='same'))

model.add(Flatten())

model.add(Dense(512))
```

```
model.add(Activation('relu'))
model.add(Dropout(0.5, seed=1000))

model.add(Dense(10))
model.add(Activation('softmax'))
```

The next step is, as usual, compiling the model. We have chosen the `Adam` optimizer with a learning rate `lr=0.001`. The `decay` parameter is responsible for a progressive reduction (per epoch) of the learning rate according to the following formula:

$$\eta^{(t+1)} = \frac{\eta^{(t)}}{1 + decay}$$

In this way, during the first iterations, the weight updates are stronger, while, when approaching the minimum, they become smaller and smaller, so that more precise adaptations (fine-tuning) are possible:

```
from keras.optimizers import Adam

model.compile(optimizer=Adam(lr=0.001, decay=1e-5),
              loss='categorical_crossentropy',
              metrics=['accuracy'])
```

Once the model has been compiled, it's possible to start the training process (we have decided to set `epochs=200` and `batch_size=256`):

```
history = model.fit(X_train, Y_train,
                    epochs=200,
                    batch_size=256,
                    validation_data=(X_test, Y_test))

Train on 60000 samples, validate on 10000 samples
Epoch 1/200
60000/60000 [==============================] - 18s 301us/step - loss: 0.5601 - acc: 0.8163 - val_loss: 0.1253 - val_acc: 0.9624
Epoch 2/200
60000/60000 [==============================] - 15s 254us/step - loss: 0.1846 - acc: 0.9422 - val_loss: 0.0804 - val_acc: 0.9769
Epoch 3/200
60000/60000 [==============================] - 16s 265us/step - loss: 0.1392 - acc: 0.9565 - val_loss: 0.0618 - val_acc: 0.9818
Epoch 4/200
60000/60000 [==============================] - 16s 272us/step - loss: 0.1160 - acc: 0.9645 - val_loss: 0.0484 - val_acc: 0.9848
Epoch 5/200
```

```
60000/60000 [==============================] - 17s 281us/step - loss:
0.1063 - acc: 0.9665 - val_loss: 0.0480 - val_acc: 0.9857

...

Epoch 200/200
60000/60000 [==============================] - 29s 484us/step - loss:
0.0783 - acc: 0.9923 - val_loss: 0.0141 - val_acc: 0.9943
```

The final validation accuracy is about 99.5%, which is absolutely acceptable considering that many samples are quite similar (for example, many handwritten 1s can be misclassified as 7s, even by human beings). The final validation loss confirms that the learned distribution is almost identical to the data-generating process and, assuming that the test samples are real digits written by different people, that the model is able to generalize correctly.

In this example, we have collected the output of the `fit()` function, which is a Keras `History` class instance, containing the following fields computed at each epoch:

- `history['acc']` and `history['val_acc']`, which are respectively the training and validation accuracies
- `history['loss']` and `history['val_loss']`, which are respectively the training and validation losses

The corresponding plots are shown in the following diagram:

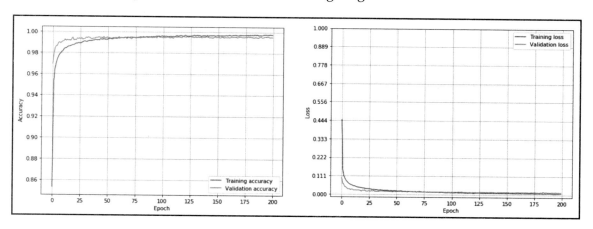

Accuracy plot (left), loss plot (right)

Advanced Deep Learning Models

As you can see, the model reaches its maximum accuracy very early, and it's not necessary to reach 200 epochs. However, the dropout layers have prevented the model to overfit; in fact, the loss functions continue to decrease and, in the end, remain almost flat. I invite the reader to repeat the exercise with different configurations (for example, fewer or more layers and tune up both the batch size and number of epochs, to maximize the final accuracy and minimize the training time). I also suggest to remove the pooling layers and observe the behavior. How is the number of parameters affected by this choice? Another exercise is to replace the dropout layers with `BatchNormalization()` layers and to observe the final effect. In this case, the batch normalization layers must be inserted between the convolutional/dense layer and its activation.

An example of an LSTM network with Keras

Even if we haven't analyzed in detail the internal dynamics of LSTM cells, we want to present a simple example of a time-series forecast using this kind of model. For this task, we have chosen a dataset of average Earth temperature anomalies (collected every month) provided by the **Global Component of Climate at a Glance (GCAG)** and available through DataHub (`https://datahub.io`).

It is possible to download the CSV files directly from `https://datahub.io/core/global-temp`; however, I suggest installing the Python package through the `pip -U install datapackage` command and using the API (as shown in the example) to get all the available datasets.

The first step is downloading and preparing the dataset:

```
from datapackage import Package

package = Package('https://datahub.io/core/global-temp/datapackage.json')

for resource in package.resources:
    if resource.descriptor['datahub']['type'] == 'derived/csv':
        data = resource.read()

data_gcag = data[0:len(data):2][::-1]
```

As the dataset contains two interleaved series (collected by different organizations), we have selected only the first one and reversed it because it's sorted by date in descending order. Each element of the `data_gcag` list contains three values: the source, the timestamp, and the actual temperature anomaly. We are interested only in the last one, so we need to extract the column. Moreover, as the values range between -0.75 and 1.25, it's preferable to normalize them in the interval `(-1.0, 1.0)` using the `MinMaxScaler` class (it's always possible to obtain the original values using the `inverse_transform()` method):

```
from sklearn.preprocessing import MinMaxScaler

Y = np.zeros(shape=(len(data_gcag), 1), dtype=np.float32)

for i, y in enumerate(data_gcag):
    Y[i - 1, 0] = y[2]

mmscaler = MinMaxScaler((-1.0, 1.0))
Y = mmscaler.fit_transform(Y)
```

A plot of the resultant time-series (containing 1,644 samples) is shown in the following graph:

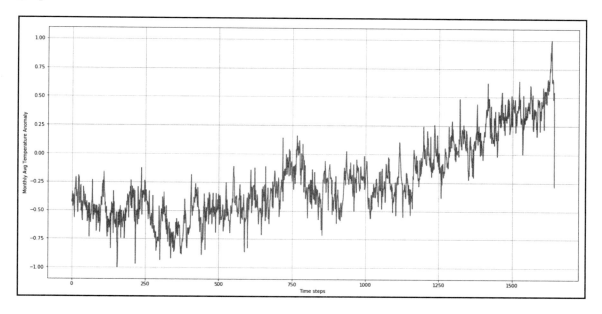

Complete plot of the time-series

Advanced Deep Learning Models

As it's possible to observe, the time-series shows a seasonality, very high-frequency small oscillations, and a trend starting from about the time step **750**. Our goal is to train a model that is able to take 20 input samples and predict the subsequent one. To train the model, we need to split the sequence into a list of fixed-length sliding blocks:

```
import numpy as np

nb_samples = 1600
nb_test_samples = 200
sequence_length = 20

X_ts = np.zeros(shape=(nb_samples - sequence_length, sequence_length, 1), dtype=np.float32)
Y_ts = np.zeros(shape=(nb_samples - sequence_length, 1), dtype=np.float32)

for i in range(0, nb_samples - sequence_length):
    X_ts[i] = Y[i:i + sequence_length]
    Y_ts[i] = Y[i + sequence_length]

X_ts_train = X_ts[0:nb_samples - nb_test_samples, :]
Y_ts_train = Y_ts[0:nb_samples - nb_test_samples]

X_ts_test = X_ts[nb_samples - nb_test_samples:, :]
Y_ts_test = Y_ts[nb_samples - nb_test_samples:]
```

We have limited the sequence to 1600 samples (1,400 for training and the last 200 for validation). As we're working with a time-series, we avoid shuffling the data in order also to exploit the internal memory of the LSTM blocks throughout the whole training process. In this way, we expect to model more efficiently both the short-term and the long-term dependencies. Considering the didactic purpose of this example, we are going to create a network with a single LSTM layer containing four cells and an output linear neuron:

```
from keras.models import Sequential
from keras.layers import LSTM, Dense, Activation

model = Sequential()

model.add(LSTM(8, stateful=True, batch_input_shape=(20, sequence_length, 1)))

model.add(Dense(1))
model.add(Activation('linear'))
```

We have set the `stateful=True` attribute because we want to force the LSTM to use the internal state (representing the memory) corresponding to the last sample of a batch as the initial state for the subsequent batch. With this choice, we must provide a `batch_input_shape` where the batch size is explicitly declared as the first element of the tuple (the remaining part is analogous to the content of `input_shape`). In this case, we have chosen to have a batch size of 20 blocks. We can compile the model using a **mean squared error** (**MSE**) loss function (`loss='mse'`) and `Adam` optimizer:

```
from keras.optimizers import Adam

model.compile(optimizer=Adam(lr=0.001, decay=0.0001),
              loss='mse',
              metrics=['mse'])
```

As a metric function, we have chosen the same loss function; hence, it's also possible to remove it. The next step is starting the training process, with `epochs=100`, `batch_size=20`, and `shuffle=False` (which is necessary when the network is stateful):

```
model.fit(X_ts_train, Y_ts_train,
          batch_size=20,
          epochs=100,
          shuffle=False,
          validation_data=(X_ts_test, Y_ts_test))

Train on 1400 samples, validate on 180 samples
Epoch 1/100
1400/1400 [==============================] - 8s 5ms/step - loss: 0.0568 - mean_squared_error: 0.0568 - val_loss: 0.0651 - val_mean_squared_error: 0.0651
Epoch 2/100
1400/1400 [==============================] - 3s 2ms/step - loss: 0.0206 - mean_squared_error: 0.0206 - val_loss: 0.0447 - val_mean_squared_error: 0.0447
Epoch 3/100
1400/1400 [==============================] - 3s 2ms/step - loss: 0.0177 - mean_squared_error: 0.0177 - val_loss: 0.0322 - val_mean_squared_error: 0.0322
Epoch 4/100
1400/1400 [==============================] - 3s 2ms/step - loss: 0.0151 - mean_squared_error: 0.0151 - val_loss: 0.0248 - val_mean_squared_error: 0.0248
Epoch 5/100
1400/1400 [==============================] - 3s 2ms/step - loss: 0.0138 - mean_squared_error: 0.0138
 - val_loss: 0.0206 - val_mean_squared_error: 0.0206
```

Advanced Deep Learning Models

```
...
Epoch 100/100
1400/1400 [==============================] - 3s 2ms/step - loss: 0.0100 -
mean_squared_error: 0.0100 - val_loss: 0.0100 - val_mean_squared_error:
0.0100
```

The final MSE is `0.010`, which corresponds to an average accuracy of about 95%. To have a confirmation, we can start plotting the training set and the relative predictions:

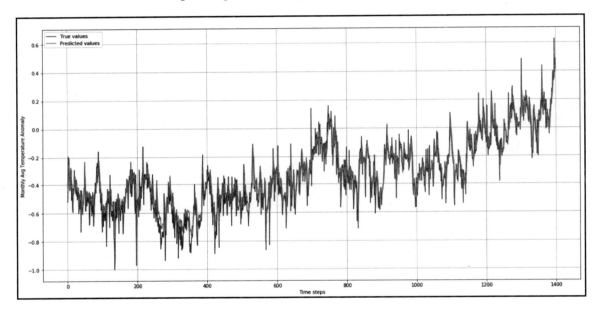

Time-series training set and relative predictions

As you can see, the model is very accurate on the training data and it has successfully learned the seasonality and the local oscillations. Let's now show the same graph for the validation data (200 samples):

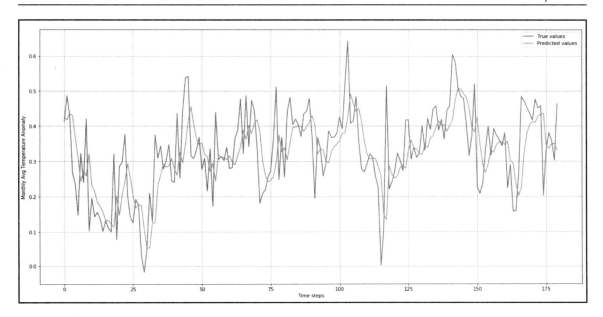

Time-series validation set and relative predictions

In this case, the larger scale allows observing some imprecisions, which, however, are limited to the small local oscillations. The global trend and the seasonality are predicted correctly (with a minimum delay), confirming the ability of LSTM cells to learn short- and long-term dependencies. I invite the reader to repeat the example with different configurations, remembering that when an LSTM is followed by another LSTM layer, the first one must have the `return_sequences=True` parameter. In this way, the whole output sequence is fed into the subsequent layer. Instead, the last one outputs only the final value, which is processed further by the dense layers.

A brief introduction to TensorFlow

TensorFlow is a computational framework created by Google and has become one of the most diffused deep learning toolkits. It can work with both CPUs and GPUs and already implements most of the operations and structures required to build and train a complex model. TensorFlow can be installed as a Python package on Linux, macOS, and Windows (with or without GPU support). However, I suggest you follow the instructions provided on the website (the link can be found in the info box at the end of this chapter) to avoid common mistakes and install it in the best way considering every specific environment.

The main concept behind TensorFlow is the computational graph or a set of subsequent operations that transform an input batch into the desired output. In the following diagram, there's a schematic representation of a graph:

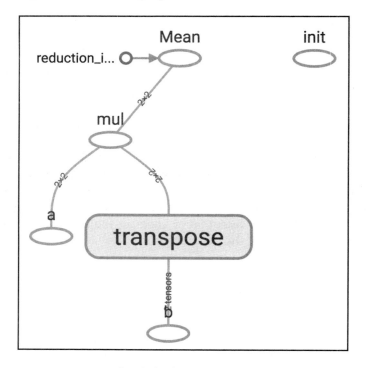

Example of simple computational graph

Starting from the bottom, we have two input nodes (**a** and **b**), a transpose operation (that works on **b**), a matrix multiplication, and a mean reduction. The **init** block is a separate operation, which is formally part of the graph, but it's not directly connected to any other node; therefore, it's autonomous (indeed, it's a global initializer).

As this one is only a brief introduction, it's useful to list all of the most important strategic elements needed to work with TensorFlow so as to be able to build a few simple examples that can show the enormous potential of this framework:

- **Graph**: This represents the computational structure that connects a generic input batch with the output tensors through a directed network made of operations. It's defined as a `tf.Graph()` instance and normally used with a Python context manager.
- **Placeholder**: This is a reference to an external variable, which must be explicitly supplied when it's requested for the output of an operation that uses it directly or indirectly. For example, a placeholder can represent an x variable, which is first transformed into its squared value and then summed to a constant value. The output is then x^2+c, which is materialized by passing a concrete value for x. It's defined as a `tf.placeholder()` instance.
- **Variable**: An internal variable used to store values that are updated by the algorithm. For example, a variable can be a vector containing the weights of a logistic regression. It's normally initialized before a training process and automatically modified by the built-in optimizers. It's defined as a `tf.Variable()` instance. A variable can also be used to store elements that must not be considered during training processes; in this case, it must be declared with the `trainable=False` parameter.
- **Constant**: A constant value defined as a `tf.constant()` instance.
- **Operation**: A mathematical operation that can work with placeholders, variables, and constants. For example, the multiplication of two matrices is an operation defined as `tf.matmul(A, B)`. Among all operations, gradient calculation is one of the most important. TensorFlow allows determining the gradients starting from a determined point in the computational graph, until the origin or another point that must be logically before it. We're going to see an example of this operation.

- **Session**: This is a sort of wrapper-interface between TensorFlow and our working environment (for example, Python or C++). When the evaluation of a graph is needed, this macro-operation will be managed by a session, which must be fed with all placeholder values and will produce the required outputs using the requested devices. For our purposes, it's not necessary to go deeper into this concept; however, I invite the reader to retrieve further information from the website or from one of the resources listed at the end of this chapter. It's declared as an instance of `tf.Session()` or, as we're going to do, an instance of `tf.InteractiveSession()`. This type of session is particularly useful when working with notebooks or shell commands because it places itself automatically as the default one.
- **Device**: A physical computational device, such as a CPU or a GPU. It's declared explicitly through an instance of the class `tf.device()` and used with a context manager. When the architecture contains more computational devices, it's possible to split the jobs so as to parallelize many operations. If no device is specified, TensorFlow will use the default one (which is the main CPU or a suitable GPU if all the necessary components are installed).

We can now analyze some simple examples using these concepts.

Computing gradients

The option to compute the gradients of all output tensors with respect to any connected input or node is one of the most interesting features of TensorFlow, because it allows us to create learning algorithms without worrying about the complexity of all transformations. In this example, we first define a linear dataset representing the function $f(x) = x$ in the range $(-100, 100)$:

```
import numpy as np

nb_points = 100
X = np.linspace(-nb_points, nb_points, 200, dtype=np.float32)
```

The corresponding plot is shown in the following graph:

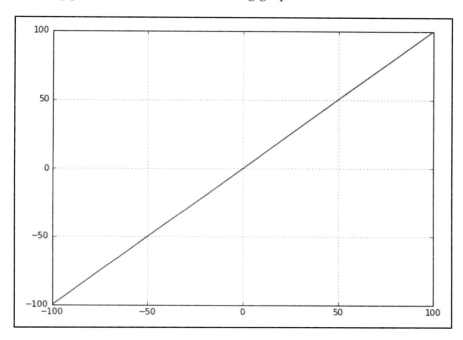

Input data for gradient computation

Now we want to use TensorFlow to compute the following:

$$\begin{cases} g(x) = x^3 \\ \dfrac{\partial g}{\partial x} = 3x^2 \\ \dfrac{\partial^2 g}{\partial x^2} = 6x \end{cases}$$

The first step is defining a graph variable:

```
import tensorflow as tf

graph = tf.Graph()
```

Within the context of this graph, we can define our input placeholder and other operations:

```
with graph.as_default():
    Xt = tf.placeholder(tf.float32, shape=(None, 1), name='x')
```

Advanced Deep Learning Models

```
Y = tf.pow(Xt, 3.0, name='x_3')
Yd = tf.gradients(Y, Xt, name='dx')
Yd2 = tf.gradients(Yd, Xt, name='d2x')
```

A placeholder is generally defined with a type (first parameter), a shape, and an optional name. We've decided to use a `tf.float32` type because this is the only type also supported by GPUs. Selecting `shape=(None, 1)` means that it's possible to use any bidimensional vectors with the second dimension equal to 1.

The first operation computes the third power if `Xt` is working on all elements. The second operation computes all the gradients of `Y` with respect to the input placeholder `Xt`. The last operation will repeat the gradient computation, but in this case, it uses `Yd`, which is the output of the first gradient operation.

We can now pass some concrete data to see the results. The first thing to do is create a session connected to this graph:

```
session = tf.InteractiveSession(graph=graph)
```

By using this session, we ask any computation using the `run()` method. All the input parameters must be supplied through a feed-dictionary, where the key is the placeholder, while the value is the actual array:

```
X2, dX, d2X = session.run([Y, Yd, Yd2], feed_dict={Xt: X.reshape((nb_points*2, 1))})
```

We needed to reshape our array to be compliant with the placeholder. The first argument of `run()` is a list of tensors that we want to be computed. In this case, we need all operation outputs. The plot of each of them is shown in the following graphs:

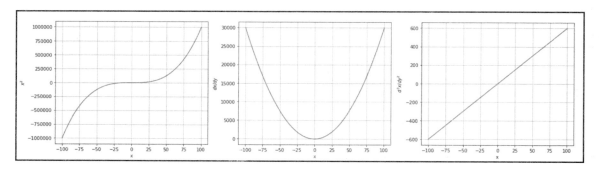

TensorFlow outputs: g(x) (left), first derivative (center), and second derivative (right)

As expected, they represent respectively: x^3, $3x^2$, and $6x$.

Logistic regression

Now we can try a more complex example implementing a logistic regression algorithm. The first step, as usual, is to create a dummy dataset:

```
from sklearn.datasets import make_classification

nb_samples = 500
X, Y = make_classification(n_samples=nb_samples, n_features=2,
n_redundant=0, n_classes=2)
```

The dataset is shown in the following graph:

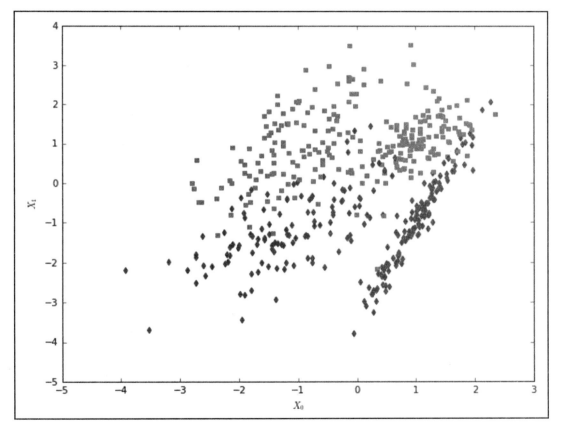

Dataset employed for a logistic regression with TensorFlow

Advanced Deep Learning Models

At this point, we can create the graph and all placeholders, variables, and operations:

```
import tensorflow as tf

graph = tf.Graph()

with graph.as_default():
    Xt = tf.placeholder(tf.float32, shape=(None, 2), name='points')
    Yt = tf.placeholder(tf.float32, shape=(None, 1), name='classes')

    W = tf.Variable(tf.zeros((2, 1)), name='weights')
    bias = tf.Variable(tf.zeros((1, 1)), name='bias')

    Ye = tf.matmul(Xt, W) + bias
    Yc = tf.round(tf.sigmoid(Ye))

    loss = tf.reduce_mean(tf.nn.sigmoid_cross_entropy_with_logits(logits=Ye, labels=Yt))
    training_step = tf.train.GradientDescentOptimizer(0.025).minimize(loss)
```

The `Xt` placeholder is needed for the points, while `Yt` represents the labels. At this point, we need to involve a couple of variables: if you remember, they store values that are updated by the training algorithm. In this case, we need a `W` weight vector (with two elements) and a single `bias`. When a variable is declared, its initial value must be provided; we've decided to set both to zero using the `tf.zeros()` function, which accepts as an argument the shape of the desired tensor.

Now we can compute the output (if you don't remember how a logistic regression works, please step back to Chapter 5, *Logistic Regression*) in two steps: first, the sigmoid exponent `Ye`, and then the actual binary output `Yc`, which is obtained by rounding the sigmoid value. The training algorithm for a logistic regression minimizes the negative log-likelihood, which corresponds to the cross-entropy between the real distribution `Y` and `Yc`. It's easy to implement this loss function; however, the `tf.log()` function is numerically unstable (when its value becomes close to zero, it tends to negative infinity and yields a `NaN` value); therefore, TensorFlow has implemented a more robust function, `tf.nn.sigmoid_cross_entropy_with_logits()`, which computes the cross-entropy assuming the output is produced by a sigmoid. It takes two parameters, the `logits` (which corresponds to the exponent `Ye`) and the target `labels`, which are stored in `Yt`.

Now we can work with one of the most powerful TensorFlow features: the training optimizers. After defining a loss function, it will be dependent on placeholders, constants, and variables. A training optimizer (such as `tf.train.GradientDescentOptimizer()`), through its `minimize()` method, accepts the loss function to optimize. Internally, according to every specific algorithm, it will compute the gradients of the loss function with respect to all trainable variables and will apply the corresponding corrections to the values. The parameter passed to the optimizer is the learning rate.

Therefore, we have defined an extra operation called `training_step`, which corresponds to a single stateful update step. It doesn't matter how complex the graph is; all trainable variables involved in a loss function will be optimized with a single instruction.

Now it's time to train our logistic regression. The first thing to do is to ask TensorFlow to initialize all variables so that they are ready when the operations have to work with them:

```
session = tf.InteractiveSession(graph=graph)
tf.global_variables_initializer().run()
```

At this point, we can create a simple training loop (it should be stopped when the Loss stops decreasing; however, we have a fixed number of iterations):

```
feed_dict = {
    Xt: X,
    Yt: Y.reshape((nb_samples, 1))
}

for i in range(5000):
    loss_value, _ = session.run([loss, training_step], feed_dict=feed_dict)
    if i % 100 == 0:
        print('Step %d, Loss: %.3f' % (i, loss_value))

Step 0, Loss: 0.269
Step 100, Loss: 0.267
Step 200, Loss: 0.265
Step 300, Loss: 0.264
Step 400, Loss: 0.263
Step 500, Loss: 0.262
Step 600, Loss: 0.261
Step 700, Loss: 0.260
Step 800, Loss: 0.260
Step 900, Loss: 0.259
...
Step 9900, Loss: 0.256
```

As you can see, at each iteration, we ask TensorFlow to compute the loss function and a training step, and we always pass the same dictionary containing X and Y. At the end of this loop, the loss function is stable, and we can check the quality of this logistic regression by plotting the separating hyperplane:

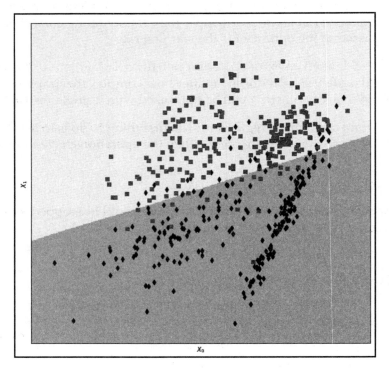

Classification result with the separating line

The result is approximately equivalent to the one obtained with the scikit-learn implementation. If we want to know the values of both coefficients (weights) and intercept (bias), we can ask TensorFlow to retrieve them by calling the `eval()` method on each variable:

```
Wc, Wb = W.eval(), bias.eval()

print(Wc)
[[-1.16501403]
 [ 3.10014033]]

print(Wb)
[[-0.12583369]]
```

Classification with a multilayer perceptron

We can now build an architecture with two dense layers and train a classifier for a more complex dataset. Let's start by creating it:

```
from sklearn.datasets import make_classification

nb_samples = 1000
nb_features = 3

X, Y = make_classification(n_samples=nb_samples, n_features=nb_features,
                           n_informative=3, n_redundant=0, n_classes=2,
                           n_clusters_per_class=3)
```

Even if we have only two classes, the dataset has three features and three clusters per class; therefore, it's almost impossible that a linear classifier can separate it with very high accuracy. A plot of the dataset is shown in the following graph:

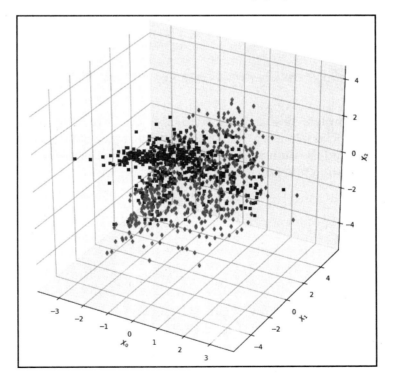

Three-dimensional representation of the dataset

For benchmarking purposes, it's useful to test a logistic regression:

```
from sklearn.model_selection import train_test_split
from sklearn.linear_model import LogisticRegression

X_train, X_test, Y_train, Y_test = train_test_split(X, Y, test_size=0.2)

lr = LogisticRegression()
lr.fit(X_train, Y_train)

print('Score: %.3f' % lr.score(X_test, Y_test))
Score: 0.715
```

The score computed on the test set is about 71%, which is not really bad but below an acceptable threshold. Let's try with an MLP with 50 hidden neurons (with hyperbolic tangent activation) and one sigmoid output neuron. The hyperbolic tangent is defined as follows:

$$tanh(x) = \frac{e^x - e^{-x}}{e^x + e^{-x}}$$

And it's bounded asymptotically between -1.0 and 1.0, where the gradients tend to saturate to 0.0.

We are not going to implement each layer manually, but we're using the built-in `tf.layers.dense()` class. It accepts the input tensor or placeholder as the first argument (`inputs` parameter) and the number of layer-output neurons as the second one (`units` parameter). The activation function can be specified using the `activation` attribute:

```
import tensorflow as tf

graph = tf.Graph()

with graph.as_default():
    Xt = tf.placeholder(tf.float32, shape=(None, nb_features), name='X')
    Yt = tf.placeholder(tf.float32, shape=(None, 1), name='Y')

    layer_1 = tf.layers.dense(input=Xt, units=50, activation=tf.nn.tanh)
    layer_2 = tf.layers.dense(input=layer_1, units=1,
activation=tf.nn.sigmoid)

    Yo = tf.round(layer_2)

    loss = tf.nn.l2_loss(layer_2 - Yt)
    training_step = tf.train.GradientDescentOptimizer(0.025).minimize(loss)
```

As in the previous example, we have defined two placeholders, Xt and Yt, and two fully connected layers. The first one accepts as input Xt and has 50 output neurons (with `tanh` activation), while the second accepts as input the output of the previous layer (`layer_1`) and has only one sigmoid neuron, representing the class. The rounded output is provided by Yo, while the loss function is the total squared error, and it's implemented using the `tf.nn.l2_loss()` function computed on the difference between the output of the network (`layer_2`) and the target class placeholder Yt. The training step is implemented using a standard gradient descent optimizer, as per the logistic regression example.

We can now implement a training loop, splitting our dataset into a fixed number of batches (the number of samples is defined in the `batch_size` variable) and repeating a complete cycle for `nb_epochs` epochs:

```
session = tf.InteractiveSession(graph=graph)
tf.global_variables_initializer().run()

nb_epochs = 200
batch_size = 50

for e in range(nb_epochs):
    total_loss = 0.0
    Xb = np.ndarray(shape=(batch_size, nb_features), dtype=np.float32)
    Yb = np.ndarray(shape=(batch_size, 1), dtype=np.float32)

    for i in range(0, X_train.shape[0]-batch_size, batch_size):
        Xb[:, :] = X_train[i:i+batch_size, :]
        Yb[:, 0] = Y_train[i:i+batch_size]

        loss_value, _ = session.run([loss, training_step],
                            feed_dict={Xt: Xb, Yt: Yb})
        total_loss += loss_value

        Y_predicted = session.run([Yo],
                            feed_dict={Xt: X_test.reshape((X_test.shape[0], nb_features))})
        accuracy = 1.0 - (np.sum(np.abs(np.array(Y_predicted[0]).squeeze(axis=1) - \
                    Y_test)) / float(Y_test.shape[0]))

        print('Epoch %d) Total loss: %.2f - Accuracy: %.2f' % (e, total_loss, accuracy))

Epoch 0) Total loss: 78.19 - Accuracy: 0.66
Epoch 1) Total loss: 75.02 - Accuracy: 0.67
Epoch 2) Total loss: 72.28 - Accuracy: 0.68
```

```
Epoch 3) Total loss: 68.52 - Accuracy: 0.71
Epoch 4) Total loss: 63.50 - Accuracy: 0.79
Epoch 5) Total loss: 57.51 - Accuracy: 0.84
...
Epoch 195) Total loss: 15.34 - Accuracy: 0.94
Epoch 196) Total loss: 15.32 - Accuracy: 0.94
Epoch 197) Total loss: 15.31 - Accuracy: 0.94
Epoch 198) Total loss: 15.29 - Accuracy: 0.94
Epoch 199) Total loss: 15.28 - Accuracy: 0.94
```

As you can see, without paying particular attention to all the details, the accuracy computed on the test set is 94%. This is an acceptable value, considering the structure of the dataset. In the aforementioned book, the reader will find details of many important concepts that can still improve the performance and speed up the convergence process.

Image convolution

Even if we're not building a complete deep learning model, we can test how convolution works with a simple example. The input image we're using is already provided by SciPy:

```
from scipy.misc import face

img = face(gray=True)
```

The original picture is shown as follows:

Sample picture to test the convolution features

We're going to apply a Laplacian filter (that is, it is equivalent to a discrete Laplacian operator), which emphasizes the boundary of each shape:

```
import numpy as np

kernel = np.array(
    [[0, 1, 0],
     [1, -4, 0],
     [0, 1, 0]],
     dtype=np.float32)

cfilter = np.zeros((3, 3, 1, 1), dtype=np.float32)
cfilter[:, :, 0, 0] = kernel
```

The `kernel` must be repeated twice because the TensorFlow `tf.nn.conv2d` convolution function expects an input and an output filter. We can now build the `graph` and test it:

```
import tensorflow as tf

graph = tf.Graph()

with graph.as_default():
    x = tf.placeholder(tf.float32, shape=(None, 768, 1024, 1), name='image')
    f = tf.constant(cfilter)
    y = tf.nn.conv2d(x, f, strides=[1, 1, 1, 1], padding='same')

session = tf.InteractiveSession(graph=graph)

c_img = session.run([y], feed_dict={x: img.reshape((1, 768, 1024, 1))})
n_img = np.array(c_img).reshape((768, 1024))
```

The `strides` parameter is a four-dimensional vector (each value corresponds to the input dimensions, so the first is the batch and the last one is the number of channels) that specifies how many pixels the sliding window must shift. In this case, we don't want to skip pixels while shifting, so the parameter is set equal to 1 for all dimensions. The `padding` parameter determines how the new dimensions must be computed and whether it's necessary to apply a zero padding. In our case, we're using the value `'same'`, which computes the dimensions by rounding off to the next integer the original dimensions divided by the corresponding strides value (as the latter are both 1.0, the resultant image size will be exactly the same as the original one).

The output is shown here:

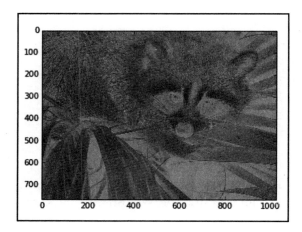

Convolution result

 The installation instructions of TensorFlow for every operating system (with and without GPU support) can be found at https://www.tensorflow.org/install/.

Summary

In this chapter, we have briefly presented the most important deep learning layers, and we have discussed two concrete examples based on Keras. We have seen how to model a deep convolutional network to classify images and how an LSTM model can be easily employed when it's necessary to learn short- and long-term dependencies in a time-series.

We also saw how TensorFlow computes the gradients of an output tensor with respect to any previous connected layer, and therefore how it's possible to implement the standard back propagation strategy seamlessly to deep architectures. We haven't discussed the actual deep learning problems and methods in detail because they require much more space. However, the reader can easily find many valid resources to continue their exploration of this fascinating field. At the same time, it's possible to modify the examples and continue a *learning-by-doing* process.

In the next chapter, Chapter 17, *Advanced Deep Learning Models* we're going to summarize many of the concepts previously discussed, to create complex machine learning architectures.

17
Creating a Machine Learning Architecture

In this chapter, we're going to summarize many of the concepts discussed in the book with the purpose of defining a complete machine learning architecture that is able to preprocess the input data, decompose/augment it, classify/cluster it, and, eventually, show the results, using graphical tools. We're also going to show how scikit-learn manages complex pipelines and how it's possible to fit them and search for the optimal parameters in the global context of a complete architecture.

In particular, we are going to discuss the following:

- Data collection, preprocessing, and augmentation
- Normalization, regularization, and dimensionality reduction
- Vectorized computation and GPU support
- Distributed architectures
- Pipelines and feature unions

Machine learning architectures

Until now, we have discussed single methods that could be employed to solve specific problems. However, in real contexts, it's very unlikely to have well-defined datasets that can be immediately fed into a standard classifier or clustering algorithm. A machine learning engineer often has to design a full architecture that a layman would consider to be like a black box, where the raw data enters and the outcomes are automatically produced. All the steps necessary to achieve the final goal must be correctly organized and seamlessly joined together in a processing chain similar to a computational graph (indeed, it's very often a direct acyclic graph). Unfortunately, this is an unconventional process, as every real-life problem has its own peculiarities. However, there are some common steps that are normally included in almost any ML pipeline.

Creating a Machine Learning Architecture

In the following diagram, there's a schematic representation of this process:

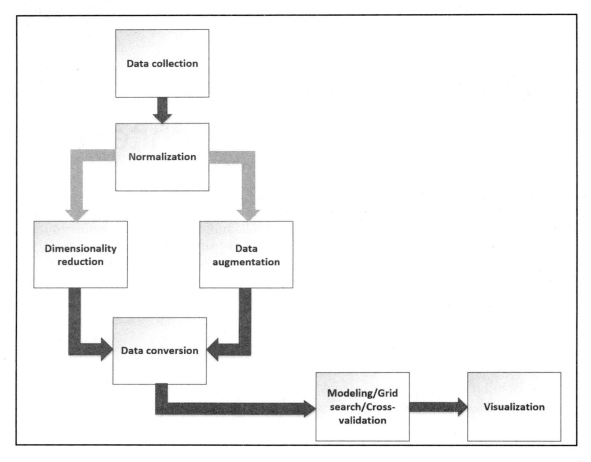

Structure of a machine learning pipeline

Now we will briefly explain the details of each phase with some possible solutions.

Data collection

The first step is always the most generic because it depends on every single context. However, before working with any dataset, it's necessary to collect the data from all the sources where it is stored. The ideal situation is to have a **Comma-Separated Values** (**CSV**) (or another suitable format) dump that can be immediately loaded, but, more often, the engineer has to look for all the database tables, define the right SQL query to collect all the pieces of information, and manage data type conversion and encoding.

We're not going to discuss this topic (which is more related to data engineering), but it's important not to under-evaluate this stage, because it can be much more difficult than expected. I suggest, whenever possible, to extract flattened tables, where all the fields are placed on the same row, because it's easier to manipulate a large amount of data using a **Database Management System** (**DBMS**) or a big data tool, but it can be very time and memory-consuming if done on a normal PC directly with Python tools. Moreover, it's important to use a standard character encoding for all text fields. The most common choice is UTF-8, but it's also possible to find database tables encoded with other charsets, and normally it's good practice to convert all the documents before starting with the other operations.

A very famous and powerful Python library for data manipulation is **pandas** (part of SciPy). It's mainly based on the concept of **DataFrame** (an abstraction of SQL table) and implements many methods that allow the selection, joining, grouping, and statistical processing of datasets that can fit in memory. In the majority of data science tasks, this is probably one of the best choices, as it can quickly provide the user with a complete insight of the data, and it allows manipulating the fields or the sources in real time. In *Learning pandas - Python Data Discovery and Analysis Made Easy, Heydt M, Packt Publishing*, the reader can find all the information needed to use this library to solve many real-life scenarios. Two common problems that must be managed during this phase are imputing the missing features and managing categorical ones. In `Chapter 3`, *Feature Selection and Feature Engineering*, we discussed some practical methods that can be employed automatically before starting the following steps. Even if scikit-learn provides many powerful tools, this step can be also carried out using pandas and, hence, assigned to data scientists/engineers without a full machine learning knowledge.

Normalization and regularization

Normalizing a numeric dataset is one of the most important steps, particularly when different features have different scales. In Chapter 3, *Feature Selection and Feature Engineering*, we discussed several methods that can be employed to solve this problem. Very often, it's enough to use StandardScaler to whiten the data, but sometimes it's better to consider the impact of noisy features on the global trend and use RobustScaler to filter them out without the risk of conditioning the remaining features. Moreover, there are many algorithms that can benefit from whitened datasets; therefore, I suggest using a grid search (we're going to discuss how to merge different transformations later in this chapter) and pick the optimal combination.

The reader can easily verify the different performances of the same classifier (in particular, **Support Vector Machines** (**SVMs**) and neural networks) when working with normalized and unnormalized datasets. As we're going to see in the next section, it's possible to include the normalization step in the processing pipeline as one of the first actions and include the C parameter in grid search to impose an *L1/L2* regularization during the training phase (see the importance of regularization in Chapter 4, *Regression Algorithms*, when discussing Ridge, Lasso and ElasticNet).

Dimensionality reduction

This step is not always mandatory, but, in many cases, it can be a good solution to avoid memory leaks or long computational times. When the dataset has many features, the probability of some hidden correlation is relatively high. For example, the final price of a product is directly influenced by the price of all materials, and, if we remove one secondary element, the value changes slightly (generally speaking, we can say that the total variance is almost preserved). If you remember how **Principal Component Analysis** (**PCA**) works, you know that this process decorrelates the input data too. Therefore, it's useful to check whether a PCA or a Kernel PCA (for non-linear datasets) can remove some components while keeping the explained variance close to 100 percent (this is equivalent to compressing the data with minimum information loss).

There are also other methods discussed in `Chapter 3`, *Feature Selection and Feature Engineering*, such as `NMF` or `SelectKBest`, that can be useful for selecting only the best features according to various criteria (such as **Analysis of Variance (ANOVA)** or chi-squared). Testing the impact of each factor during the initial phases of the project can save time that can be useful when it's necessary to evaluate slower and more complex algorithms. **Independent Component Analysis (ICA)** and dictionary learning can be efficiently employed whenever it's preferable to have an internal encoding based on a sparse combination of features. For example, an image-processing model can work more efficiently with a synthetic input (a feature vector) representing the dominant elements extracted from a sample.

As every scenario has its own peculiarities, it's not easy to define a rule of thumb regarding dimensionality reduction; however, I suggest to consider its usage as a trade-off between complexity and accuracy. The main evaluation criterion is based on the explained variance, which is always lower when PCA is applied. As the information is proportional to the variance, the loss must be justified by a consistent complexity reduction (for example, if the explained variance is 90% with a model that requires one-tenth of the parameters and yields an accuracy comparable to the original one, PCA should be employed).

Data augmentation

Sometimes, the original dataset has only a few non-linear features, and it's quite difficult for a standard classifier to capture the dynamics. Moreover, forcing an algorithm on a complex dataset can result in overfitting the model because all the capacity is exhausted in trying to minimize the error considering only the training set, and without taking into account the generalization ability. For this reason, it's sometimes useful to enrich the dataset with derived features that are obtained through some functions of the existing ones.

`PolynomialFeatures` is an example of data augmentation that can really improve the performances of standard algorithms and avoid overfitting. In other cases, it can be useful to introduce trigonometric functions (such as *sin(x)* or *cos(x)*) or correlating features (such as $x_1 x_2$). The former allows a simpler management of radial datasets, while the latter can provide the classifier with information about the cross-correlation between two features. Another common usage of data augmentation is with image classifiers, where the datasets are often too small considering the number of possibilities required in real-life applications. For example, we can have a dataset of traffic signs that must be recognized by an autonomous vehicle. However, the camera captures rotated or partially occluded images that often don't match the training distribution. For this reason, it's useful to enlarge the dataset with variants based on rotations, rescaling, noise-additions, shearing, and so on. All these samples, even if synthetic, represent potential samples that the car can encounter; hence, data augmentation allows training larger and more efficient models, without the need of millions of contextual variants.

In general, a feature-based data augmentation can be employed before trying a more complex algorithm. For example, a logistic regression (which is a linear method) can be successfully applied to augmented non-linear datasets (we saw a similar situation in `Chapter 4`, *Regression Algorithms*, when we discussed the polynomial regression). The choice to employ a more complex (with higher capacity) model or to try to augment the dataset is up to the engineer and must be considered carefully, taking into account both the pros and the cons. A very important criterion to consider is to evaluate the model on a validation set sampled from the true data-generating process. In fact, data augmentation can produce quite different distributions that don't match the one considered in production. As the goal is to maximize the validation accuracy, it's extremely important to be sure that the model is tested against real examples (for example, images with the same resolution, feature quality, and so forth).

In many cases, for example, it's preferable not to modify the original dataset (which could be quite large), but to create a scikit-learn interface to augment the data in real time. In other cases, a neural model can provide faster and more accurate results without the need for data augmentation. Together with parameter selection, this is more of an art than a real science, and the experiments are the only way to gather useful knowledge.

Data conversion

This step is probably the simplest and, at the same time, the most important when handling categorical data. We have discussed several methods to encode labels using numerical vectors, and it's not necessary to repeat the concepts already explained. A general rule concerns the usage of integer or binary values (one-hot encoding). The latter is probably the best choice when the output of the classifier is the value itself, because, as discussed in Chapter 3, *Feature Selection and Feature Engineering*, it's much more robust to noise and prediction errors. On the other hand, one-hot encoding is quite memory-consuming. Therefore, whenever it's necessary to work with probability distributions (as in NLP), an integer label (representing a dictionary entry or a frequency/count value) can be much more efficient.

Modeling/grid search/cross-validation

Modeling implies the choice of the classification/clustering algorithm that best suits every specific task. We have discussed different methods, and the reader should be able to understand when a set of algorithms is a reasonable candidate, and when it's better to look for another strategy. However, the success of a machine learning technique often depends on the right choice for each parameter involved in the model as well. As already discussed, when talking about data augmentation, it's very difficult to find a precise method to determine the optimal values to assign, and the best approach is always based on a grid search. scikit-learn provides a very flexible mechanism to investigate the performance of a model with different parameter combinations, together with cross-validation (which allows a robust validation without reducing the number of training samples), and this is indeed a more reasonable approach, even for expert engineers. Moreover, when performing different transformations, the effect of a choice can impact the whole pipeline, and, therefore, (we're going to see a few examples in the next section) I always suggest for application of the grid search to all components at the same time, to be able to evaluate the cross-influence of each possible choice.

Visualization

Sometimes, it's useful/necessary to visualize the results of intermediate and final steps. In this book, we have always shown plots and diagrams using matplotlib, which is part of SciPy and provides a flexible and powerful graphics infrastructure. Even if it's not part of the book, the reader can easily modify the code to get different results. For a deeper understanding, refer to *Mastering matplotlib, McGreggor D, Packt Publishing*. As this is an evolving sector, many new projects are being developed, offering new and more stylish plotting functions. One of them is **Bokeh** (http://bokeh.pydata.org), which works using some JavaScript code to create interactive graphs that can be embedded into web pages too. I highly suggest that the reader learn the main matplotlib commands and use this library directly from Jupyter notebooks. The possibility to embed plots and recompute them easily is extremely helpful during the whole analytic and modeling process.

GPU support

In the previous two chapters, Chapter 15, *Introducing Neural Networks*, and Chapter 16, *Advanced Deep Learning Models*, we have discussed the basic concepts of deep learning models, highlighting how large the number of parameters that characterizes such architectures is. If we consider also the dimensionality of the dataset, it's easy to understand that the computational complexity can become extremely large. Frameworks such as scikit-learn are mainly based on NumPy, which contains highly optimized native code that runs on multi-core CPUs. Moreover, NumPy works with **Single Instruction Multiple Data** (**SIMD**) commands and exposes vectorized primitives. This feature, in many cases, allows getting rid of explicit loops, speeding up at the same time the execution time. Let's consider a simple example based on the multiplication of two 500 × 500 matrices. Without any vectorization, the code looks as follows:

```
import numpy as np
import time

A1 = np.random.normal(0.0, 2.0, size=(size, size)).astype(np.float32)
A2 = np.random.normal(0.0, 2.0, size=(size, size)).astype(np.float32)

size = 500

D = np.zeros(shape=(size, size)).astype(np.float32)

start_time = time.time()

for i in range(size):
    for j in range(size):
        d = 0.0
```

```
        for k in range(size):
            d += A1[i, k] * A2[k, j]
        D[i, j] = d
end_time = time.time()
elapsed = end_time - start_time

print(elapsed)
115.05700373649597
```

As each element is obtained as the dot product between a row and a column, the algorithm has a complexity $O(n^3)$, corresponding to 125 million operations, and the execution time on a modern CPU (Intel Core i7) is 115 seconds. Let's now repeat the example using the NumPy `dot()` function:

```
import numpy as np
import time

start_time = time.time()

D = np.dot(A1, A2)

end_time = time.time()
elapsed = end_time - start_time

print(elapsed)
0.003000020980834961
```

The execution time is now about 3 ms! It should be clear that using NumPy is almost essential to almost any machine learning project, and the reader has to take care of employing the vectorization whenever possible.

The execution times are influenced by the CPU, global architecture, and Python and NumPy versions; however, the discrepancy between the two versions is always extremely large. In the vectorized version, the start/end time measures can impact the final value. Therefore, I suggest using large matrices to increase the computational time.

However, the majority of machine learning and deep learning tasks are quite repetitive, and, in many cases, the matrix products and related operations are executed 1 million times. Unfortunately, even if the CPU SIMD instructions are very powerful, the number of cores is normally very small *(4, 8, 16)*. In fact, CPUs are designed to be general-purpose processors, and every core must be able to perform the most complex operations. Conversely, GPUs are very specialized processors, with many more cores (for example, 2,000 or more) that can execute parallel computations with limited complexity. In a deep-learning context, even the most complex models are mainly based on matrix manipulations where the main obstacle is represented by the dimensions. For this reason, specific low-level libraries (for example, NVIDIA Cuda) have been designed and freely distributed. Frameworks such as TensorFlow can easily interface with one or more GPUs, allowing the training of large models in reasonable times (in this very period, large companies such as Google have started offering even more specialized processors; these are called **Tensor Processing Units** (**TPUs**).

One of the main limitations of GPU usage is memory. In fact, while a CPU can access all the available RAM, a GPU has its own dedicated memory (**VRAM**), which is normally not very large (in particular, in common-usage devices). When a matrix needs to be processed, the first step is moving it into the VRAM and then starting the GPU computation. Modern frameworks are optimized to minimize the number of swaps between CPU and GPU; however, it can happen that very large blocks don't fit into the VRAM. Nowadays, the majority of common deep architectures can be trained on medium-level GPUs when the input data is not too large. A mitigating solution is reducing the batch size, but this choice will negatively impact the performances, because more memory swaps are needed. When working with a single GPU, I invite the reader to check what the memory limit is and test different configurations, to try to find the most appropriate trade-off.

TensorFlow is an extremely powerful choice for the majority of tasks; however, sometimes, the code has already been written using NumPy, and its conversion can be expensive. A very powerful solution is offered by **CuPy** (https://cupy.chainer.org/), which is a free framework that implements an interface quite similar to NumPy (some functions are not available, but the majority of operations have the exactly same syntax) but works with NVIDIA GPU support. A complete discussion is beyond the scope of this book (I recommend the official documentation, which is extremely clear), but I want to show the performance improvement that is possible to obtain using a medium-level GTX 960M.

Suppose we have two random 5,000 × 5,000 matrices, A/B_1 and A/B_2 (they must be defined as `float32` because GPUs don't support 64-bit floats). Let's perform 100 sequential dot products (that is, the result of a dot product is multiplied times A_2) to simulate a complex processing pipeline. The vectorized NumPy code is as follows:

```
import numpy as np
import time

size = 5000

A1 = np.random.normal(0.0, 2.0, size=(size, size)).astype(np.float32)
A2 = np.random.normal(0.0, 2.0, size=(size, size)).astype(np.float32)

Ad = A1.copy()

start_time = time.time()

for _ in range(100):
    Ad = np.dot(Ad, A2)

end_time = time.time()
elapsed = end_time - start_time

print(elapsed)
236.12225341796875
```

NumPy requires about four minutes to complete the task. Let's now employ CuPy:

```
import cupy as cp
import time

B1 = cp.random.normal(0.0, 2.0, size=(size, size))
B2 = cp.random.normal(0.0, 2.0, size=(size, size))

Bd = B1.copy()

start_time = time.time()

for _ in range(100):
    Bd = cp.dot(Bd, B2)

end_time = time.time()
elapsed = end_time - start_time

print(elapsed)
0.3540008068084717
```

The computational time is now 0.35 seconds! Moreover, the only external difference in the code is the `cp` namespace instead of `np`. Of course, not all NumPy functions can be implemented on a GPU, but the majority of machine learning tasks can be efficiently converted without particular limitations. Behind the scenes, CuPy is, however, a little bit different than NumPy, in fact, when the two matrices are created, they are transferred to the VRAM (if there's not enough free space, an exception is raised), and the results are taken into the GPU memory until they are required for further CPU computations. For this reason, once `B1` and `B2` are created (together with the support matrix `Bd`), the `dot()` command will run without any other communication with the CPU, as the referenced data blocks are immediately available. The result is an extremely fast parallel processing that shows better performances (with respect to a modern CPU) in the majority of cases.

CuPy (`https://cupy.chainer.org/`) can be installed using the `pip install -U cupy` command; however, it requires the CUDA libraries already installed in the system. I invite the reader to check out `https://docs-cupy.chainer.org/en/stable/install.html`, where there are detailed instructions and links for all the required components.

A brief introduction to distributed architectures

Many real-life problems can be solved using single machines, with enough computational power; however, in some cases, the amount of data is so large that it is impossible to perform in-memory operations. We have seen an example of such a scenario in Chapter 12, *Introducing Recommendation Systems*, when discussing the **Alternating Least Squares** (**ALS**) strategy using Spark. In that case, the user-product matrix can become extremely large, and the only way to solve the problem is to employ distributed architectures. A generic schema is shown in the following diagram:

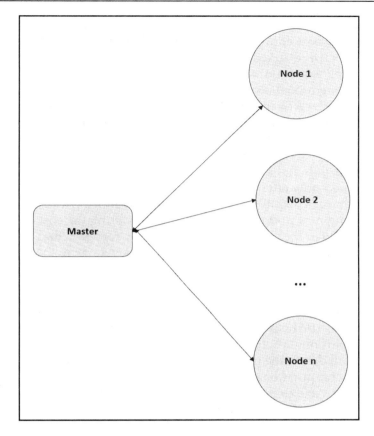

Generic distributed architecture

In this kind of architecture, there are generally two main components: a master (the name can change, but its role is generally the same) and some computational nodes. Once the model has been designed and deployed, the master starts analyzing it to determine the optimal physical distribution of the jobs. Normally, the first step is loading the training and validation datasets, which we assume to be too large to fit into the memory of a single node. Hence, the master will inform each node to load only a part and work with it until a new command is received. At this point, each node is ready to start its computation, which is always local (that is, the node A doesn't know anything about its peers). Once finished, the results are sent back to the **Master**, which will run further operations (for example, shuffling and aggregating the blocks). If the process is iterative (as it often happens in machine learning), the master is also responsible to check for the target value (accuracy, loss, or any other indicator), share the new parameters to all the nodes, and restart the computations.

According to the nature of the problem, the strategy can be different. In some cases, it's possible to employ the MapReduce paradigm (which requires the functions to be commutative and associative), but sometimes it's preferable to train the same model on each node using a small subset with a periodic synchronization with the master. For example, a deep learning model can be locally trained on every single node accumulating the gradient updates. After a fixed number of steps, the node will send its update contribution to the master and wait for the updated weights. In this case, the role of the master is to cumulate all the gradient updates, apply them to a central model, perform the validation (which is generally lightweight), and synchronize with the nodes. Clearly, each node will always work with a slightly suboptimal model, because the update is often partial and is performed only after a few epochs. However, the final result is generally extremely accurate, and it's possible to perform more iterations without an impact on the training time by adding new nodes to the cluster. Considering the cost of medium-level cloud machines (very often they can be provisioned only for the actual training phase), this approach allows solving efficiently many complex problems where the amount of data is prohibitively large.

A detailed analysis of these methods is beyond the scope of this book; however, I invite the reader to check out Spark MLlib (https://spark.apache.org/mllib/) as an alternative to scikit-learn for big data. The main advice is to accurately evaluate the nature of the problem and avoid complex solutions when they are not really needed. In fact, despite its power, Spark has an overhead, due to the management of the cluster. When a single machine with enough memory and computational power can solve the problem, it's generally preferable to a distributed system. An exception is when the dataset is subject to continuous growth. For example, a start-up company that needs a recommendation system can start working with a limited amount of data with a single machine. However, the number of users increases constantly; hence, the dataset becomes too large for the provisioned machine. In this case, instead of looking for a more powerful solution, it's preferable to switch to a distributed one, starting with a small cluster, and adding new resources when needed. In fact, the effort of a scaling is normally extremely low (in general, when working with cloud environments, it's only necessary to update the configuration), while rewriting the code when it's already in production (and it's not possible to process the whole dataset anymore) is harder and riskier.

Scikit-learn tools for machine learning architectures

Now we're going to present two very important scikit-learn classes that can help the machine learning engineer to create complex processing structures, including all the steps needed to generate the desired outcomes from the raw datasets.

Pipelines

Scikit-learn provides a flexible mechanism for creating pipelines made up of subsequent processing steps. This is possible thanks to a standard interface implemented by the majority of classes; therefore, most of the components (both data processors/transformers and classifiers/clustering tools) can be exchanged seamlessly. The `Pipeline` class accepts a single `steps` parameter, which is a list of tuples in the form (name of the component—instance), and creates a complex object with the standard fit/transform interface. For example, if we need to apply a PCA, a standard scaling, and then we want to classify using an SVM, we could create a pipeline in the following way:

```
from sklearn.decomposition import PCA
from sklearn.pipeline import Pipeline
from sklearn.preprocessing import StandardScaler
from sklearn.svm import SVC

pca = PCA(n_components=10)
scaler = StandardScaler()
svc = SVC(kernel='poly', gamma=3)

steps = [
    ('pca', pca),
    ('scaler', scaler),
    ('classifier', svc)
]

pipeline = Pipeline(steps)
```

At this point, the pipeline can be fitted like a single classifier (using the standard `fit()` and `fit_transform()` methods), even if the input samples are first passed to the `PCA` instance, the reduced dataset is normalized by the `StandardScaler` instance, and, finally, the resultant samples are passed to the classifier.

Creating a Machine Learning Architecture

A `pipeline` is also very useful together with `GridSearchCV`, to evaluate different combinations of parameters, not limited to a single step but considering the whole process. Considering the previous example, we can create a dummy dataset and try to find the optimal parameters:

```
from sklearn.datasets import make_classification

nb_samples = 500
X, Y = make_classification(n_samples=nb_samples, n_informative=15,
n_redundant=5, n_classes=2)
```

The dataset is redundant. Therefore, we need to find the optimal number of components for PCA and the best kernel for the SVM. When working with a pipeline, the name of the parameter must be specified using the component ID followed by a double underscore and then the actual name, for example, `classifier__kernel` (if you want to check all the acceptable parameters with the right name, it's enough to execute `print(pipeline.get_params().keys())`. Therefore, we can perform a grid search with the following parameter dictionary:

```
from sklearn.model_selection import GridSearchCV

param_grid = {
    'pca__n_components': [5, 10, 12, 15, 18, 20],
    'classifier__kernel': ['rbf', 'poly'],
    'classifier__gamma': [0.05, 0.1, 0.2, 0.5],
    'classifier__degree': [2, 3, 5]
}

gs = GridSearchCV(pipeline, param_grid)
gs.fit(X, Y)
```

As expected, the best estimator (which is a complete pipeline) has 15 principal components (which means they are uncorrelated) and an RBF SVM with a relatively high `gamma` value (`0.2`):

```
print(gs.best_estimator_)
Pipeline(steps=[('pca', PCA(copy=True, iterated_power='auto',
n_components=15, random_state=None,
   svd_solver='auto', tol=0.0, whiten=False)), ('scaler',
StandardScaler(copy=True, with_mean=True, with_std=True)), ('classifier',
SVC(C=1.0, cache_size=200, class_weight=None, coef0=0.0,
   decision_function_shape=None, degree=2, gamma=0.2, kernel='rbf',
   max_iter=-1, probability=False, random_state=None, shrinking=True,
   tol=0.001, verbose=False))])
```

The corresponding score is as follows:

```
print(gs.best_score_)
0.96
```

It's also possible to use `pipeline` together with `GridSearchCV` to evaluate different combinations. For example, it can be useful to compare some decomposition methods, mixed with various classifiers:

```
from sklearn.datasets import load_digits
from sklearn.decomposition import NMF
from sklearn.feature_selection import SelectKBest, f_classif
from sklearn.linear_model import LogisticRegression

digits = load_digits()

pca = PCA()
nmf = NMF()
kbest = SelectKBest(f_classif)
lr = LogisticRegression()

pipeline_steps = [
    ('dimensionality_reduction', pca),
    ('normalization', scaler),
    ('classification', lr)
]

pipeline = Pipeline(pipeline_steps)
```

We want to compare `PCA`, `NMF`, and `kbest` feature selection based on the ANOVA criterion, together with logistic regression and kernelized SVM:

```
pca_nmf_components = [10, 20, 30]

param_grid = [
    {
        'dimensionality_reduction': [pca],
        'dimensionality_reduction__n_components': pca_nmf_components,
        'classification': [lr],
        'classification__C': [1, 5, 10, 20]
    },
    {
        'dimensionality_reduction': [pca],
        'dimensionality_reduction__n_components': pca_nmf_components,
        'classification': [svc],
        'classification__kernel': ['rbf', 'poly'],
        'classification__gamma': [0.05, 0.1, 0.2, 0.5, 1.0],
        'classification__degree': [2, 3, 5],
```

```
            'classification__C': [1, 5, 10, 20]
    },
    {
        'dimensionality_reduction': [nmf],
        'dimensionality_reduction__n_components': pca_nmf_components,
        'classification': [lr],
        'classification__C': [1, 5, 10, 20]
    },
    {
        'dimensionality_reduction': [nmf],
        'dimensionality_reduction__n_components': pca_nmf_components,
        'classification': [svc],
        'classification__kernel': ['rbf', 'poly'],
        'classification__gamma': [0.05, 0.1, 0.2, 0.5, 1.0],
        'classification__degree': [2, 3, 5],
        'classification__C': [1, 5, 10, 20]
    },
    {
        'dimensionality_reduction': [kbest],
        'classification': [svc],
        'classification__kernel': ['rbf', 'poly'],
        'classification__gamma': [0.05, 0.1, 0.2, 0.5, 1.0],
        'classification__degree': [2, 3, 5],
        'classification__C': [1, 5, 10, 20]
    },
]

gs = GridSearchCV(pipeline, param_grid)
gs.fit(digits.data, digits.target)
```

Performing a grid search, we get the pipeline made up of PCA with 20 components (the original dataset has 64 features) and an RBF SVM with a very small `gamma` value (0.05) and a medium (5.0) *L2* penalty parameter `C`:

```
print(gs.best_estimator_)
Pipeline(steps=[('dimensionality_reduction', PCA(copy=True,
   iterated_power='auto', n_components=20, random_state=None,
    svd_solver='auto', tol=0.0, whiten=False)), ('normalization',
   StandardScaler(copy=True, with_mean=True, with_std=True)),
   ('classification', SVC(C=5.0, cache_size=200, class_weight=None, coef0=0.0,
    decision_function_shape=None, degree=2, gamma=0.05, kernel='rbf',
    max_iter=-1, probability=False, random_state=None, shrinking=True,
    tol=0.001, verbose=False))])
```

Considering the need to capture small details in the digit representations, these values are an optimal choice. The score for this pipeline is indeed very high:

```
print(gs.best_score_)
0.968836950473
```

Feature unions

Another interesting class provided by scikit-learn is the `FeatureUnion` class, which allows concatenating different feature transformations into a single output matrix. The main difference with a pipeline (which can also include a feature union) is that the pipeline selects from alternative scenarios, while a feature union creates a unified dataset where different preprocessing outcomes are joined together. For example, considering the previous results, we could try to optimize our dataset by performing a PCA with 10 components joined with the selection of the best 5 features chosen according to the ANOVA metric. In this way, the dimensionality is reduced to 15 instead of 20:

```
from sklearn.pipeline import FeatureUnion

steps_fu = [
  ('pca', PCA(n_components=10)),
  ('kbest', SelectKBest(f_classif, k=5)),
]

fu = FeatureUnion(steps_fu)

svc = SVC(kernel='rbf', C=5.0, gamma=0.05)

pipeline_steps = [
  ('fu', fu),
  ('scaler', scaler),
  ('classifier', svc)
]

pipeline = Pipeline(pipeline_steps)
```

We already know that an RBF SVM is a good choice, and, therefore, we keep the remaining part of the architecture without modifications. Performing a cross-validation, we get the following:

```
from sklearn.model_selection import cross_val_score

print(cross_val_score(pipeline, digits.data, digits.target, cv=10).mean())
0.965464333604
```

The score is slightly lower than before (< 0.002), but the number of features has been considerably reduced, and therefore also the computational time. Joining the outputs of different data preprocessors is a form of data augmentation, and it must always be taken into account when the original number of features is too high or redundant/noisy and a single decomposition method doesn't succeed in capturing all the dynamics.

Summary

In this final chapter, we discussed the main elements of machine learning architecture, considering some common scenarios and the procedures that are normally employed to prevent issues and improve the global performance. None of these steps should be discarded without a careful evaluation because the success of a model is determined by the joint action of many parameters, and hyperparameters, and finding the optimal final configuration starts with considering all possible preprocessing steps.

We saw that a grid search is a powerful investigation tool and that it's often a good idea to use it together with a complete set of alternative pipelines (with or without feature unions), so as to find the best solution in the context of a global scenario. Modern personal computers are fast enough to test hundreds of combinations in a few hours, and when the datasets are too large, it's possible to provision a cloud server using one of the existing providers.

Finally, I'd like to repeat that until now (also considering the research in the deep learning field), creating an up-and-running machine learning architecture needs a continuous analysis of alternative solutions and configurations, and there's no silver bullet for any but the simplest cases. This is a science that still keeps an artistic heart!

Other Books You May Enjoy

If you enjoyed this book, you may be interested in these other books by Packt:

Mastering Machine Learning Algorithms
Giuseppe Bonaccorso

ISBN: 978-1-78862-111-3

- Explore how a ML model can be trained, optimized, and evaluated
- Understand how to create and learn static and dynamic probabilistic models
- Successfully cluster high-dimensional data and evaluate model accuracy
- Discover how artificial neural networks work and how to train, optimize, and validate them
- Work with Autoencoders and Generative Adversarial Networks
- Apply label spreading and propagation to large datasets
- Explore the most important Reinforcement Learning techniques

Python Deep Learning

Valentino Zocca, Gianmario Spacagna, Daniel Slater, Peter Roelants

ISBN: 978-1-78646-445-3

- Get a practical deep dive into deep learning algorithms
- Explore deep learning further with Theano, Caffe, Keras, and TensorFlow
- Learn about two of the most powerful techniques at the core of many practical deep learning implementations: Auto-Encoders and Restricted Boltzmann Machines
- Dive into Deep Belief Nets and Deep Neural Networks
- Discover more deep learning algorithms with Dropout and Convolutional Neural Networks
- Get to know device strategies so you can use deep learning algorithms and libraries in the real world

Leave a review - let other readers know what you think

Please share your thoughts on this book with others by leaving a review on the site that you bought it from. If you purchased the book from Amazon, please leave us an honest review on this book's Amazon page. This is vital so that other potential readers can see and use your unbiased opinion to make purchasing decisions, we can understand what our customers think about our products, and our authors can see your feedback on the title that they have worked with Packt to create. It will only take a few minutes of your time, but is valuable to other potential customers, our authors, and Packt. Thank you!

Index

A

AdaBoost 273, 274, 275, 276, 277
AdaBoost.M1 273
adaptive system
  interaction diagram 10
adjusted rand index 321
affinity 344
agglomerative clustering
  about 343, 344
  connectivity constraints 354, 355, 357, 359
  dendrograms 346
  in scikit-learn 349, 350, 351, 353, 354
Akaike Information Criterion (AIC) 298
Alternating Least Squares (ALS)
  about 373, 488
  with Apache Spark MLlib 374, 377
Analysis of Variance (ANOVA) 80, 481
Apriori 45, 188
Area Under the Curve (AUC) 184, 203
artificial neural network (ANN) 430, 431
artificial neuron
  generic structure 431
atom extraction 100

B

back propagation 432
Bag-of-Words strategy 382, 383
bagging 268
Balanced Iterative Reducing and Clustering using Hierarchies (BIRCH) 334
ball tree
  about 290
  example 291
batch normalization layers 451
Bayes' theorem 188, 189
Bayesian Information Criterion (BIC) 299
Bayesian regression 130, 131, 132, 133
Bernoulli Naive Bayes 191, 192, 193, 194
Best Linear Unbiased Estimator (BLUE) 110
Between-Cluster-Dispersion (BCD) 314
biclustering 337, 338, 340
bidimensional dataset
  example 107, 108, 109, 110
big data 27
bigrams 391
Binary Decision Tree 246, 247
binary decisions 247, 249
bits 57
Bokeh
  reference 484
boosting 268
bootstrapping 268

C

C-Support Vector Machines 214
Calinski-Harabasz index 314, 315
categorical data
  managing 69, 71
categorical distribution 414
categories 15
Characteristic-Feature Tree 334
circular dummy dataset 354
class balancing
  about 51
  resampling with replacement 52
  SMOTE resampling 54
classical system
  generic representation 8
classification 15, 30
Classification and Regression Trees (CART) 246
classification metrics
  about 170, 171
  Cohen's Kappa 178, 179

confusion matrix 172, 173, 175
  F-Beta 177
  global classification report 180
  learning curve 180, 182
  precision 176
  recall 176
classifier 31
cluster instability 316, 318, 336
Clustering-Feature Tree (CF-Tree) 334
clustering
  basics 285, 286
  online clustering 331
  spectral clustering 328, 330
clusters
  about 286
  inertia, optimizing 309
coefficient of determination 116
Cohen's Kappa 178, 179
cold-startup 367
Comma-Separated Values (CSV) 479
completeness 320
computational neuroscience 23, 24
conditional entropy 60
conditional independence 189
confusion matrix 172, 173, 175
conjugate prior 131
constant 463
content-based systems 365, 367
convolutional layers 449
corpora, NLTK
  reference 382
cosine distance 345
cot function 40
count vectorizing 389, 390
covariate shift 451
Crab
  about 368, 369
  reference 369
cross-entropy 59
cross-entropy impurity index 250
cross-validation 39
CuPy
  reference 486, 488
curse of dimensionality 43
custom kernels 219

D

data augmentation 481
data formats 30, 31, 32
data generating process 30
data scaling 74
Database Management System (DBMS) 479
DataFrame 479
decision tree classification
  with scikit-learn 252, 254, 255, 257, 259
Decision Tree regression
  about 260
  with Concrete Compressive Strength dataset 261, 262, 263, 264, 266
deep convolutional network
  with Keras 452, 454, 455
deep learning
  about 24, 429
  applications 25
  artificial neural network (ANN) 430
deep model layers
  about 447
  fully connected layers 448
Deep Neural Networks (DNN) 421
dendrograms
  about 346
  computing 347, 349
dense layer 448
Density-Based Spatial Clustering of Applications with Noise (DBSCAN)
  about 324
  half-moons 325, 326, 327
descriptive analysis 11
device 464
diagnostic model 11
dictionary learning 99, 101
dimensionality reduction 480
discriminant analysis 203, 205, 206, 207
distributed architectures 488, 490
Divisive Analysis (DIANA) 343
divisive clustering 343
document-term matrix 400
dropout layers 450

E

ElasticNet 124
ensemble learning
  about 267
  bagging 268
  boosting 268
  stacking 268
ensemble methods
  AdaBoost 273, 274, 276, 277
  Gradient Tree Boosting 277, 280
  Random Forests 268, 270
  voting classifier 280, 281, 283, 284
entropy 57, 58
error measures 39, 40
Euclidean distance 344
evaluation methods, based on ground truth
  about 319
  adjusted rand index 321
  completeness 320
  homogeneity 319
Expectation-Maximization (EM) 46, 408

F

F-Beta 177
false negative (FN) 171
false positive (FP) 171
false positive rate (FPR) 183
Fast Independent Component Analysis (FastICA) 97
feature importance
  about 252
  in Random Forests 271, 272
feature selection 78, 79
feature unions 495
filtering 78, 79
Forward Stage-wise Additive Modeling 277
fully connected layers
  about 448
  batch normalization layers 451
  convolutional layers 449
  dropout layers 450
  recurrent neural networks 451
fuzzy clustering 287

G

Gated Recurrent Unit (GRU) 451
Gauss-Markov theorem 110
Gaussian mixture
  about 294, 295, 296, 297, 298
  optimal number of components, finding 298
Gaussian Naive Bayes 199, 201, 202, 203
Generative Adversarial Networks (GANs) 62
generic dataset 74
generic likelihood expression 47
Gensim
  reference 421
Geo-political entity (GPE) 395
Gini importance 252
Gini impurity index 250
global classification report 180
Global Component of Climate at a Glance (GCAG) 456
Gradient Tree Boosting 277, 280
graph 463
Graphviz
  reference 253
grid search
  optimal hyperparameters, finding through 167

H

Hamming distance 366
hard clustering 287
hard voting 281
Hierarchical Clustering algorithm 335
hierarchical clustering
  about 343
  agglomerative clustering 343
  divisive clustering 343
high-dimensional datasets
  visualizing, t-SNE used 102, 103
homogeneity 319
Huber loss 128
Huber regression 128, 130
Hughes phenomenon 43

I

imbalanced-learn
  reference 56

impurity measures
  about 250
  cross-entropy impurity index 251
  Gini impurity index 250
  misclassification impurity index 252
independent and identically distributed (i.i.d) 30, 108, 203
Independent Component Analysis (ICA) 95, 96, 97, 99, 481
inference 13
information theory
  elements 57
  entropy 57, 58
instance-based learning 31
instance-based learning methods 288
Inverse Document Frequency (IDF) 392
ISO 639-1 codes
  reference 387
isotonic regression 138, 139, 140
Iterative Dichotomizer 3 (ID3) 246

J

Jaccard distances 366
Jaccard similarity coefficient 171
Jensen-Shannon divergence 62
Jupyter
  reference 186

K

K-fold cross-validation 114, 115
K-means++ 302
K-means
  about 301, 302, 303, 305, 306, 307
  disadvantages 308
k-Nearest Neighbors (k-NN) 328, 354
K-NN 288
KD tree 289
Keras
  about 434
  deep convolutional network 452, 454, 455
  interfacing, to scikit-learn 443, 444, 445
  LSTM network 456, 458, 459, 460, 461
  reference 434
kernel PCA 92, 94
kernel support vector machines 106

kernel trick 218
kernel-based classification
  about 217
  custom kernels 219
  polynomial kernel 219
  Radial Basis Function (RBF) 218
  sigmoid kernel 219
kernel-based support vector machines 31
kernels, in scikit-learn
  reference 95
Kullback-Leibler divergence 61
Kurtosis 96

L

Lagrange multipliers 217
Laplace 194
Lasso regressor 122, 123
Latent Dirichlet Allocation 413, 414, 415, 417
latent factors 370
Latent Semantic Analysis 400, 401, 402, 404, 405
learnability 34, 35, 36
learning 13
learning curve 180, 182
Lidstone factor 195
Lidstone smoothing factor 194
likelihood 45
linear classification 144, 145, 146, 215, 216
Linear Discriminant Analysis (LDA) 204
linear models
  about 105
  for regression 105, 106
linear regression
  with higher dimensionality 112
  with scikit-learn 112, 114
linear SVM 209, 210, 211, 212, 213, 214
linearly separable 145
linkage
  average linkage 346
  complete linkage 346
  Ward's linkage 346
Lloyd's algorithm 302
log-likelihood 47
logistic regression 147, 148, 149
LogisticRegression class

   implementing 150, 151, 152, 153
Long Short-Term Memory (LSTM) 451
loss function 40
LSTM network
   with Keras 456, 457, 458, 459, 460, 461

M

machine 8
machine learning architectures
   about 477
   data augmentation 481
   data collection 479
   data conversion 483
   dimensionality reduction 480
   GPU support 484
   modeling/grid search/cross-validation 483
   normalization 480
   regularization 480
   scikit-learn tools 491
   visualization 484
Machine Learning Library (MLlib) 374
machine learning
   about 9, 13
   big data 26
   reinforcement learning 21
   semi-supervised learning 19
   supervised learning 14, 15
   unsupervised learning 17, 18
Manhattan 344
manifold assumption 81
MAP learning 45
matplotlib
   reference 186
maximum-likelihood learning 46, 50
mean absolute error (MAE) 260
Mean Squared Error (MSE) 376
mean squared error (MSE) 260
Mercer's theorem 218
Microsoft Cognitive Toolkit (CNTK) 434
mini-batch gradient descent 153
mini-batch k-means 332
minimum description length (MDL) 61
Minkowski distance 366
misclassification impurity index 252
missing features

   managing 72
MLPs
   structure 431
   with Keras 434, 435, 436, 438, 440, 442
model evidence 132
model-based collaborative filtering
   about 370
   Alternating Least Squares (ALS) 373
   Singular Value Decomposition strategy 371, 373
model-free collaborative filtering 367, 368, 369
Multi-layer Perceptron (MLP) 448
multiclass strategies
   about 33
   one-vs-all strategy 33
   one-vs-one strategy 34
   reference 34
multinomial Naive Bayes
   about 194
   for text classification 196, 198
MurmurHash 3
   reference 71

N

n-grams 391
Naive Bayes classifier 190, 191
Naive Bayes, in scikit-learn
   about 191
   Bernoulli Naive Bayes 191, 192
naive Bayes, in scikit-learn
   Bernoulli Naive Bayes 193, 194
Naive Bayes, in scikit-learn
   Gaussian Naive Bayes 199, 201, 202, 203
   multinomial Naive Bayes 194
Naive user-based systems
   about 362
   implementing, with scikit-learn 363, 364
Named Entity Recognition (NER) 395
nats 58
Natural Language Processing (NLP) 9, 379
Natural Language Toolkit (NLTK)
   about 380
   Corpora examples 381, 382
   downloader interface 380
   reference 381
nearest neighbor

of noisy sample 293
neural networks 106
NLopt
  reference 242
non-linearly separable 145
Non-Negative Matrix Factorization (NMF) 406
non-negative matrix factorization (NNMF)
  about 88, 89
  reference 90
non-parametric learning 31
normalization 74, 76, 480
normalized-cuts 329
numerical outputs
  examples 30
NumPy random number generation
  reference 68

O

Occam's razor 51, 133
occurrence matrix 400
one-hot encoding 70
online clustering
  about 331
  BIRCH 334
  mini-batch k-means 332
operation 463
optimal hyperparameters
  finding, through grid search 167
optimizations 150, 151, 152
Ordinary Least Squares (OLS) 108
overfitting 14, 36, 37

P

PAC learning 43
Pandas 114
pandas 479
parameteric learning 31
passive-aggressive algorithms 157, 158, 160, 161, 162
passive-aggressive regression 163, 164, 165, 166
Penn Treebank POS corpus 394
perplexity 58
pipelines 491
placeholder 463

plate notation
  reference 407
polynomial kernel 219
polynomial regression 106, 134, 136, 137, 138
pooling layers
  average pooling 450
  max pooling 450
POS Tagging 394
positive predictive value 176
posteriori 189
Posteriori 45
posteriori probability 45
precision 176
prediction 13
predictive analysis 12
predictor 41, 42
principal component analysis
  about 81, 82, 83, 85, 86
  kernel PCA 92, 93, 94
  non-negative matrix factorization (NNMF) 88, 89
  sparse PCA 90
Probabilistic Latent Semantic Analysis 407, 408, 409, 412
probably approximately correct (PAC) 43
PySpark
  reference 378

Q

Quadratic Discriminant Analysis (QDA) 204

R

R2 score 116
Radial Basis Function (RBF) 218, 328
Random Forests
  about 268, 270
  feature importance 271, 272
RANSAC 126, 127
recall 176
Receiver Operating Characteristic (ROC) 258
Rectified Linear Unit (ReLU) 438
Recurrent Neural Networks (RNNs) 451
regression 15, 30
regressor 31
regressor analytic expression 118, 119
regularization 480

reinforcement learning 21
representational capacity 36
resilient distributed dataset (RDD) 376
reward 21
ridge regression 119, 121
robust regression
   about 125
   Huber regression 128, 130
   RANSAC 126, 127
ROC curve 182, 183, 185

S

sample text classifier
   based on Reuters corpus 396, 397
scikit-learn implementation
   about 214
   linear classification 215, 216
   non-linear examples 220, 221, 223, 224
scikit-learn score functions
   reference 80
scikit-learn tools, for machine learning architectures
   about 491
   feature unions 495
   pipelines 491
scikit-learn toy datasets
   about 66
   reference 66
scikit-learn
   Keras, interfacing to 443, 444, 445
   user-based system, implementing 363, 364
SciPy sparse matrices
   reference 91
SciPy
   reference 51
semi-supervised learning 19
semi-supervised Support Vector Machines (S3VM) 236, 238, 240, 241
sensitivity 176
sentence tokenizing 384
sentiment analysis
   about 422, 423, 424
   VADER sentiment analysis with NLTK 426
Sequential Least SQuares Programming (SLSQP) 238
session 464

Shi-Malik algorithm 329
sigmoid kernel 219
sign indeterminacy 406
silhouette plots 336
silhouette score 310, 311, 312, 313
Single Instruction Multiple Data (SIMD) 484
Singular Value Decomposition (SVD) 88, 338, 371, 373, 400
slack variables 214
soft clustering 287
soft K-means 294
soft voting 281
Spark MLlib
   reference 490
sparse matrices, SciPy
   reference 72
sparse PCA 90
spectral biclustering 338
spectral clustering 328, 330
stacking 268
statistical learning approaches
   about 44
   MAP learning 45
   maximum-likelihood learning 46, 47, 48, 50
stemming 388
Stochastic Gradient Descent (SGD) 434
stochastic gradient descent (SGD)
   reference 124
stochastic gradient descent algorithms 153, 154, 155, 156, 157
stopword 384
stopword removal
   about 386
   language detection 387
strategic elements, required to work with TensorFlow
   constant 463
   device 464
   graph 463
   operation 463
   placeholder 463
   session 464
   variable 463
strong learners 267
summation 47

supervised learning
  about 14
  applications 16
Support Vector Machines (SVM) 19, 421, 480
support vector regression
  about 228, 229, 230, 231
  Airfoil Self-Noise dataset 233
  with Airfoil Self-Noise dataset 232
support vectors 211
symmetric affinity matrix 328
symmetric Dirichlet 413
Synthetic Minority Over-sampling Technique (SMOTE) 54, 55, 56

T

t-Distributed Stochastic Neighbor Embedding (t-SNE)
  about 421
  high-dimensional datasets, visualizing 102, 103
Tensor Processing Units (TPUs) 486
TensorFlow
  about 462
  classification, with multi-layer perceptron 471, 472
  gradients, computing 464, 465
  image convolution 474, 475
  logistic regression, implementing 467, 468
Term Frequency (TF) 391
test sets
  creating 67, 68
TF-IDF vectorizing 391, 393
Tikhonov regularization 119
tokenizing
  sentence tokenizing 384
  word tokenizing 385
tokens 384
topic modeling
  about 399
  Latent Dirichlet Allocation 413, 414, 415, 417
  Latent Semantic Analysis 401, 402, 404, 405
  Probabilistic Latent Semantic Analysis 407, 408, 409, 412
topics 401

training sets
  creating 67, 68
trigrams 391
true negative (TN) 171
true positive (TP) 171
true positive rate (TPR) 183

U

UCI Machine-Learning Repository
  reference 134
underfitting 36, 37
unigrams 391
unsupervised learning
  about 17
  applications 19

V

v-Support Vector Machines 225, 226, 227, 228
VADER sentiment analysis
  with NLTK 426
Valence Aware Dictionary and sEntiment Reasoner (VADER) 426
variable 463
variance 117
vectorizing
  about 389
  count vectorizing 389, 390
  TF-IDF vectorizing 391, 393
voting classifier 280, 282, 283, 284

W

Ward's linkage 346
weak learners 267
whitening 76
whitening matrix 77
Within-Cluster-Dispersion (WCD) 314
word tokenizing 385
word2vec
  about 418
  reference 419

Z

zero-one-loss function 40